AF332903

Biochalcogen Chemistry:
The Biological Chemistry of Sulfur, Selenium, and Tellurium

ACS SYMPOSIUM SERIES **1152**

Biochalcogen Chemistry: The Biological Chemistry of Sulfur, Selenium, and Tellurium

Craig A. Bayse, Editor
Old Dominion University
Norfolk, Virginia

Julia L. Brumaghim, Editor
Clemson University
Clemson, South Carolina

Sponsored by the
ACS Division of Inorganic Chemistry, Inc.
Society of Biological Inorganic Chemistry

American Chemical Society, Washington, DC

Distributed in print by Oxford University Press

Library of Congress Cataloging-in-Publication Data

Biochalcogen chemistry : the biological chemistry of sulfur, selenium, and tellurium / Craig A. Bayse, editor, Old Dominion University, Norfolk, Virginia, Julia L. Brumaghim, editor, Clemson University, Clemson, South Carolina ; sponsored by the ACS Division of Inorganic Chemistry, Inc., Society of Biological Inorganic Chemistry.
    pages cm. -- (ACS symposium series ; 1152)
  Includes bibliographical references and index.
  ISBN 978-0-8412-2903-7 (alk. paper)
  1. Chalcogens--Congresses. 2. Sulfur--Congresses. 3. Selenium--Congresses. 4. Tellurium--Congresses. I. Bayse, Craig A., editor of compilation. II. Brumaghim, Julia L., editor of compilation. III. American Chemical Society. Division of Inorganic Chemistry, sponsoring body. IV. Society of Biological Inorganic Chemistry, sponsoring body.
  TP245.O9B56 2013
  546'.72--dc23

                     2013041540

The paper used in this publication meets the minimum requirements of American National Standard for Information Sciences—Permanence of Paper for Printed Library Materials, ANSI Z39.48n1984.

Foreword

The ACS Symposium Series was first published in 1974 to provide a mechanism for publishing symposia quickly in book form. The purpose of the series is to publish timely, comprehensive books developed from the ACS sponsored symposia based on current scientific research. Occasionally, books are developed from symposia sponsored by other organizations when the topic is of keen interest to the chemistry audience.

Before agreeing to publish a book, the proposed table of contents is reviewed for appropriate and comprehensive coverage and for interest to the audience. Some papers may be excluded to better focus the book; others may be added to provide comprehensiveness. When appropriate, overview or introductory chapters are added. Drafts of chapters are peer-reviewed prior to final acceptance or rejection, and manuscripts are prepared in camera-ready format.

As a rule, only original research papers and original review papers are included in the volumes. Verbatim reproductions of previous published papers are not accepted.

ACS Books Department

Contents

Indexes

Preface

The redox activity of the heavier chalcogens, sulfur, selenium and tellurium, has long been a focus of biological and medical interest. Thiols and selenoproteins, in particular, play a critical role in maintaining healthy states by scavenging excess oxidants that contribute to increased risk of cancer, cardiovascular disease and other oxidative-stress-related illnesses. To this end, natural sulfur and selenium compounds found in many foods and a number of small synthetic organosulfur, -selenium and -tellurium compounds have been explored for their potential role as chemopreventives. Sulfur, especially in the form of cysteine, is a biomarker for oxidative stress as well as a ligand in the active site of numerous metalloproteins, notably iron hydrogenase and transcription factors, where the conversion of thiolates to disulfides is an important redox switch. Thus, while the chemistry of ubiquitous oxygen is distinct and often studied separately, much of the biological chemistry of the heavier chalcogens are defined by their interaction with this lightest member of the group. Highlighting both the potential value and the pitfalls of chalcogens in biology and medicine, the National Institutes of Health has spent over \$250,000,000 in the past decade on selenium-supplementation clinical trials alone, leading to mixed results and demonstrating the clear need for further basic research.

This book highlights the biological uses of heavy chalcogens as a key area of focus in bioinorganic chemistry and a unifying theme for research in a wide variety of disciplines. Recent achievements in these multidisciplinary efforts are presented that discuss the subtle, yet important roles of biochalcogens in living systems as sulfur- and selenium-containing metabolic intermediates and products (Chapters 1 and 10) and in their oxidation when coordinated to metals (Chapters 3 and 4). Chemical and instrumental tools for detecting sulfur and selenium species and their functionalities are also discussed (Chapters 2 and 6), as are new directions in biochalcogen applications to redox scavenging, both in terms of synthesis (Chapters 7 and 8) and mechanistic modeling (Chapter 9). Tellurium, with no natural biological function, is represented together with sulfur and selenium as a phasing agent in nucleic acid crystallography and for other biological studies (Chapter 5).

This book will serve as a useful collection of reviews and research results in this diverse field, encompassing research in bioinorganic chemistry, organic synthesis, computational approaches, and biochemistry; as an inspiration for researchers wishing to enter the variety of fields that encompass these multidisciplinary research efforts; and as a useful resource for undergraduate or graduate courses focusing on main group and transition element biochemistry. We hope that a wide audience finds this book a helpful resource for this rapidly expanding field.

We thank the American Chemical Society's Division of Inorganic Chemistry and the Society for Biological Inorganic Chemistry for their generous support of the 'Biochalcogen Chemistry' symposium at the 2012 National ACS Meeting in Philadelphia.

Craig A. Bayse
Department of Chemistry and Biochemistry
Old Dominion University
Hampton Boulevard
Norfolk, Virginia 23529, U.S.A.
cbayse@odu.edu (e-mail)

Julia L. Brumaghim
Chemistry Department
Clemson University
Clemson, South Carolina 29634-0973, U.S.A.
brumagh@clemson.edu (e-mail)

Chapter 1

Smelling Sulfur: Discovery of a Sulfur-Sensing Olfactory Receptor that Requires Copper

Eric Block[*,1] and Hanyi Zhuang[2,3]

[1]Department of Chemistry, University at Albany, State University of New York, Albany, New York 12222, U.S.A.
[2]Department of Pathophysiology, Shanghai Jiaotong University School of Medicine, Shanghai 200025, P. R. China
[3]Institute of Health Sciences, Shanghai Jiaotong University School of Medicine/Shanghai Institutes for Biological Sciences of Chinese Academy of Sciences, Shanghai 200025, P. R. China
[*]E-mail: eblock@albany.edu.

Olfactory receptors (ORs), located in olfactory sensory neurons (OSNs), mediate detection of odorants. Volatile sulfur compounds (VSCs), e.g., thiols and thioethers, are potent odorants. A mouse OR, MOR244-3, has been identified as robustly responding to strong-smelling (methylthio)methanethiol (MTMT) in heterologous cells. MTMT is a male mouse urine semiochemical attracting female mice. Proximate thiol and thioether groups in MTMT suggest a chelated metal complex in the activation of MOR244-3. Metal ion involvement in interaction of thiols with ORs was previously proposed but unproven. Recent work shows that Cu is required for activation of MOR244-3 toward ppb levels of MTMT, related sulfur compounds, and other metal-coordinating odorants, such as odorous *trans*-cyclooctene, among >125 compounds tested. Use of a Cu-chelator (TEPA) abolishes the response of MOR244-3 to MTMT. An olfactory discrimination assay showed that mice injected with TEPA failed to discriminate MTMT. The above work establishes for the first time the role of copper in detection of sulfur-containing odorants by ORs.

Introduction

Humans, and other animals, have an exquisitely sensitive sense of smell toward low-valent, volatile sulfur compounds (VSCs). In 1887, Emil Fischer wrote that concentrations of ethanethiol as low as 0.05 parts per billion (ppb) are "clearly perceptible to the sense of smell" (*1*). Spider monkeys are yet more sensitive, detecting 0.001 ppb ethanethiol (*2*), and chiral 3-methyl-3-sulfanylhexan-1-ol, present in onions and in armpit odor (*3*) can be perceived at levels as low as 0.001 ng/L (~0.001 parts per trillion) (*4*). Thiols with very low odor thresholds are also present in grapefruit (*5*), skunk scent (*6*), skunky-smelling beer (*7*), male mouse urine (*8*), and in the aromas of durian (Figure 1) (*9*) and bell peppers (Figure 2) (*10*), among other sources, as well as in scent markers, e.g., Chevron's Scentinel® (*11*), for detection of otherwise odorless natural gas.

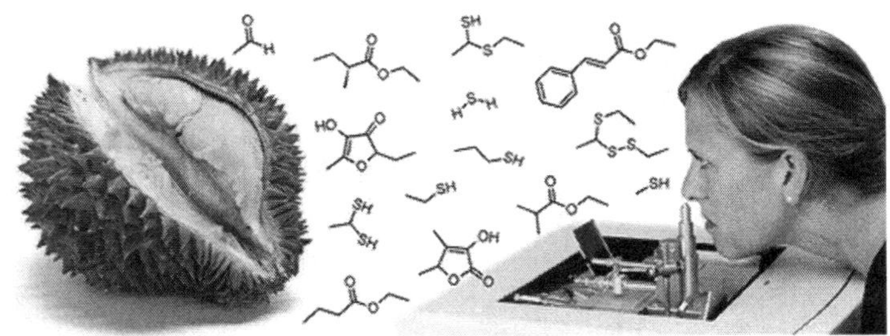

Figure 1. Strong smelling VSCs in Thai durian identified by headspace GC-olfactometry (9). Copyright 2012, American Chemical Society.

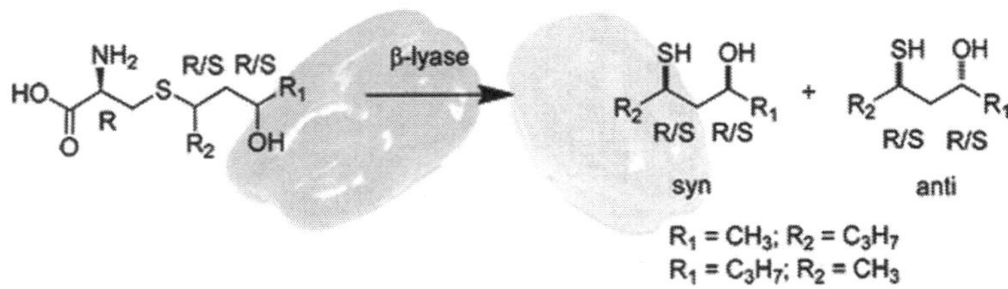

Figure 2. Cysteine-S-conjugate origin of bell pepper VSCs (10). Copyright 2011, American Chemical Society.

Strong-smelling heterocyclic thioethers, thietane and thiolane, also used as gas odorants, are found derivatized in anal scent glands of musteloid (weasels, etc.) species who use them as trail markers (*12*). Malodorous VSCs and amines are protein degradation products found in putrid food, and H_2S is present in oxygen-depleted air, hence the need for animals to have heightened sensitivity to these compounds to avoid intoxication (*12–15*). It should be noted that the sensory perception of VSCs can vary with concentration, with lower concentrations being perceived as favorable and higher concentrations as unpleasant, e.g., as in the case of 3-methyl-2-butene-1-thiol in beer (*7*), and dimethyl sulfide in wine, which at trace levels is perceived as fruity, whereas in higher concentrations it is described as skunky (*16, 17*).

Little is known about perception of low molecular weight VSCs by the sense of smell, and why there is such a striking difference in smell between the structurally similar molecules ethanol and ethanethiol (Figure 3). For example, ethanol "is only perceptible in air in a concentration of 0.4 % wt./wt., whilst ethyl mercaptan is perceptible at 0.3×10^{-8} % wt./wt.; our perception of it is one hundred million times more delicate" (*18*). This chapter describes recent efforts to understand the molecular basis for sensitive olfactory detection of VSCs, complementing recent publications by the author on occurrence and analysis of VSCs, including those from genus *Allium* plants (garlic, onions, etc.) (*19–21*).

Figure 3. Left: space-filling model of ethanol. Right: space-filling model of ethanethiol. Both structures are from Wikipedia.

Possible Role of Metals in Olfaction

The alternative name for ethanethiol, ethyl mercaptan, provides a clue about the possible role of metals in olfaction: "mercaptan" comes from the Latin *mercurium captans* ("capturing mercury"). Over the past 40 years several researchers have proposed that transition metals such as Zn^{2+}, Ni^{2+}, Cu^{2+}, or Cu^+ (generally in the form of metalloproteins) may mediate taste or odor perception of thiols and amines. In 1969, Henkin and Bradley (*22*) suggested that the physiology of taste involved copper. In 1978, Crabtree (*23*) proposed that H_2S, thiols and sulfides and other strong-smelling small molecules "bind chemically to a nasal receptor. . . containing a transition metal at the active site," that Cu(I) is "the most likely candidate for a metallo-receptor site in olfaction," and that "the Cu(I) centre would be stabilized by coordination, perhaps to a protein thiolato-group, and . . . two or three additional protein S or N neutral donor groups." Crabtree provided support for his hypothesis by noting that, compared with unstrained, mild-smelling olefins, strained, strong-smelling *trans*-cyclooctene gives "much more stable [metal] complexes, e.g., [Cu₂Cl₂-(*trans*-cyclooctene)₃]." In 1978, Day (*24*) argued that "possibly a transition metal serves in the olfaction of certain functional groups, based on the absence of a "lutidine-like" odor for purified, sterically hindered 2,6-di-*tert*-butylpyridine." Furthermore, hindered *o*-trimethylsilylbenzenethiol is reported to have a greatly reduced odor compared to the parent benzenethiol (*25*). In 1996, Turin (*26*) proposed a central role for zinc in olfaction. In 2003, Suslick et al. (*27, 28*) reported that synthetic pentapeptide HACKE, corresponding to a conserved sequence in the extracellular loop of olfactory receptors (ORs), could effectively bind to metal ions and therefore may form the basis for sensitive activation of ORs by thiols. In 2012, we reported compelling evidence for the central role of copper in discrimination of thiols and other metal-coordinating odorants by MOR244-3 in the mouse (*29*). Details relating to our work will be presented here.

Identification of (Methylthio)methanethiol (MTMT) as a Mouse Social-Signaling Compound

In 2005, in collaboration with Dayu Lin and (the late) Larry Katz at Duke University we discovered that male mouse urine contains (methylthio)methanethiol (MTMT; MeSCH₂SH), a semiochemical (signaling compound) with a powerful garlic-like odor, which is highly attractive to female mice (*8*). Our work is significant because MTMT is a novel sex-specific chemical cue, able to initiate a defined innate behavior and that acts through the main olfactory system. As described in Figure 4, solid phase microextraction (SPME) was used to collect mouse urinary volatiles, which were separated by GC. The effluent from the GC was split, with one stream going to a flame ionization detector (FID) or mass selective detector (MSD) and the other directed at the mouse's nose. In this manner, individual peaks from the GC were correlated with their ability to induce an electrophysiological neural response in the mouse, recording electrically from the main olfactory bulb mitral cells, which received direct excitatory inputs from olfactory sensory neurons (OSNs). When OSNs responsive to the urine were tested with individual, separated urine components, 33% were activated by a single compound, present in male but neither in female mouse nor castrated male mouse urine (*8*).

Based on the MS fragmentation pattern, and comparison of retention times and fragmentation patterns with that of an authentic sample, the male-specific compound was identified as MTMT. Mouse OSNs are highly, and specifically sensitive to MTMT, responding at a threshold of 10 ppb, yet not responding to any of the more than 100 other volatiles present in mouse urine. Importantly, MTMT was shown to elicit a specific behavioral response in female mice. Females are more interested in urine produced by intact rather than castrated males. Addition of synthetic MTMT to castrated male urine increased the attractiveness of the urine to female mice (*8*).

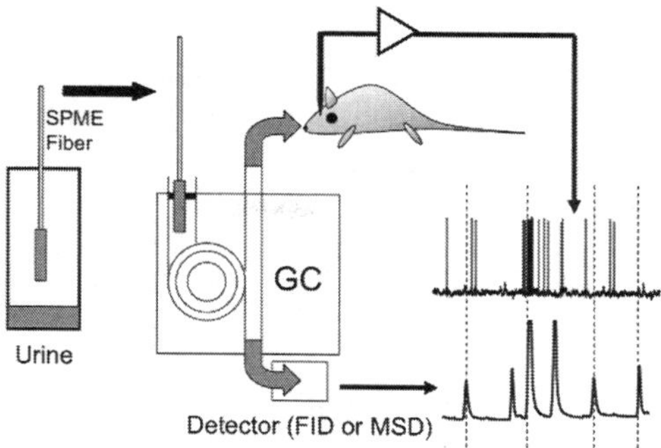

Figure 4. SPME collection of mouse urine volatiles. GC linked with single-unit electrophysiology to correlate activation of olfactory bulb mitral cells (upper trace) with specific urine volatiles detected by GC (lower trace) using a flame ionization detector (FID) or mass selective detector (MSD) to identify MTMT (CH_3SCH_2SH). With permission from ref. (21).

Overview of Olfactory Receptors

The sense of smell – olfaction – is mediated by specialized sensory cells of the nasal cavity of vertebrates. In these cells, olfactory receptors (ORs) (*30, 31*), expressed on the cell-surface membranes of olfactory sensory (receptor) neurons (OSNs [ORNs]), mediate detection of volatile odorants, and are members of the superfamily of G protein-coupled receptors (GPCRs) (*32–34*). GPCRs are transmembrane proteins that pass seven times through the plasma membrane. They comprise a large protein family of receptors sensing exogenous chemical ligands (e.g., odorants), activating signal transduction pathways and, ultimately, delivering a message to the inside of the cell. Humans and mice have 387 and 1035 ORs, respectively. At the same time, humans have many millions of OSNs, so there are a large number of replicates of each OSN expressing a certain type of OR. The genes that code for the ORs are the largest family of genes in humans, and animals in general. In vertebrates, ORs are located in the cilia of the OSNs, which are in turn located in the olfactory epithelium in the nasal cavity. In insects the ORs are mainly located on the antennae. An odorant will dissolve in the mucus of the olfactory epithelium and then bind to an OR. Rather than binding specific ligands, ORs display affinity for a range of odorants, and, conversely, a single odorant molecule may bind to a number of ORs with varying affinities. This difference in affinities causes differences in activation patterns, combinations, and permutations of which result in unique profiles for practically an infinite number of odorant molecules.

Once the odorant has bound to the OR, the latter undergoes structural changes, and it binds and activates the olfactory-type G protein on the inside of the cell membrane. This in turn activates the olfactory adenylyl cyclase, converting ATP and releasing cAMP in the process. Serving as a second messenger, cAMP then binds to the olfactory cyclic nucleotide-gated channel, leading to an influx of Ca^{2+}, effectively depolarizing the neuron. In the meantime, Ca^{2+} binds a Ca^{2+}-gated Cl^- channel and the resulting Cl^- current further depolarizes the neuron. This transduction mechanism enables the rapid detection of odorants within hundreds of milliseconds (*35, 36*). Molecular structures of ORs remain unknown due to difficulties expressing them in sufficient quantities and in crystallizing them. At the present time structural information on ORs is obtained by biochemical studies combined with computational modeling techniques, for example using the human M2 muscarinic receptor as a template (*37*).

The Role of Copper in Detection of (Methylthio)methanethiol by Mouse Olfactory Receptor MOR244-3

A key question presented by our work was how mice detect the very low concentrations of the thioether-thiol MTMT present in male mouse urine. Hiroaki Matsunami at Duke University and one of the authors (HZ) were able to isolate the specific mouse olfactory receptor (MOR) responsive to MTMT, termed MOR244-3, and, at the other author's (EB) suggestion, based on the possible chelating ability of MTMT, explored the possibility that Cu or Zn ions might be involved in the detection of MTMT by this OR. It was found that Cu, but not Zn ions or other common transition metal ions, specifically activated MOR244-3 toward MTMT as well as toward a panel of other organosulfur compounds that were structurally related to MTMT (Figure 5) (*29*). Among the compounds showing high activity were (methylseleno)methanethiol, disulfides $MeSCH_2SSMe$ and $MeSCH_2SSCH_2SMe$, and methyl dithioformate. It is was separately established that epithelial mucus taken from the mouse and analyzed by inductively coupled plasma mass spectrometry (ICP-MS) shows the presence of levels of inorganic copper similar to those used in testing MOR244-3 *in vitro*.

Behavioral Study in Mice Involving MTMT and Copper

An important part of our study was to associate a behavioral effect in the mouse, for example, olfactory recognition of MTMT, in the presence or absence of copper. Thus, mice were trained to associate either eugenol or MTMT with sugar reward. On the test day, they received bilateral nasal cavity injection of the copper chelator TEPA. Mice trained to associate MTMT with sugar reward spent significantly less time investigating the odor, whereas time spent investigating the nonsulfurous odorant was unaltered. With metabolic clearance of TEPA, the mouse group trained to recognize MTMT regained olfactory discrimination ability two days after TEPA injection. The results from the behavioral experiment indicate that copper is required for the olfactory detection of MTMT (*29*).

Selectivity of the Copper Ion Enhancement Effect

The panel of analogs tested (Figure 5) by measuring dose-response curves under *in vitro* conditions (Figure 6) were isomeric with MTMT, or differed by the addition of one or two carbon atoms with associated hydrogens or two oxygens (a sulfone), by the deletion of two hydrogen atoms (dithioformate), or by substitution of the thioether sulfur by selenium. Isomerization or atom addition, deletion, or substitution could alter the number of thiol and thioether groups, change the steric crowding at the sulfur atoms, modify the ligand "bite angle," alter the S–H acidity, and change the availability of thioether electron pairs, in turn, potentially modifying the coordinating ability of the copper complex with the functional groups of the transmembrane receptor protein. Testing of the analogs in Figure 5 was performed with addition of 30 µM Cu (as $CuSO_4$), with no added copper, and with 30 µM of copper chelator TEPA. The latter addition is important to eliminate the effect of background levels of copper we found in the medium used to culture MOR244-3 *in vitro*. The results for MTMT, shown in Figure 6 (top left graph), illustrate the dramatic difference between the response with added 30 µM Cu (top, blue trace, showing a limiting detection concentration of 10^{-8} M and an estimated concentration for 50% effect, EC_{50}, of 10^{-6} M), with no added Cu (middle, magenta trace, showing a limiting detection concentration of 10^{-6} M and EC_{50} of 10^{-5} M), and with 30 µM TEPA (bottom, green trace, showing a limiting detection concentration of 10^{-5} M and an EC_{50} of 5×10^{-4} M). The double asterisk indicates a significant difference between the curves with and without added Cu.

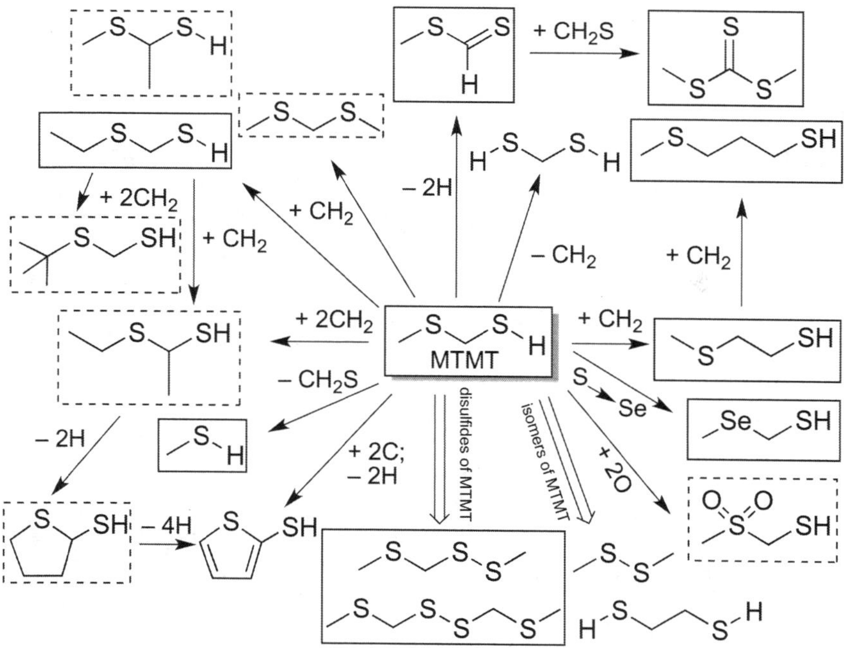

Figure 5. Structural relationship between MTMT and its analogs. Odorants boxed with solid lines are those with prominent responses in the presence of 30 µM Cu^{2+}, and odorants boxed with dashed lines are those with less prominent responses, as defined by a more than 50% reduction in efficacy compared with MTMT. Unboxed odorants did not elicit MOR244-3 response.

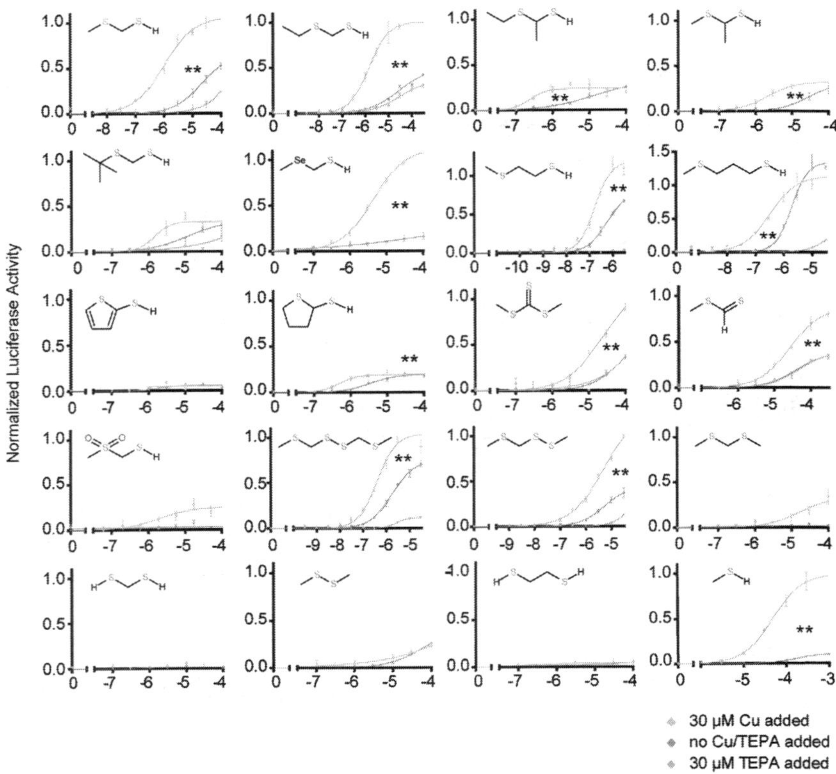

Figure 6. Dose-response curves of MOR244-3 to representative sulfur-containing compounds with and without 30 µM exogenous copper ion. The horizontal scale shows the exponent of the odorant molar concentration (e.g., $^{-6} = 10^{-6}$ M), and the vertical axis shows the normalized luciferase activity, an indirect measure of the response of the receptor to substrates. For odors with a significant response in the absence of exogenous copper ion, as defined arbitrarily by a top value greater than 0.32, dose–response curves with 30 µM of TEPA are also shown. F-tests were used to compare the pairs of dose–response curves with or without copper ion; asterisks shown represent significance of p-values after Bonferroni corrections. Adapted with permission from (29). Copyright 2012, PNAS. (see color insert)

We found that replacing the thioether methyl group in MTMT by ethyl (EtSCH$_2$SH) has no effect on activity (modest steric effect), whereas addition of a methyl group on the carbon between the sulfur atoms (MeSCHMeSH), addition of both the ethyl and methyl groups (EtSCHMeSH), or replacing the ethyl with a *tert*-butyl group (*t*-BuSCH$_2$SH) diminishes the activity (more significant steric effects). Cyclization to 2-mercaptothiolane, with removal of two hydrogens, results in modest loss of activity (shape change). Further removal of four hydrogens from 2-mercaptothiolane giving 2-thiophenethiol results in almost complete loss of OR activity, presumably due to the very poorly nucleophilic character of the thiophene exocyclic lone pair. Similarly, conversion of MTMT to $MeSO_2CH_2SH$ results in partial loss of OR activity due to the inability of the sulfone group to coordinate to copper, even though the acidity of the thiol group is increased. On the other hand, activity is retained when the thioether sulfur is replaced by selenium ($MeSeCH_2SH$) and upon removal of two hydrogens ($MeSCH=S$); good activity also remains with $MeSC(S)SMe$.

Even higher OR activity than MTMT is shown by both $MeS(CH_2)_2SH$ and $MeS(CH_2)_3SH$, where copper could form 5- and 6-membered ring chelates. On the other hand, loss of OR activity is seen upon methylation of the thiol group of MTMT in $CH_2(SMe)_2$, loss of the thioether methyl group in $CH_2(SH)_2$ or isomerization to MeSSMe or $HSCH_2CH_2SH$. 2,3,5-Trithiahexane ($MeSCH_2SSMe$), found in male mouse urine, and $MeSCH_2SSCH_2SMe$, an oxidation product of MTMT, elicited strong responses by the receptor. For these two compounds, there is also a dramatic reduction in the response by basal level of copper ion in the medium when 30µ M TEPA is added. These observations are consistent with literature reports that the ability of a neighboring electron donor in disulfides when ligated to Cu(I) "enhances the electron transfer from Cu(I) to the disulfide leading to S–S bond scission" (*38*), for example, the methylthio groups in 2,3,5-trithiahexane and 2,4,5,7-tetrathiaoctane, since dimethyl disulfide is apparently not reduced under these same conditions.

To explain the reactivity of the above disulfides, we assume that both Cu(I) and Cu(II) species could be available at the interface of receptor-ligand interaction, consistent with the known reducing environment within cells. Whether one or both of the two copper species is predominant at the active site for interaction with certain sulfurous ligands has yet to be explored. Most of our *in vitro* OR experiments in HEK293T cells were done using Cu(II) under aerobic conditions. Under the same conditions, addition of Cu(I), kept reduced by ascorbic acid prior to cell culture addition, gave similar results compared to Cu(II) (*39*).

Recently, in collaboration with Professor Victor Batista's lab at Yale University, QM/MM geometry optimizations were used to examine the different active site models, in which the oxidation state of the Cu center varies with the protonation state of the thiol group of the cysteine 109 residue (*40*).

Taken together, the data suggest that the most active complexes involve sulfur compounds of type $RX(CH_2)_nS$ (X = S or Se; n = 1–3; R = Me or Et), with one terminal thiolate ($C-S^-$) or thiocarbonyl (C=S) sulfur (*29*). The crude proposed model, shown in Figure 7, of copper simultaneously binding to OR protein residues and thiol/thiolate as well as thioether or selenoether groups in the series of odorants $MeX(CH_2)_nSH$ is analogous to the binding that occurs in blue copper proteins, such as azurin, where copper(II) is coordinated by one cysteine (Cys112) by a short bond (~2.1 Å) and by two histidines (His46 and His117) in a trigonal plane (Figure 8). A weak axial ligand, Met121, is present at ~2.9 Å approximately perpendicular to the plane, while the carbonyl oxygen of Gly45 functions as a second weakly coordinated axial ligand (*41, 42*). The Yale QM/MM studies of MOR244-3 indicate that the metal-binding site, lying in the middle of a long aqueous channel, consists of copper-II coordinated to H105, C109 (as thiolate) and N202 residues, which easily binds to both sulfur atoms of MTMT or its analogs (*40*). However, it is unknown whether the receptor binds copper to induce the subsequent ligand-binding event or the copper-ligand complex binds to the receptor. Future studies involving chemical and protein crystallization experiments may help to explore complex formation events and distinguish between the different mechanisms.

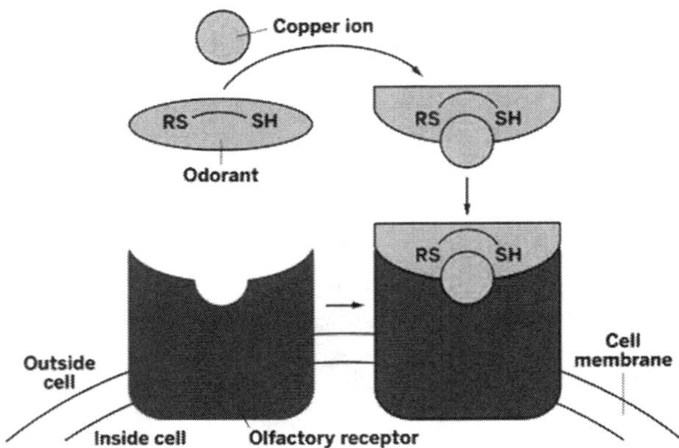

Figure 7. Schematic showing docking of copper-coordinated odorants with odorant receptor, for example, MOR244-3. The copper ion binds to the ligand followed by binding of the copper ion-odor complex to the OR. Adapted with permission from Chemical & Engineering News, 90(7), p. 9, February 13, 2012. Copyright 2012, American Chemical Society. (see color insert)

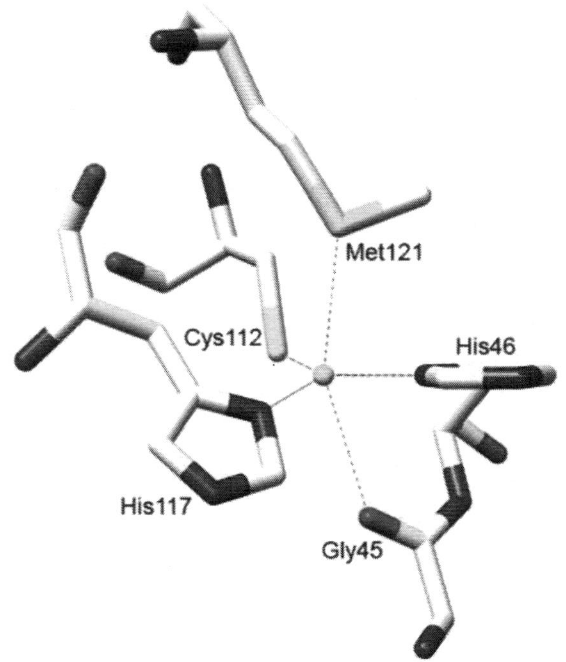

Figure 8. Type 1 copper site in Pseudomonas aeruginosa azurin. Copyright 2010, American Chemical Society (41). (see color insert)

In Vivo Studies of MOR244-3 in the Mouse Septal Organ

The mammalian nose contains several distinct chemosensory organs, including the main olfactory epithelium, the vomeronasal organ, and the septal organ. The septal organ is a small patch of olfactory neuroepithelium at the ventral base of the nasal septum found in many mammals that expresses olfactory receptors SR1, MOR244-3, and a few other ORs in high abundance (*43*). Perforated patch-clamp recordings were performed on dissected septal organs from SR1-IRES-tauGFP mice, and MTMT responses among the non-SR1 cells were examined. In fact, of 132 cells recorded, 76 cells (58%) responded to MTMT in a dose-dependent manner (*29*). This percentage is higher than that of MOR244-3 cells (27%) in the non-SR1 cells of the septal organ, suggesting that additional ORs in the septal organ are responsive to MTMT. *In vitro* screening was used to test this possibility and it was found that another OR, MOR256-17, showed robust responses to MTMT. Interestingly, the MTMT responses of MOR256-17 were not modulated by copper addition.

Mutational Studies of MOR244-3

Given the facts that amino acid residues histidine, cysteine, and methionine frequently coordinate copper in cuproenzymes and that the amino acids distant from each other in the primary structure may be closely interacting in actual spatial arrangement, a series of single-site mutants were constructed, changing all methionine residues to alanines; all histidines to arginines, lysines, tyrosines, leucines, valines, phenylalanines, asparagine, and/or alanines; and all cysteines to serines, valines, and/or phenylalanines in MOR244-3 in order to answer the question of how the respective mutations affect MOR244-3 activation by MTMT in the luciferase assay.

Some of the mutations did not significantly affect the Cu^{2+}-induced enhancement when stimulated by three concentrations of MTMT, excluding these sites as copper- and/or ligand-binding, whereas some of the sites reduced the response of MOR244-3 both with and without Cu^{2+}. The rest of the sites, including C97, H105, H155, C169, C179, and H243, abolished responses to MTMT completely when mutated, regardless of Cu^{2+} presence. Figure 9 shows a serpentine model of MOR244-3 color-coded for the methionine (M, green), histidine (H, red), and cysteine (C, yellow) residues that were subjected to mutagenesis analysis. Residues circled in blue are those that exhibited complete loss-of-function phenotypes in the luciferase assay. Of these, mutants C97S, H155R, C169S, C179S, and H243R (S = serine, R = arginine), but not H105K (K = lysine), have little or no cell-surface expression, suggesting that these first five mutants may have lost their functions as a result of defects in receptor folding/trafficking. Notably, the H105K mutant retains the ability to respond to some of MOR244-3's nonsulfurous ligands, such as cineole, and to ligands with no copper effect, such as dimethyl sulfide, indicating that the mutant receptor is intact. Presumably the H105K loss-of-function mutation disrupts copper/ligand binding, making H105–C109 the most likely location of the MTMT-copper binding active site. This possibility is the subject of a current computational study (*40*).

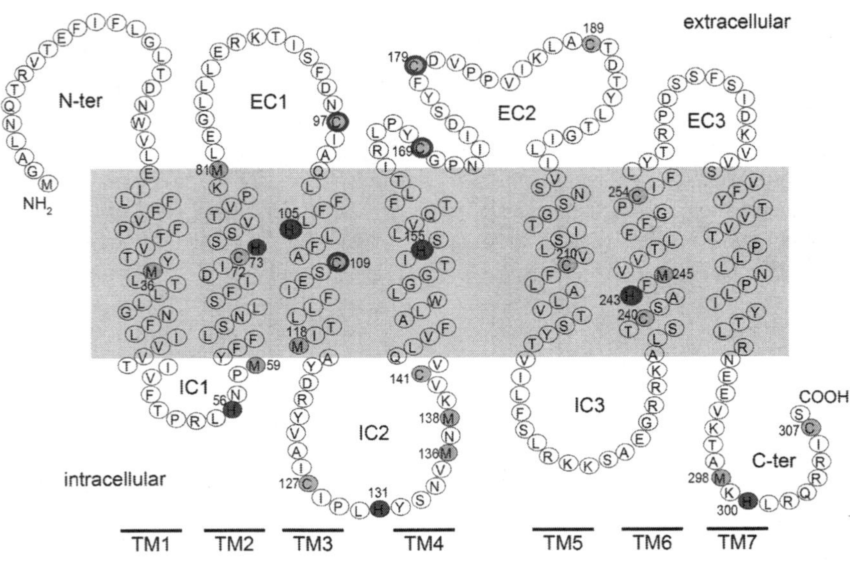

Figure 9. A serpentine model of the MOR244-3 receptor color-coded for the methionine (green), histidine (red), and cysteine (yellow) residues that were subjected to mutagenesis analysis. Residues circled in blue are those that exhibited complete loss-of-function phenotypes in the luciferase assay. Transmembrane domains, as predicted by the TMHMM server, are indicated by "TM". Adapted with permission from (29). Copyright 2012, PNAS. (see color insert)

Acknowledgments

This work was supported in part by NSF (CHE-0744578 & CHE-1265679), the University at Albany, State University of New York (all to EB), and by the National Basic Research Program of China (2012CB910400), National Natural Science Foundation of China Grants (30970981 & 31070972), the Shanghai Pujiang Program (09PJ1406900), the Program for Innovative Research Team of Shanghai Municipal Education Commission, the Chen Guang Project from Shanghai Municipal Education Commission and Shanghai Education Development Foundation (2009CG15), and the Program for Professor of Special Appointment (Eastern Scholar) at Shanghai Institutions of Higher Learning (J50201) (all to HZ).

References

1. Fischer, E.; Penzoldt, F. *Justus Liebigs Ann. Chem.* **1887**, *239*, 131–135.
2. Laska, M.; Bautista, R. M.; Hofelmann, D.; Sterlemann, V.; Salazar, L. T. *J. Exp. Biol.* **2007**, *210*, 4169–4178.
3. Natsch, A.; Schmid, J.; Flachsmann, F. *Chem. Biodiversity* **2004**, *1*, 1058–1072.
4. Kraft, P.; Mannschreck, A. *J. Chem. Educ.* **2010**, *87*, 598–603.
5. Demole, E.; Enggist, P.; Ohloff, G. *Helv. Chim. Acta* **1982**, *65*, 1785–1794.
6. Wood, W. F. *Chem. Educ.* **1999**, *4*, 44–50.
7. Callemien, D.; Dasnoy, S.; Collin, S. *J. Agric. Food Chem.* **2006**, *54*, 1409–1413.
8. Lin, D. Y.; Zhang, S.-Z.; Block, E.; Katz, L. C. *Nature* **2005**, *434*, 470–477.
9. Li, J.-X.; Schieberle, P.; Steinhaus, M. *J. Agric. Food Chem.* **2012**, *60*, 11253–11262.
10. Starkenmann, C.; Niclass, Y. *J. Agric. Food Chem.* **2011**, *59*, 3358–3365.
11. http://www.cpchem.com/bl/specchem/en-us/Pages/Odorants.aspx.
12. Block, E. *Reactions of Organosulfur Compounds*; Academic Press: New York, 1978.
13. Block, E. In *Sulfur in Pesticide Action and Metabolism*; Rosen, J. D., Magee, P. S., Casida, J. E., Eds.; ACS Symposium Series 158; American Chemical Society: Washington, DC, 1981; pp 1–16.
14. Laska, M.; Bautista, R. M.; Höfelmann, D.; Sterlemann, V.; Salazar, L. T. *J. Exp. Biol.* **2007**, *210*, 4169–4178.
15. Zarzo, M. *Sensors* **2011**, *11*, 3667–3686.
16. Gürbüz, O.; Rouseff, J.; Talcott, S. T.; Rouseff, R. *J. Agric. Food Chem.* **2013**, *61*, 532–539.
17. Siebert, T. E.; Solomon, M. R.; Pollnitz, A. P; Jeffery, D. W. *J. Agric. Food Chem.* **2010**, *58*, 9454–9462.
18. Moncrieff, R. W. *The Chemical Senses*; Leonard Hill: London, 1967; p 143.
19. Block, E. *Garlic and Other Alliums: The Lore and the Science*; Royal Society of Chemistry: Cambridge, UK, 2009 (hardback), 2010 (paperback).
20. Block, E. In *Volatile Sulfur Compounds in Food*; Qian, M. C., Fan, X., Manhattanatawee, K., Ed.; ACS Symposium Series 1068, American Chemical Society: Washington, DC, 2011; pp 35–63.
21. Block, E. *J. Sulfur Chem.* **2013**, *34*, 158–207.
22. Henkin, R. I.; Bradley, D. F. *Proc. Natl. Acad. Sci. U.S.A.* **1969**, *62*, 30–37.
23. Crabtree, R. H. *J. Inorg. Nucl. Chem.* **1978**, *40*, 1453.
24. Day, J. C. *J. Org. Chem.* **1978**, *43*, 3646–3649.
25. Nishide, K.; Miyamoto, T.; Kumar, K.; Ohsugi, S.; Node, M. *Tetrahedron Lett.* **2002**, *43*, 8569–8573.
26. Turin, L. *Chem. Senses* **1996**, *21*, 773–791.
27. Wang, J.; Luthey-Schulten, Z. A.; Suslick, K. S. *Proc. Natl. Acad. Sci. U.S.A.* **2003**, *100*, 3035–3039.
28. Zarzo, M. *Biol. Rev.* **2007**, *82*, 455–479.
29. Duan, X.; Block, E.; Li, Z.; Connelly, T.; Zhang, J.; Huang, Z.; Su, X.; Pan, Y.; Wu, L.; Chi, Q.; Thomas, S.; Zhang, S.; Ma, M.; Matsunami, H.;

Chen, G.-Q.; Zhuang, H. *Proc. Natl. Acad. Sci. U.S.A.* **2012**, *109*, 3492–3497.

30. Axel, R. *Angew. Chem., Int. Ed.* **2005**, *44*, 6110–6127.
31. Buck, L. *Angew. Chem., Int. Ed.* **2005**, *44*, 6128–6140.
32. Lefkowitz, R. J. *Angew., Chem. Int. Ed.* **2013**, *52*, 6366–6378.
33. Kobilka, B. *Angew. Chem., Int. Ed.* **2013**, *52*, 6380–6388.
34. Zarzo, M. *Sensors* **2012**, *12*, 4105–4112.
35. Carey, R. M.; Verhagen, J. V.; Wesson, D. W.; Pirez, N.; Wachowiak, M. *J. Neurophysiol.* **2009**, *101*, 1073–1088.
36. Wesson, D. W.; Verhagen, J. V.; Wachowiak, M. *J. Neurophysiol.* **2009**, *101*, 1089–1102.
37. Haga, K.; Kruse, A. C.; Asada, H.; Yurugi-Kobayashi, T.; Shiroishi, M.; Zhang, C.; Weis, W. I.; Okada, T.; Kobilka, B. K.; Haga, T.; Kobayashi, T. *Nature* **2012**, *482*, 547–551.
38. Desbenoit, N.; Galardon, E.; Frapart, Y.; Tomas, A.; Artaud, I. *Inorg. Chem.* **2010**, *49*, 8637–8644.
39. Zhuang, H.; Matsunami, H., unpublished data.
40. Sekharan, S.; Ertem, M. Z.; Zhang, R.; Wei, J. N.; Pan, Y.; Zhuang, H.; Matsunami, H.; Block, E.; Batista, V. S., manuscript submitted.
41. Clark, K. M.; Yu, Y.; Marshall, N. M.; Sieracki, N. A.; Nilges, M. J.; Blackburn, N. J.; van der Donk, W. A.; Lu, Y. *J. Am. Chem. Soc.* **2010**, *132*, 10093–10101.
42. Paraskevopoulos, K.; Sundararajan, M.; Surendran, R.; Hough, M. A.; Eady, R. R.; Hillier, I. H.; Hasnain, S. S. *Dalton. Trans.* **2006**, *7*, 3067–3076.
43. Tian, H.; Ma, M. *J. Neurosci.* **2004**, *24*, 8383–8390.

Chapter 2

Chemical Tools for Studying Biological Hydrogen Sulfide

Michael D. Pluth,* T. Spencer Bailey, Matthew D. Hammers, and Leticia A. Montoya

Department of Chemistry and Biochemistry, Institute of Molecular Biology, University of Oregon, Eugene, Oregon 97403-1253
*E-mail: pluth@uoregon.edu.

Although hydrogen sulfide (H_2S) has historically been recognized as a toxic gas, recent studies have established H_2S as an important, endogenously-produced signaling molecule active in a wide array of (patho)physiological roles in biological systems. As the multifaceted biological roles of H_2S continue to emerge, tools to modulate, measure, and detect biological H_2S are needed. Here we highlight recent biocompatible advances in reaction-based H_2S detection methods and their associated benefits and pitfalls.

Biological Relevance of Hydrogen Sulfide

The study of signal-transducing gasotransmitters has evolved over the past twenty years based on the discovery that gaseous molecules can be both biorelevant and be produced endogenously. After the discovery in 1987 that biosynthetic nitric oxide (NO) was the endothelium-derived relaxing factor (EDRF) (*1*), two other endogenous gases, namely carbon monoxide (CO) and hydrogen sulfide (H_2S) (*2, 3*), have garnered interest in the biomedical community. Hydrogen sulfide has emerged as the most recent biosynthetic gasotransmitter and is now accepted as an important signaling molecule with prominent (patho)physiological roles (*4–6*). Despite this interest, real-time detection methods compatible with biological systems are only beginning to emerge. As the field of H_2S detection progresses, development and implementation of new, sensitive, and robust H_2S detection and quantification methods are likely to greatly impact our current understanding of the basic science of biological H_2S as well as open avenues toward diagnostic techniques for different (patho)physiological conditions. Reaction-based methods of H_2S detection, which have emerged rapidly in the last two years, offer the first-generation solutions to this important problem. As the multifaceted biological roles of H_2S continue to emerge, the need for tools that modulate, measure, and detect its presence is paramount. To address these needs, chemists have devised a variety of H_2S delivery mechanisms using small molecule compounds that release H_2S at controlled rates, thereby mimicking enzymatic H_2S production more closely than adding exogenous sulfide sources such as H_2S, SH^-, or S^{2-} directly. Similarly, new biocompatible reaction-based chemical methods are emerging to detect and quantify H_2S. This review will focus on emerging strategies for H_2S detection, highlighting different sensing strategies as well as their associated benefits and pitfalls.

Hydrogen sulfide, much like NO and CO, meets the requirements of a gasotransmitter. It is a small, gaseous molecule that is produced enzymatically, and its production and metabolism are tightly regulated. Like NO, H_2S is readily oxidized, thus disfavoring long-range transport under normoxic conditions, and also suggesting the need for endogenous storage mechanisms, such as the formation of thiol hydropersulfides (RS-SH), which constitute a direct parallel to NO storage as nitrosothiols (RS-NO). Hydrogen sulfide is a weak acid (pK_{a1}: 6.76, pK_{a2}: 19.6) that exists primarily as SH^- (82%) rather than H_2S (18%) or S^{2-} (< 0.1%) under physiological conditions. This hydrosulfide anion is a more potent nucleophile than Cys or reduced glutathione (GSH) under physiological conditions due to the higher pK_a of these endogenous thiols by comparison to H_2S. Furthermore, the diprotic nature of H_2S allows for modulation between H_2S and SH^-, thus allowing for modulation of the water solubility, lipophilicity, and redox potential based on the local cellular environment.

Hydrogen-sulfide-generating enzymes produce the majority of H_2S in mammalian cells, although non-enzymatic H_2S production is also possible. Enzymes involved in transsulfuration pathways, such as cystathionine β-synthase (CBS) and cystathionine γ-lyase (CSE), are the main H_2S-producing enzymes. Additionally, 3-mercaptopyruvate sulfurtransferase (3-MST) has recently been identified as an H_2S-generating enzyme in the mitochondria (Figure 1). Production of H_2S from CBS primarily arises from conversion of L-cysteine (Cys) to L-serine with concomitant release of H_2S. Similarly, CSE can also convert L-cysteine to H_2S directly. Alternatively, CSE reacts with L-cystine to generate thiocysteine, which, upon further reaction with a thiol, generates H_2S. CBS and CSE can also work in concert; for example, CBS-mediated condensation of homocysteine (Hcy) and L-serine forms L-cystathionine, which is a substrate for subsequent CSE-mediated H_2S production. In addition to CSE and CBS, 3-MST converts 3-mercaptopyruvate, which is generated from L-cysteine by cysteine aminotransferase (CAT), to H_2S. Non-enzymatic H_2S production pathways include the conversion of thiosulfate to H_2S under reducing conditions, typically by GSH, with concomitant formation of sulfate and oxidation of the thiol reducing agent to the corresponding disulfide. Once generated, H_2S can react with a variety of organic and inorganic biological targets, including heme irons, thiols, and other reactive oxygen/nitrogen species. Although the basic H_2S-producing pathways are known, the exact intercellular interplay between these enzymes, as well as crosstalk with other signaling molecules, remains an emerging arena.

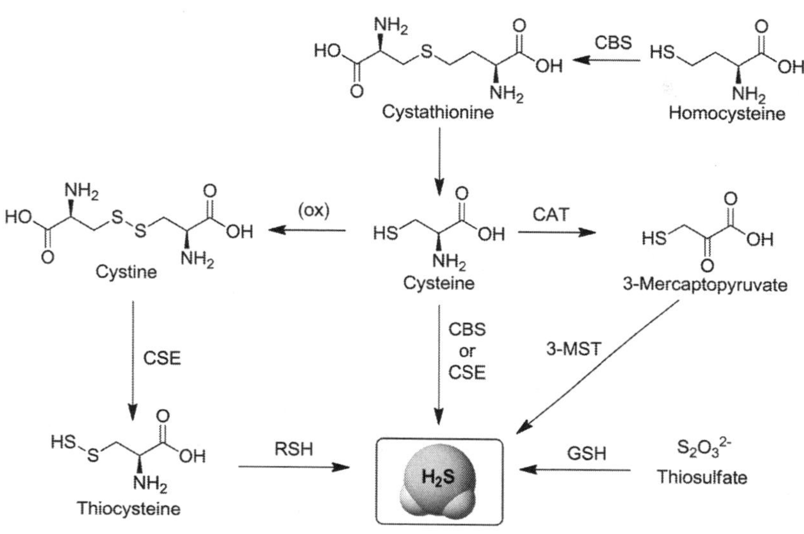

Figure 1. Biosynthetic pathways for H_2S formation in mammalian cells. CBS: cystathionine β-synthase; CSE: cystathionine γ-lyase; 3-MST: 3-mercaptopyruvate sulfurtransferase; CAT: cysteine aminotransferase.

In addition to H_2S generation, storage of biological H_2S is an important, yet still poorly understood, aspect of H_2S homeostasis. Drawing parallels to other important bioinorganic analytes that exist in both free and bound pools, such as NO and Zn(II) (*7*), different H_2S storage mechanisms likely play important roles in releasing H_2S under different physiological conditions. For example, iron-sulfur clusters can be a source of acid-labile H_2S, although release of H_2S is only efficient under acidic conditions. Although these conditions are significantly removed from normal physiological pH, such acidities are accessible in different cellular locales, such as lysosomes. A likely more important pool of stored biological H_2S is sulfane-bound sulfur, resulting from reaction of H_2S with a thiol under oxidizing conditions (*8, 9*). Such sulfane-bound sulfur sources include hydropersulfides (RS-SH) and polysulfides (RS-$S_{(n>1)}$-R), which release H_2S under reducing conditions or after transulfurization reactions with other reduced thiols. A variety of stored sulfur pools are likely required for ensuring H_2S homeostasis, but the release of biologically-stored sulfur from these sources complicates H_2S detection. For example, many classical methods of H_2S measurement require sample acidification or disruption of the pre-established redox balance prior to analysis (*vide infra*).

Emerging biological and physiological roles of H_2S clearly establish that H_2S plays important and multifaceted roles in the cardiovascular, nervous, endocrine, and immune systems (6). One major challenge in establishing and understanding such roles is that the physiological response to H_2S is often dependent upon the method of H_2S administration or modulation. Despite these complications, the role of H_2S has been established in numerous biological processes. For example, H_2S functions as a vasorelaxant in the cardiovascular system with EC_{50} levels for induced vasorelaxation that correlate well with measured plasma H_2S levels (10), although the detection limit of the H_2S measurement method used in these studies has subsequently been revised (11). Additionally, high expression levels of CBS in the hippocampus and cerebellum, as well as the interaction with *N*-methyl-D-aspartate (NMDA) receptors, suggest important roles for H_2S in the central nervous system (CNS) in the modulation of neurotransmission and long term potentiation (LTP) (12, 13). Furthermore, the existence of H_2S has been implicated in the endocrine system by influencing glucose metabolism homeostasis in islets through action on K_{ATP} channels in beta cells (14). Hydrogen sulfide plays important function in the immune system, displaying both pro- and anti-inflammatory effects depending on the mode and concentration of H_2S administration (15–18). Taken together, H_2S clearly plays diverse and important roles in various physiological systems. Although a complete description of the diverse biological roles of H_2S is beyond the scope of this review, the interested reader is referred to a recent, comprehensive summary of H_2S in biology (6). As the field continues to grow, revision and refinement of many of the current paradigms is likely as better tools for selectively delivering and measuring biological H_2S levels continue to emerge.

H₂S Detection Strategies

Classical instrumental methods of H_2S quantification include gas chromatography (GC), polarography, and sulfide-selective electrodes (19–22). For these techniques, samples are typically homogenized, and either the resultant solution or the gaseous headspace is analyzed. Polarographic and GC methods detect H_2S gas released from the solution and therefore require accurate pH measurements to correct for H_2S speciation under physiological conditions. Based on what is now known about acid-labile endogenous sulfur pools, such sample acidification may result in releasing bound sulfur, thereby resulting in total, rather than free, sulfide measurements. Most sulfide-selective electrodes also require sample homogenization followed by treatment of a sulfide antioxidant buffer containing sodium salicylate, ascorbic acid, and sodium hydroxide. Because the electrodes only measure S^{2-}, the least prevalent species in the H_2S acid-base equilibria, sulfide-sensitive electrodes are quite sensitive to small changes in sample pH. Additionally, commonly-used additives contain redox-active species, thereby increasing the possibility of perturbing the redox homeostasis of the sample and allowing for either release of or additional storage by sulfane-bound sulfur. Driven by the drawbacks of the instrumental methods of H_2S detection and quantification, biocompatible chemical analyses for H_2S are beginning to emerge.

One of the most commonly-used chemical methods for H_2S quantification is the methylene blue assay (Scheme 1) (*23, 24*). In this assay, H_2S from the desired sample is typically trapped initially with $Zn(OAc)_2$ to form ZnS. Sample acidification releases the trapped H_2S, and heating with *N,N*-dimethyl-*p*-phenylenediamine (**1**) and addition of $FeCl_3$ generates the methylene blue dye (**2**). After removal of precipitated proteins by centrifugation, the characteristic methylene blue absorbance at 670 nm is then measured and compared to a background sample and calibration curve. Despite the wide use of this quantification method, the revised detection limit (2 μM) is much less sensitive than the initially indicated detection limit (~10 nM) (*11*). Furthermore, although the methylene blue method has been used to detect and measure endogenously-produced H_2S, recent studies have demonstrated its inability to detect differential H_2S levels in mice deficient in H_2S-producing enzymes. Taking these limitations into account, many of the measured levels of H_2S may soon come under increased scrutiny as new, improved methods for H_2S measurement are developed.

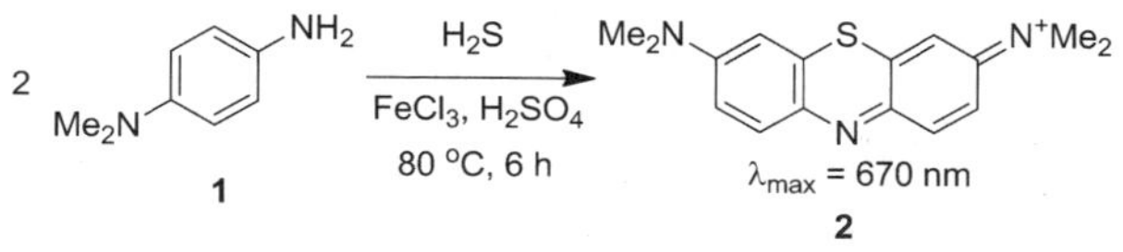

Scheme 1. Formation of methylene blue (2) to trap H_2S

Common features of the H_2S detection and quantification methods described above include the requirement of sample homogenization and/or acidification prior to analysis. Such pre-treatments, as well as the time required to perform them, complicates the actual H_2S levels reported. Because H_2S is tightly regulated, lengthy or caustic workups, which often require pH changes, or incubation with metal salts, likely remove sulfur from other endogenous stores such as persulfides or iron sulfur clusters, thereby complicating detection and quantification. Such considerations highlight the highly controversial reported levels of H_2S in different tissues or cellular locales, partially because it is unclear whether free or total sulfide levels are being measured (*25*). To address the need for contemporary H_2S detection methods compatible with living tissues, numerous research groups, including our group, are investigating new methods of H_2S detection and quantification by utilizing the innate chemical reactivity of H_2S.

Although H_2S is often grouped with NO and CO as a gasotransmitter, its chemical reactivity is much different than either of these important gasses. Hydrogen sulfide is a weak reductant and is a weaker reducing agent than Cys or GSH (*26*). Similarly, H_2S, particularly SH^-, is a potent nucleophile able to react with electrophilic targets. Notably, the solubility of many metal-sulfur compounds is quite low, resulting in the use metal-sulfide precipitation as a classical gravimetric method for H_2S-mediated metal determination. These three properties have been exploited by chemists in the last few years to generate new detection methods for H_2S based on utilization of its physical and reactive properties. Consequently, reaction-based methods of H_2S detection have focused on exploiting the physical properties of H_2S to result in chemical sensing. These strategies can be broken down into three categories: (1) H_2S as a reductant, (2) H_2S as a nucleophile, and (3) H_2S as a metal precipitant. Recent advances using these strategies for H_2S detection are highlighted below.

H_2S Detection Methods Based on Chemical Reduction

Thiols play an important role in cellular redox chemistry, with GSH playing a central role in maintaining cellular redox homeostasis. Drawing parallels to biologically-relevant thiols, H_2S is also a reducing agent, although its reduction potential is lower than GSH or Cys (*26*). Although the reduction of azide and nitro groups by H_2S has been known for over 30 years (*27–30*), it has only recently been applied to reaction-based chemical methods for H_2S detection (Scheme 2). Empirically, even though H_2S is a weaker reductant than GSH, it reduces azides and other oxidized nitrogen species faster than GSH or other thiols, resulting in a viable H_2S detection method. This reduction-based strategy has emerged as a general method for H_2S detection, with various fluorescent probes being reported in the last two years. The selectivity of these probes for H_2S over GSH is typically moderate to good, although variable conditions and concentrations under which the different probes were evaluated make direct comparison difficult. For example, although cellular levels of GSH are typically in the millimolar concentration range, many of the reported azide-based H_2S probes only test micromolar levels of GSH to compare with H_2S, thereby limiting the relevance to typical biological conditions. Despite these limitations, a palette of different excitation and emission wavelengths are possible with currently developed probes, ranging from standard blue, green, yellow, and red channels (Figure 2). One important note on reduction-based probes is that the reaction product from H_2S- and thiol-mediated reduction is identical, thus precluding accurate quantification of H_2S unless exact thiol concentrations are known for the sample of interest. Furthermore, because the reaction rates may depend on pH, ionic strength, or local protein environment, drawing definitive quantitative conclusions remains difficult. Although many of the developed probes demonstrate a highly-linear response to H_2S, caution should be taken when using these probes to quantify H_2S unless adequate control experiments are performed. Recent advances in reduction-based probes for H_2S will be outlined below, although not all probe features are discussed in depth.

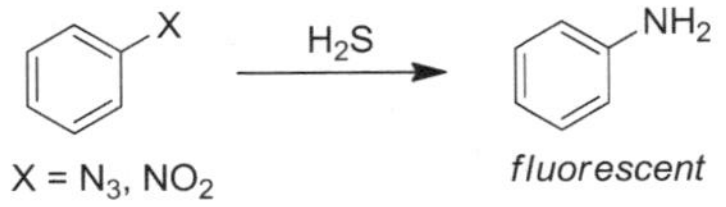

Scheme 2. Reduction of an oxidized amine with H_2S regenerates the parent amine. Translation of this detection mechanism to fluorophores with fluorogenic amines provides a viable strategy for H_2S detection.

A flurry of fluorescent probes using H_2S-mediated azide reduction as the turn-on mechanism were reported in late 2011 and early 2012 by a variety of research groups. For example, Chang and co-workers initially reported functionalized rhodamine azides SF1 (**3**) and SF2 (**4**) in which reduction of the azide to the parent amine resulted in regeneration of the rhodamine platform and concomitant fluorescence turn-on (*31*). When tested in buffer, SF1 and SF2 demonstrated a 6-8 fold turn-on with H_2S after 60 min with 2-5 fold selectivity over other reactive sulfur, oxygen, and nitrogen species (RSONS). Furthermore, both SF1 and SF2 were used to detect exogenous H_2S in HEK293T cells (*31*). A diazido rhodamine-based H_2S sensing platform was later reported by Chang and co-workers to generate SF4, and SF5-AM to SF7-AM (**5-8**), which feature different ester functionalization to impart cell trappability (*32*). These later probes offer higher fluorescence turn-on than that of SF1 and SF2; for example SF7-AM results in a 20-fold fluorescence turn-on after 60 min with 8-10 fold selectivity over other RSONS. SF-7-AM was used to detect endogenous H_2S produced in human umbilical vein endothelial cells treated with vascular endothelium growth factor (VEGF), establishing a role for cellular crosstalk between H_2S and H_2O_2 (*32*).

Azide reduction has also been used for ratiometric H_2S detection as reported by Han and co-workers (*33*). Using the infrared chromophore cyanine to generate Cy-N_3 (**9**), which upon reduction to the parent Cy-NH_2 amine by H_2S generates new absorption and emission bands centered at 660 nm and 750 nm, respectively. Ratiometric H_2S detection based on the fluorescent signals of the Cy-N_3 and Cy-NH_2 signals (F_{750}/F_{710}) resulted in a ratio of 2.0 for H_2S and signals of less than 0.7 for other RSONS. The response of Cy-N_3 probe for both H_2S and ADT-OH, a common organic-based H_2S donor, was also demonstrated. Additionally, the Cy-N_3 probe was used to image exogenous H_2S in Raw 264.7 cells using ratiometric imaging (*33*).

Our group reported the naphthalimide-based azide HSN-2 (**11**) that reacts with H_2S to afford the parent naphthalimide amine (*34*). This sensing scaffold benefits from a large Stokes shift of the naphthalimide dye and a large fluorescence turn-on. Reaction of the azide-based HSN-2 with H_2S results in a 60-fold turn-on while maintaining high selectivity at large (2,000-fold, 10 mM) excesses of Cys and GSH. In addition to azide reduction, H_2S-reduction of NO_2 groups was demonstrated for H_2S detection in HSN-1 (**10**). Although HSN-1 is selective for H_2S over equimolar RSONS, the selectivity eroded when challenged with high (10 mM) GSH concentrations (*34*). Additionally, H_2S-mediated reduction of the azide is faster than reduction of the nitro group. Nevertheless, both HSN-1 and HSN-2 were demonstrated to detect exogenous H_2S in live HeLa cells.

Use of naphthalimide-based azide and nitro compounds for H_2S detection has been further elaborated by the Zhang and Wu groups, which have reported micelle-based and carbon-dot-based probes, respectively. By appending hydrophobic groups to the diimide of the naphthalimide azide scaffold, the highly-hydrophobic probes could be encapsulated in CTAB micelles, which were used to detect H_2S in buffer and in fetal bovine serum (FBS) (*35*). Alternatively, appending a naphthalimide azide to carbon dots allows for coupling of the naphthalimide and carbon dot absorptions/emission profiles, resulting in ratiometric H_2S detection (*36*). The resultant scaffold was used to detect exogenous H_2S in HeLa and L929 cells. Additionally, Wang and co-workers reported a naphthalimide hydroxylamine for H_2S sensing, since hydroxylamines are proposed intermediates in NO_2 group reduction (*37*). The hydroxylamine naphthalimide scaffold resulted in a ~9-fold turn-on with H_2S, a 3-fold selectivity over other reactive biological species including Cys and GSH, and was used to image exogenous H_2S in astrocyte cells.

Although the majority of azides used for H_2S detection have been aryl azides, Wang and co-workers reported the use of dansyl azide (Ds-N_3, **12**), a sulfonyl azide, for H_2S detection (*38*). The dansyl azide reacts quickly with H_2S to afford the fluorescent dansyl amine product. This probe is selective for SH^- over other anions although only moderate selectivity over low concentrations of thiols such as thiophenol, benzylthiol, or Cys is reported. The Ds-N_3 probe showed linear H_2S detection both in Tween buffer and in FBS. Furthermore, Ds-N_3 was also used to detect and quantify H_2S in mouse blood, providing concentrations (32 ± 9 μM) that agreed well with previously-determined concentrations (*38*).

Coumarin-derived azides for H_2S detection have been reported by both the Tang and Li groups. Tang reported the use of both 6- and 7-azido coumarin for H_2S detection; however, only the 7-isomer (C7-Az, **13**) is reactive toward H_2S (*39*). Subsequent DFT calculations demonstrated that the orbital density on the azide is much lower for the non-reactive 6-azidocoumarin than for the 7-isomer, thereby suggesting that differences in electronic structure may be responsible for the different reactivity. The 7-azido coumarin reacts selectively with H_2S over other anions and reactive biological species and displays a linear response to H_2S both in buffer and in FBS (*39*). The resultant probe was used to visualize exogenous H_2S in HeLa cells using two-photon laser scanning confocal imaging. Similarly, Li reported ethylamino coumarin azide (**14**) for H_2S sensing (*40*). The resultant probe demonstrated ~40-fold turn-on with H_2S but only 2-fold selectivity for H_2S over GSH. Despite this low selectivity, the resultant probe was used to detect H_2S in rabbit plasma as well as exogenous H_2S in PC-3 cells.

Other azide/nitro group reduction chromophores have also been used for H_2S detection, many of which address specific applications for common fluorescent probes. For example, Ai and co-workers reported the genetically-encoded azide in cpGFP-Tyr66*p*AzF (**15**) to allow for direct incorporation of the azide-reduction method in biomolecules (*41*). Reaction of the resultant azide with NaSH results in a 0.6-fold turn-on with H_2S and was used to detect exogenous H_2S in HeLa cells. For applications in which high water solubility is important, Hartman and co-workers prepared 8-azidopyrene-1,3,6-trisulfonic acid (N_3-PTS, **16**) which, due to the anionic sulfonate groups, maintains water solubility of ~100 mM (*42*). The trianionic probe reacts with NaSH to afford the parent amine, resulting in a linear fluorescent response to H_2S in buffer and FBS as well as moderate selectivity over other RSONS. Aliphatic azides have received significantly less attention for H_2S sensing than aryl azides, although Han and co-workers reported the use of the *o*-azidomethylbenzoyl group appended to coumarin in AzMB-Coumarin (**17**) (*43*). When treated with equimolar thiols or H_2S, AzMB-Coumarin resulted in 6-fold selectivity for H_2S over other thiols. Although this probe was demonstrated to detect exogenous H_2S in HeLa cells, very high probe (1 mM) and NaSH (1 mM) concentrations were required, thus limiting the potential biological applications of this method. Additionally, H_2S probes based on phenanthroimidazole (**18**) (*44*), resorufamine (**19**) (*45*), nitrobenzofurazan (**20**) (*46*), benzathiazole fluorine (**21**) (*47*), cresyl violet (**22**) (*48*), as well as constructs based on cleavage of dinitrophenyl esters (*49*) or modulation of nitrophenyl fluorescence quenching groups (*50*), have also been reported.

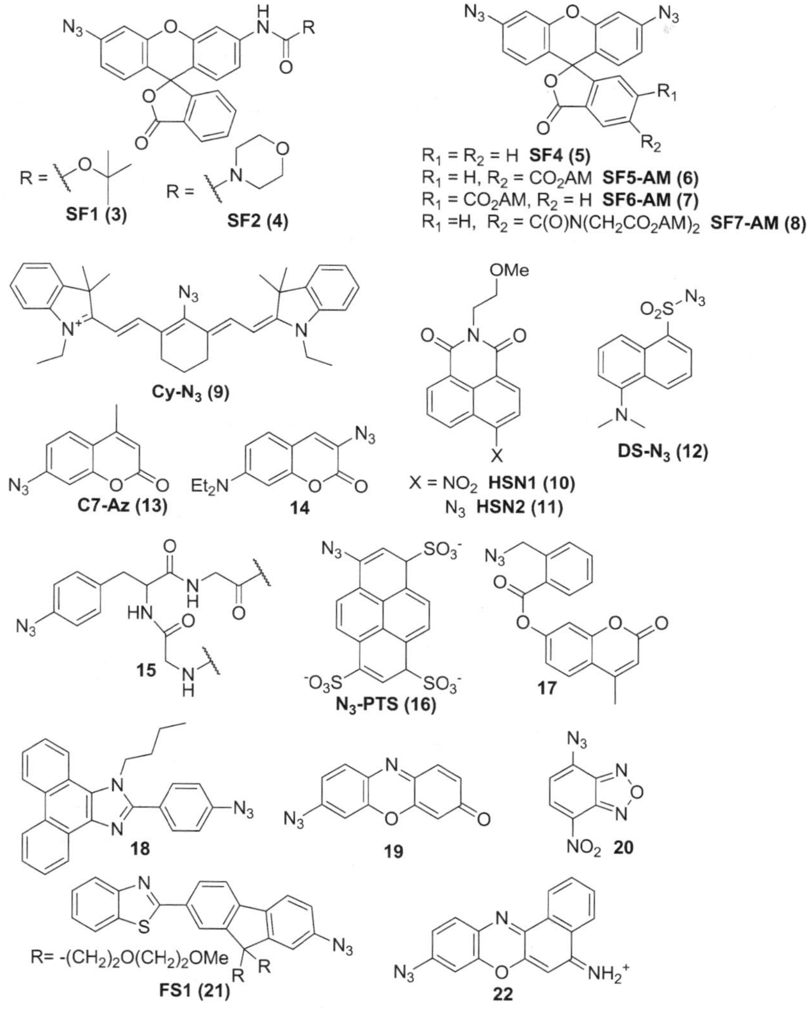

Figure 2. Highlighted reduction-based methods of H_2S detection.

Taken together, reduction-based H_2S detection methods offer a rapidly-evolving class of H_2S detection methods amenable to a wide array of fluorophores. Based on the available probes, many absorption/emission wavelengths are accessible with many probe alternatives available. One major challenge in comparing the potential biological efficacy of currently-available probes is that most of the probes have been investigated under significantly different conditions, including probe and analyte concentration as well as solvent/buffer composition. Similarly, because H_2S is well known to undergo redox chemistry with molecular oxygen, different procedures in handling the probe and H_2S under an ambient or inert atmosphere may influence the observed response. Although most current probes display moderate to good selectivity for H_2S over other reactive species, the fact that the final product after reaction with H_2S is identical to that generated from unwanted side reactions with thiols makes quantification difficult unless the exact thiol concentrations are known in the sample of interest. Despite these current limitations, azide/nitro-based H_2S detection methods have emerged as a viable method for observation of biological H_2S and will likely serve as a robust platform for future probe development and refinement.

H_2S Detection Methods Based on Nucleophilic Attack

In addition to its redox chemistry, H_2S is also a potent nucleophile. Hydrogen sulfide primarily exists as SH^- at physiological pH, whereas thiols are primarily in the neutral RSH form. This difference in protonation state results in increased nucleophilicity of H_2S by comparison to thiols under typical physiological conditions. The nucleophilic nature of H_2S, as well as its diprotic nature and ability to undergo two sequential nucleophilic attacks, has been exploited to both detect and quantify H_2S. For example, upon reaction with two equivalents of monobromobimane (**23**), H_2S can be trapped as the fluorescent thioether product (**24**) (*11, 51*) (Scheme 3). For this system, separation by HPLC is required to determine H_2S content because both the H_2S-derived thioether product, as well as unwanted side reactions of monobromobimane with thiols to form bimane-SR, produce fluorescent products. Although the monobromobimane method cannot be used to detect H_2S in real-time, it has emerged as one of the preferred methods for H_2S quantification with detection limits as low as 2 nM (*11*). Recent advances in the use of H_2S as a nucleophile for H_2S detection have focused on attack by H_2S on an activated electrophile to generate a thiol followed by a second intramolecular nucleophilic attack on a second electrophile to generate a fluorescent response (Figure 3).

Scheme 3. Monobromobimane can trap H_2S to form fluorescent bimane thioether (24), allowing for H_2S quantification by HPLC

Building on the doubly-nucleophilic nature of H_2S, He and co-workers reported H_2S fluorescent probes SFP-1 (**25**) and SFP-2 (**26**), in which initial attack of H_2S on an aldehyde results in formation of a thiol that undergoes a second attack on an α-β-unsaturated olefin (*52*). Upon addition of H$_2$S to the olefin, the fluorescence quenching mechanism is abrogated, thereby resulting in fluorescence turn-on. This detection strategy was used in SFP-1 to generate a 16-fold fluorescence enhancement upon H$_2$S addition with a 3-4 fold selectivity over Cys and GSH. Translation of the same sensing strategy to a BODIPY scaffold generated SFP-2, which was used to detect H$_2$S generated from purified CBS as well as exogenous H$_2$S in live HeLa cells (*52*). Although the rate of fluorescence turn-on was slow and required 3-4 hours for complete reaction, increasing the electrophilicity of the olefin in second generation scaffolds resulted in a more active probe. The more highly-reactive SFP-3 (**27**) exhibited complete reaction with H$_2$S within 30 minutes, thereby greatly enhancing the temporal resolution of the sensing platform (*53*). This probe was further used to detect and measure H$_2$S levels in blood plasma and brain tissue of mice.

Similarly, electrophilic olefins have been used for H$_2$S detection. For example, Xian and co-workers reported two probes for H$_2$S detection exploiting acrylates substituted with either nitrile or ester groups to enhance their electrophilicity (**28**) (*54*). These probes rely on initial addition of H$_2$S to the activated olefin, followed by intramolecular attack on an ester linkage to the fluorophore. This strategy allows for fluorophore liberation after the double nucleophilic attack by H$_2$S and a fluorescence turn on-of ~160-fold. Additionally, these Michael acceptor olefins can react with thiol nucleophiles but may do so reversibly, thereby enhancing the selectivity of these probes for H$_2$S over other biological nucleophiles (*54*).

Electrophilic disulfides have also been used to generate H$_2$S-selective fluorescent probes. By appending such a disulfide on a fluorophore protecting group, Xian and co-workers reported the generation of an esterified methylfluorescein platform (**29**) which, upon reaction with H$_2$S, generated a nucleophilic bound hydropersulfide that cleaves the ester bond ligating the protecting group to the fluorophore (*55*). This strategy was selective for H$_2$S over equimolar Cys or GSH and was used to detect exogenous H$_2$S in COS7 cells. A similar disulfide exchange reaction was later used by Qian and co-workers by appending a pyridyl disulfide onto a benzathiazole platform to generate the H$_2$S probe E1 (**30**) (*56*). The resultant platform reacted quickly with H$_2$S, was selective for H$_2$S over Cys, Hcy, and GSH, and was used to detect exogenous H$_2$S in HeLa cells.

In addition to scaffolds exploiting the doubly-nucleophilic nature of H$_2$S highlighted above, the high nucleophilicity of H$_2$S has also been utilized to disrupt the π-system of a conjugated fluorophore, resulting in modulation of the fluorescent properties of the dye. For example, Guo and He reported the development of CouMC (**31**), which contains a diethylaminocoumarin fluorophore appended to an indolenium group (*57*). Nucleophilic attack by H$_2$S on the electrophilic carbon of the indolenium ring breaks the conjugation of the system and changes the emission characteristics of the probe. This chemistry results in a ratiometric response to H$_2$S with high selectivity for H$_2$S over other RSONS. The reaction of CouMC with H$_2$S occurs within minutes, and ratiometric imaging of exogenous H$_2$S in MCF-7 cells was demonstrated (*57*). Similarly, Guo and co-workers synthesized a flavylium derivative (**32**), in which attack of H$_2$S on the electrophilic benzopyrylium moiety disrupts conjugation, thereby leading to a decrease in fluorescence in the red channel and increase in the green channel (*58*). This shift in fluorescence was used to detect exogenous H_2S in HeLa cells.

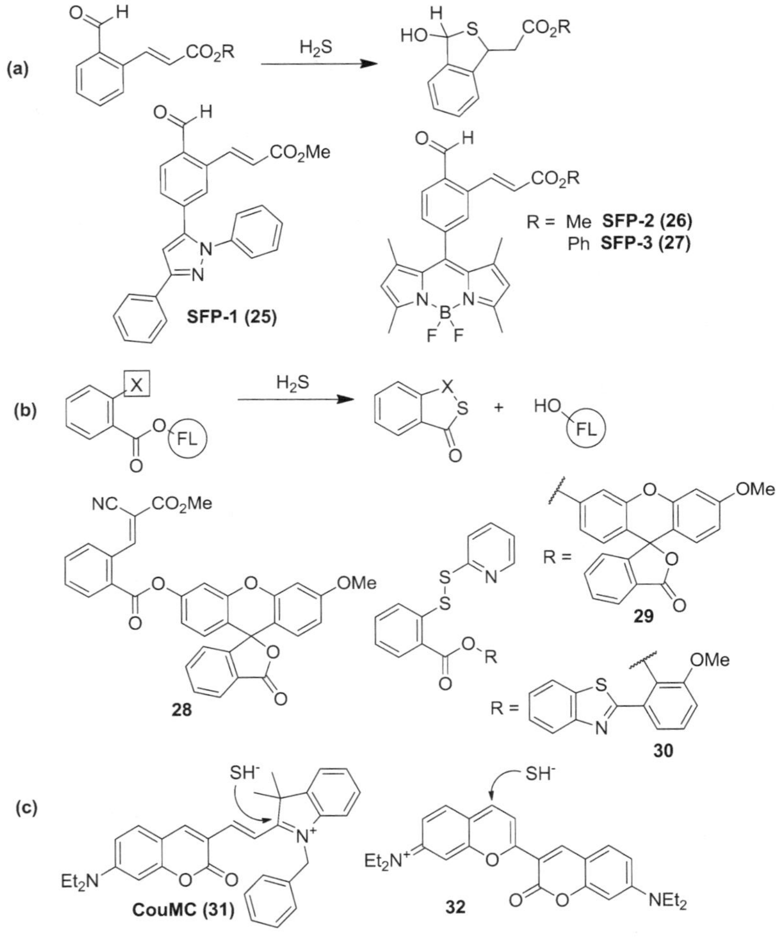

Figure 3. Selected method of H_2S detection based on nucleophilic attack. (a) Addition to an aldehyde followed by intramolecular attack on an olefin; (b) Attack on an activated electrophile followed by attack on an ester-bound fluorophore; (c) Attack on an electrophilic center to disrupt fluorophore conjugation.

Nucleophilic attack by SH$^-$ is an attractive strategy for H_2S detection that utilizes the ability of H_2S to participate in two sequential nucleophilic attacks, thus providing a clear pathway to differentiate H_2S from endogenous thiols. One major challenge, however, is ensuring that the developed constructs are not inactivated by thiols. Although attack by a thiol typically does not result in a fluorescence response, it often deactivates the probe and prevents future reaction with H_2S. Emerging strategies based on electrophiles able to react reversibly with thiols offer an attractive platform on which this problem can be addressed, but fine-tuning of the reactivity of such compounds to react reversibly with thiols and quickly with H_2S remains challenging (*59*).

H_2S Detection Methods Based on Metal Precipitation

A third strategy for reaction-based H_2S detection relies on modification of the classic gravimetric method for Cu(II) detection using H_2S to precipitate solid CuS (Scheme 4). By appending a Cu(II)-binding ligand to a fluorophore, the proximity of the unpaired electron from Cu(II) can potentially quench fluorescence from the fluorophore due to photoinduced electron transfer (PET). Such Cu(II)-mediated fluorescence quenching has been used to generate a variety of nitric oxide and nitroxyl-detecting probes (*60–63*). By judicial choice of Cu(II)-binding scaffold, H_2S can precipitate CuS without unwanted reduction of Cu(II) to Cu(I). Because Cu(II) and Cu(I) have different preferred coordination geometries, reduction to Cu(I) would likely eject the metal and result in fluorescence turn-on from the probe. Minimizing such unwanted reduction from thiols is a key requirement in maintaining high selectivity for H_2S over other biologically relevant thiols. To date, the H_2S-mediated CuS precipitation coupled with fluorescence appears to be predominantly empirical, with small modifications of the ligand resulting in large changes in thiol versus H_2S selectivity (Figure 4).

$$R-N-Cu^{2+} \xrightarrow{\ H_2S\ } R-N \quad + \quad CuS$$

Scheme 4. Treatment of a Cu(II)-bound ligand with H_2S can result in precipitation of the insoluble CuS thus releasing the fluorophore

The initial example of H_2S detection using the CuS precipitation strategy was reported by Chang and co-workers using a dipicolylamine-appended fluorescein derivative (**33**) (*64*). Complexation of the fluorescein construct with Cu(II) resulted in fluorescent quenching. Upon treatment with S^{2-}, and concomitant precipitation of CuS, the fluorescence of the fluorescein platform was restored. This method for sulfide detection was selective for H_2S over other anions but was not selective over biologically-relevant thiols such as GSH.

Nagano and co-workers expanded on this sulfide detection manifold by preparing a cyclen-containing fluorescein derivative (HSip-1, **34**) to bind Cu(II) (*65*). This scaffold allows for the fast detection of H_2S, thus generating a fluorescence turn-on. Additionally, the fluorescence response of HSip-1 was selective for H_2S over RSONS including potential reducing agents. In addition to detecting H_2S from NaSH and a Cys-activated H_2S donor (*66*), HSip-1 was also used to detect exogenous H_2S in HeLa cells.

Similar platforms utilizing different Cu(II) binding ligands have also been used for H_2S detection. For example, Ramesh, Das, and co-workers reported the use of a FRET-based indole-ligated rhodamine (**35**) for H_2S detection. This scaffold resulted in a turn-on NIR fluorescence response upon treatment with H_2S and was selective for sulfide over other anions or transition metals, but the probe was not tested with thiols (*67*). Similarly, Bai, Zeng, and co-workers reported a 8-hydroxyquinoline-appended fluorescein derivative (**36**) able to bind Cu(II) (*68*). Treatment with H_2S resulted in a 5-fold fluorescence turn-on and the probe maintained selectivity over other anions. Again, this complex was not tested with other thiol-containing reactive species. Both **35** and **36** were used to detect exogenous H_2S in HeLa cells. The same general CuS precipitation strategy has been used to generate colorimetric H_2S detection methods based on a Cu(II)-chelated azo dye (**37**) (*69*) and a quinoline-dipicolyl amine BODIPY platform (**38**) (*70*).

The method of CuS precipitation has generated a highly-modifiable strategy for H_2S detection. This method does have practical limitations, however, because the usable probe concentration ranges depend on the binding constant of Cu(II) to the ligand scaffold. Additionally, high probe concentrations are disfavored due to the stoichiometric precipitation of solid CuS. To demonstrate biocompatibility for future probes based on CuS precipitation, such scaffolds need to be tested with both biologically relevant thiols and reductants to ensure that reduction of Cu(II) to Cu(I) does not compete with CuS precipitation. Similarly, such scaffolds must be tested with other biologically-relevant metal ions such as Zn(II) to ensure that metal exchange is not occurring under physiological conditions.

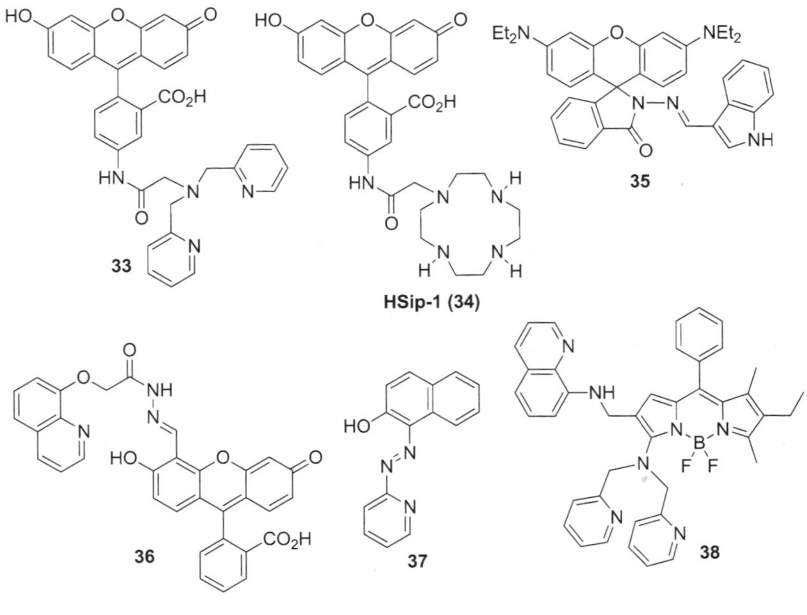

Figure 4. Highlighted metal precipitation-base methods of H₂S detection.

Conclusion

The field of H$_2$S sensing has grown rapidly in the last two years based on three general detection strategies: reduction of azide or nitro groups, nucleophilic attack on activated electrophiles, and precipitation of CuS from Cu(II)-ligated fluorophores. These detection strategies have provided an array of tools available to scientists interested in understanding the multifaceted roles of H$_2$S in biology. In addition to the above highlighted examples, other reaction-based detection methods have emerged, typically revolving around the reaction of H$_2$S with various metals or metals housed in biomimetic scaffolds (*71–74*). One of the major challenges in characterizing new probes remains drawing definitive conclusions about reaction rates or absolute selectivity over other reactive biological species because of the disparate conditions in which current probes have been tested. One clear conclusion from currently-available probes is that each of the developed strategies offers its own advantages and disadvantages, suggesting that specific properties of individual probes are likely to dictate the suitability of that platform to answer a specific biological question.

Acknowledgments

Work in our lab focusing on biological H$_2$S detection and understanding the biological roles of H$_2$S is supported by the National Institute of General Medical Sciences (R00 GM092970) and the Oregon Medical Research Foundation.

References

1. Palmer, R. M.; Ferrige, A. G.; Moncada, S. *Nature* **1987**, *237*, 524–526.
2. Benavides, G. A.; Squadrito, G. L.; Mills, R. W.; Patel, H. D.; Isbell, T. S.; Patel, R. P.; Darley-Usmar, V. M.; Doeller, J. E.; Kraus, D. W. *Proc. Natl. Acad. Sci. U.S.A.* **2007**, *104*, 17977–17982.
3. Qu, K.; Lee, S. W.; Bian, J. S.; Low, C. M.; Wong, P. T. H. *Neurochem. Int.* **2008**, *52*, 155–165.
4. Pryor, W. A.; Houk, K. N.; Foote, C. S.; Fukuto, J. M.; Ignarro, L. J.; Squadrito, G. L.; Davies, K. J. A. *Am. J. Physiol-Reg I* **2006**, *291*, R491–R511.
5. Olson, K. R. *Antioxid. Redox Signaling* **2012**, *17*, 32–44.
6. Wang, R. *Physiol. Rev.* **2012**, *92*, 791–896.
7. Pluth, M. D.; Tomat, E.; Lippard, S. J. *Annu. Rev. Biochem.* 2011, 80, 333–355.
8. Ishigami, M.; Hiraki, K.; Umemura, K.; Ogasawara, Y.; Ishii, K.; Kimura, H. *Antioxid. Redox Signaling* **2009**, *11*, 205–214.
9. Toohey, J. I. *Biochem. J.* **1989**, *264*, 625–632.
10. Hosoki, R.; Matsuki, N.; Kimura, H. *Biochem. Biophys. Res. Commun.* **1997**, *237*, 527–531.
11. Shen, X.; Pattillo, C. B.; Pardue, S.; Bir, S. C.; Wang, R.; Kevil, C. G. *Free Radical Biol. Med.* **2011**, *50*, 1021–1031.
12. Abe, K.; Kimura, H. *J. Neurosci.* **1996**, *16*, 1066–1071.

13. Kimura, H. *Mol. Neurobiol.* **2002**, *26*, 13–19.

14. Yang, W.; Yang, G. D.; Jia, X. M.; Wu, L. Y.; Wang, R. *J. Physiol. (London)* **2005**, *569*, 519–531.

15. Collin, M.; Anuar, F. B. M.; Murch, O.; Bhatia, M.; Moore, P. K.; Thiemermann, C. *Br. J. Pharmacol.* **2005**, *146*, 498–505.

16. Zhang, H. L.; Moochhala, S. M.; Bhatia, M. *J. Immunol.* **2008**, *181*, 4320–4331.

17. Elrod, J. W.; Calvert, J. W.; Morrison, J.; Doeller, J. E.; Kraus, D. W.; Tao, L.; Jiao, X. Y.; Scalia, R.; Kiss, L.; Szabo, C.; Kimura, H.; Chow, C. W.; Lefer, D. J. *Proc. Natl. Acad. Sci. U.S.A.* **2007**, *104*, 15560–15565.

18. Hu, L. F.; Wong, P. T. H.; Moore, P. K.; Bian, J. S. *J. Neurochem.* **2007**, *100*, 1121–1128.

19. Brunner, U.; Chasteen, T. G.; Ferloni, P.; Bachofen, R. *Chromatographia* **1995**, *40*, 399–403.

20. Doeller, J. E.; Isbell, T. S.; Benavides, G.; Koenitzer, J.; Patel, H.; Patel, R. P.; Lancaster, J. R.; Darley-Usmar, V. M.; Kraus, D. W. *Anal. Biochem.* **2005**, *341*, 40–51.

21. Esfandyarpour, B.; Mohajerzadeh, S.; Khodadadi, A. A.; Robertson, M. D. *IEEE Sens. J.* **2004**, *4*, 449–454.

22. Kuhl, M.; Steuckart, C.; Eickert, G.; Jeroschewski, P. *Aquat. Microb. Ecol.* **1998**, *15*, 201–209.

23. Fogo, J. K.; Popowsky, M. *Anal. Chem.* **1949**, *21*, 732–734.

24. Lawrence, N. S.; Davis, J.; Compton, R. G. *Talanta* **2000**, *52*, 771–784.

25. Shen, X. G.; Peter, E. A.; Bir, S.; Wang, R.; Kevil, C. G. *Free Radical Biol. Med.* **2012**, *52*, 2276–2283.

26. Kabil, O.; Banerjee, R. *J. Biol. Chem.* **2010**, *285*, 21903–21907.

27. Gronowitz, S.; Westerlund, C.; Hornfeldt, A. B. *Acta Chem. Scand., Ser. B* **1975**, *B29*, 224–232.

28. Huber, D.; Andermann, G.; Leclerc, G. *Tetrahedron Lett.* **1988**, *29*, 635–638.

29. Lin, Y.; Lang, S. A. *J. Heterocycl. Chem.* **1980**, *17*, 1273–1275.

30. Nickson, T. E. *J. Org. Chem.* **1986**, *51*, 3903–3904.

31. Lippert, A. R.; New, E. J.; Chang, C. J. *J. Am. Chem. Soc.* **2011**, *133*, 10078–10080.

32. Lin, V. S.; Lippert, A. R.; Chang, C. J. *Proc. Natl. Acad. Sci. U.S.A.* **2013**, *110*, 7131–7135.

33. Yu, F.; Li, P.; Song, P.; Wang, B.; Zhao, J.; Han, K. *Chem. Commun.* **2012**, *48*, 2852–2854.

34. Montoya, L. A.; Pluth, M. D. *Chem. Commun.* **2012**, *48*, 4767–4769.

35. Tian, H.; Qian, J.; Bai, H.; Sun, Q.; Zhang, L.; Zhang, W. *Anal. Chim. Acta* **2013**, *768*, 136–142.

36. Yu, C.; Li, X.; Zeng, F.; Zheng, F.; Wu, S. *Chem. Commun.* **2013**, *49*, 403–405.

37. Xuan, W.; Pan, R.; Cao, Y.; Liu, K.; Wang, W. *Chem. Commun.* **2012**, *48*, 10669–10671.

38. Peng, H.; Cheng, Y.; Dai, C.; King, A. L.; Predmore, B. L.; Lefer, D. J.; Wang, B. *Angew. Chem., Int. Ed.* **2011**, *50*, 9672–9675.

39. Chen, B.; Li, W.; Lv, C.; Zhao, M.; Jin, H.; Jin, H.; Du, J.; Zhang, L.; Tang, X. *Analyst* **2013**, *138*, 946–951.

40. Li, W. H.; Sun, W.; Yu, X. Q.; Du, L. P.; Li, M. Y. *J. Fluoresc.* **2013**, *23*, 181–186.

41. Chen, S.; Chen, Z.-J.; Ren, W.; Ai, H.-W. *J. Am. Chem. Soc.* **2012**, *134*, 9589–9592.

42. Hartman, M. C. T.; Dcona, M. M. *Analyst* **2012**, *137*, 4910–4912.

43. Wu, Z.; Li, Z.; Yang, L.; Han, J.; Han, S. *Chem. Commun.* **2012**, *48*, 10120–10122.

44. Zheng, K.; Lin, W.; Tan, L. *Org. Biomol. Chem.* **2012**, *10*, 9683–9688.

45. Chen, B.; Lv, C.; Tang, X. *Anal. Bioanal. Chem.* **2012**, *404*, 1919–1923.

46. Zhou, G.; Wang, H.; Ma, Y.; Chen, X. *Tetrahedron* **2013**, *69*, 867–870.

47. Das, S. K.; Lim, C. S.; Yang, S. Y.; Han, J. H.; Cho, B. R. *Chem. Commun.* **2012**, *48*, 8395–8397.

48. Wan, Q.; Song, Y.; Li, Z.; Gao, X.; Ma, H. *Chem. Commun.* **2013**, *49*, 502–504.

49. Cao, X.; Lin, W.; Zheng, K.; He, L. *Chem. Commun.* **2012**, *48*, 10529–10531.

50. Wang, R.; Yu, F.; Chen, L.; Chen, H.; Wang, L.; Zhang, W. *Chem. Commun.* **2012**, *48*, 11757–11759.

51. Wintner, E. A.; Deckwerth, T. L.; Langston, W.; Bengtsson, A.; Leviten, D.; Hill, P.; Insko, M. A.; Dumpit, R.; VandenEkart, E.; Toombs, C. F.; Szabo, C. *Br. J. Pharmacol.* **2010**, *160*, 941–957.

52. Qian, Y.; Karpus, J.; Kabil, O.; Zhang, S.-Y.; Zhu, H.-L.; Banerjee, R.; Zhao, J.; He, C. *Nat. Commun.* **2011**, *2*, 495.

53. Qian, Y.; Zhang, L.; Ding, S.; Deng, X.; He, C.; Zheng, X. E.; Zhu, H.-L.; Zhao, J. *Chem. Sci.* **2012**, *3*, 2920–2923.

54. Liu, C.; Peng, B.; Li, S.; Park, C.-M.; Whorton, A. R.; Xian, M. *Org. Lett.* **2012**, *14*, 2184–2187.

55. Liu, C.; Pan, J.; Li, S.; Zhao, Y.; Wu, L. Y.; Berkman, C. E.; Whorton, A. R.; Xian, M. *Angew. Chem., Int. Ed.* **2011**, *50*, 10327–10329.

56. Xu, Z.; Xu, L.; Zhou, J.; Xu, Y.; Zhu, W.; Qian, X. *Chem. Commun.* **2012**, *48*, 10871–10873.

57. Chen, Y.; Zhu, C.; Yang, Z.; Chen, J.; He, Y.; Jiao, Y.; He, W.; Qiu, L.; Cen, J.; Guo, Z. *Angew. Chem., Int. Ed.* **2013**, *52*, 1688–1691.

58. Liu, J.; Sun, Y.-Q.; Zhang, J.; Yang, T.; Cao, J.; Zhang, L.; Guo, W. *Chem.-Eur. J.* **2013**, *19*, 4717–4722.

59. Montoya, L. A.; Pearce, T. F.; Hansen, R. J.; Zakharov, L. N.; Pluth, M. D. *J. Org. Chem.* **2013**, *78*, 6550–6557.

60. Lim, M. H.; Xu, D.; Lippard, S. J. *Nat. Chem. Biol.* **2006**, *2*, 375–380.

61. McQuade, L. E.; Ma, J.; Lowe, G.; Ghatpande, A.; Gelperin, A.; Lippard, S. J. *Proc. Natl. Acad. Sci. U.S.A.* **2010**, *107*, 8525–8530.

62. Pluth, M. D.; Chan, M. R.; McQuade, L. E.; Lippard, S. J. *Inorg. Chem.* **2011**, *50*, 9385–9392.

63. Rosenthal, J.; Lippard, S. J. *J. Am. Chem. Soc.* **2010**, *132*, 5536–5537.

64. Choi, M. G.; Cha, S.; Lee, H.; Jeon, H. L.; Chang, S. K. *Chem. Commun.* **2009**, 7390–7392.

65. Sasakura, K.; Hanaoka, K.; Shibuya, N.; Mikami, Y.; Kimura, Y.; Komatsu, T.; Ueno, T.; Terai, T.; Kimura, H.; Nagano, T. *J. Am. Chem. Soc.* **2011**, *133*, 18003–18005.

66. Zhao, Y.; Wang, H.; Xian, M. *J. Am. Chem. Soc.* **2011**, *133*, 15–17.

67. Kar, C.; Adhikari, M. D.; Ramesh, A.; Das, G. *Inorg. Chem.* **2013**, *52*, 743–752.

68. Hou, F. P.; Huang, L.; Xi, P. X.; Cheng, J.; Zhao, X. F.; Xie, G. Q.; Shi, Y. J.; Cheng, F. J.; Yao, X. J.; Bai, D. C.; Zeng, Z. Z. *Inorg. Chem.* **2012**, *51*, 2454–2460.

69. Zhang, D. Q.; Jin, W. S. *Spectrochim. Acta, Part A* **2012**, *90*, 35–39.

70. Gu, X.; Liu, C.; Zhu, Y.-C.; Zhu, Y.-Z. *Tetrahedron Lett.* **2011**, *52*, 5000–5003.

71. Quek, Y.-L.; Tan, C.-H.; Bian, J.; Huang, D. *Inorg. Chem.* **2011**, *50*, 7379–7381.

72. Strianese, M.; De Martino, F.; Pellecchia, C.; Ruggiero, G.; D'Auria, S. *Protein Pept. Lett.* **2011**, *18*, 282–286.

73. Strianese, M.; Palm, G. J.; Milione, S.; Kühl, O.; Hinrichs, W.; Pellecchia, C. *Inorg. Chem.* **2012**, *51*, 11220–11222.

74. Wang, B.; Li, P.; Yu, F.; Song, P.; Sun, X.; Yang, S.; Lou, Z.; Han, K. *Chem. Commun.* **2013**, *49*, 1014–1016.

Chapter 3

Thione- and Selone-Containing Compounds, Their Late First Row Transition Metal Coordination Chemistry, and Their Biological Potential

Bradley S. Stadelman and Julia L. Brumaghim*

Chemistry Department, Clemson, South Carolina 29634-0973, United States
***E-mail: brumagh@clemson.edu.**

Thione- and selone-containing compounds and their transition metal complexes have been investigated for purposes ranging from antioxidant and anti-tumor activity to materials for optical electronics and light emitting diodes. With current investigations of sulfur-rich metalloenzyme active sites as well as the antioxidant activity of ergothioneine, selenoneine, and thione-containing anti-thyroid drugs, the coordination chemistry of thione and selone ligands with biologically common first-row transition metals has sparked considerable recent interest. This review focuses on the biological functions of thione- and selone-containing compounds, their coordination chemistry with iron, cobalt, nickel, copper, and zinc, and the biological implications of this chalcogenone-metal binding. While thione-metal coordination has been widely studied, only a few selone-metal complexes with first-row transition metals have been reported.

Introduction

Sulfur and selenium are essential elements and are incorporated into amino acids and enzymes (*1–7*). Sulfur has been very widely studied for its biological importance as well as its excess and deficiency (*8*). Although the recommended dietary allowance for selenium is low, 55-350 µg/day (*9*), selenium deficiency can lead to Keshan, Keshan-Beck, and numerous neurodegenerative diseases

(*10*). An excess of selenium, however, can also have malnutritional effects such as garlic breath, selenosis, and selenium poisoning (*10*). Humans obtain their nutritional needs for sulfur and selenium from milk, cheeses, garlic, eggs, fruits, vegetables, Brazil nuts, grains, legumes, and meats (*7*). The main biological form of selenium in humans is selenocysteine, which is synthesized in the body from dietary selenium sources (*8, 10, 11*).

Common sulfur and selenium compounds in biological systems include the amino acids cysteine, selenocysteine, methionine, and selenomethionine (Figure 1). Selenocysteine differs from cysteine by replacement of a single selenium atom in place of sulfur, and has similar chemical properties; however selenocysteine has a lower pK_a and is a stronger nucleophile, making it much more reactive (*12*). Selenoproteins, including thioredoxin reductases and glutathione peroxidases (GPx), are involved in redox reactions, electron transfer, and the synthesis of thyroid hormones; however, many of the selenoproteins' functions are not well elucidated. Cysteine and selenocysteine are often found in proteins and enzymes bound to late-transition metal centers such as iron, copper, and zinc (*12*).

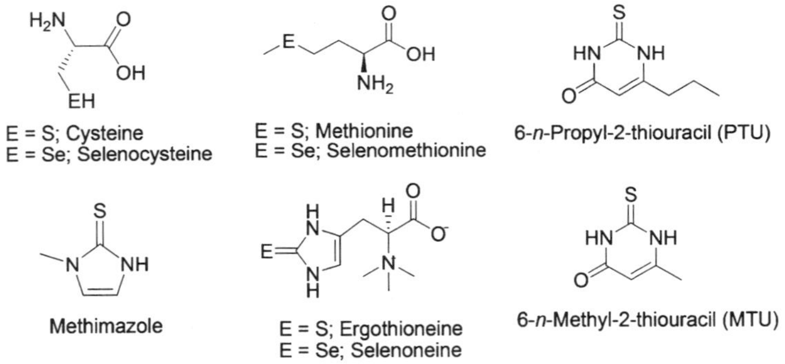

Figure 1. Common biological sulfur and selenium compounds and antithyroid drugs.

In contrast to the thiol and selenol functional groups of cysteine and selenocysteine, the amino acids ergothioneine and selenoneine (Figure 1) feature either a sulfur-carbon double bond (S=C, the thione functional group) or a selenium-carbon double bond (Se=C, the selone functional group). These thione- and selone-containing amino acids have the resonance forms shown in Figure 2, and belong to a class of pharmacologically active N,S ligands widely used in treating thyroid disorders (*13, 14*). Thione and selone compounds bind to metal ions through the sulfur or selenium atom as well as any unsubstituted NH groups of the imidazole ring. Imidazole thiones and selones have neutral and zwitterionic resonance forms; upon metal binding, spectroscopic data are consistent with stabilization of the zwitterionic resonance form with more negative charge on the sulfur or selenium atom (*13, 14*), similar to thiolate and selenolate ligands.

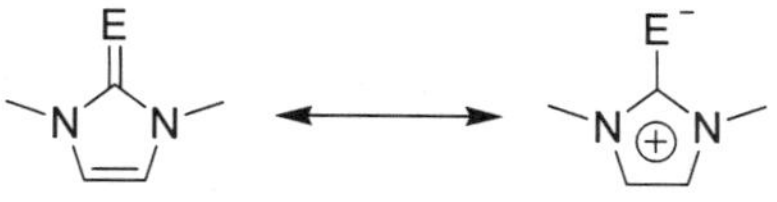

Figure 2. Resonance structures of an example imidazole thione (E = S) and selone (E = Se) compound.

Thione-metal coordination has been widely examined, but reviews do not specifically discuss late-transition-metal complexes of these ligands. General coordination chemistry of thione ligands was reviewed more than ten years ago by Raper (15) and Akrivos (16), but significant advances have since been made in this field. Reviews of multidentate (scorpionate) thione ligands and their coordination chemistry were recently published by Spicer and Reglinski (17) and Pettinari (18), but do not primarily focus on late transition metal coordination chemistry. Selone-containing compounds are scarce, and their metal complexes are correspondingly rare, so this area has not been previously reviewed. This review focuses on describing the coordination chemistry of thione- and selone-containing compounds with the late-first-row-transition metals, especially studies that have been published in the past decade, and the biological implications of this chalcogenone-metal binding.

Thione and Selone Compounds

Recently, thione- and selone-containing compounds have been primarily examined for their antioxidant behavior and their metal coordination abilities (1, 14, 19, 20). Antioxidant behavior of thione and selone compounds arises from several different mechanisms, including reactive oxygen species (ROS) scavenging, GPx-like activity to scavenge hydrogen peroxide, and binding of ROS-generating metal ions (1, 21). Because ROS damage cells and can result in disease development, it is important to understand how naturally occurring thione- and selone-containing compounds, as well as thione- and selone-containing drugs, neutralize ROS and prevent oxidative damage that leads to disease. Since the antioxidant activity of these compounds is also linked to prevention of ROS formation by metal coordination, it is essential to understand how these compounds bind metals and how they might react *in vivo*.

Thione and Selone Antioxidant Drugs

The thione-containing amino acid ergothioneine is found in plants and animals (1-3 mM (22, 23)), including humans. Ergothioneine is synthesized only in bacteria and various fungi, and it is thought that humans obtain ergothioneine in the diet (24). Although a specific transport system for ergothioneine has been identified in humans, no definite molecular pathway that involves ergothioneine has been reported (24). Franzoni, *et al.* demonstrated *in vitro* that ergothioneine scavenges peroxyl, hydroxyl, and peroxynitrite radicals more effectively than glutathione (*5*). In addition, several studies have determined that ergothioneine is a potent biological antioxidant, preventing many different types of oxidative damage, including lipid peroxidation oxidative DNA damage, and apoptosis, in cells or in animals (*25–31*). Metal coordination as an antioxidant mechanism has also been examined using density functional theory methods (*32*), and ergothioneine-copper coordination was found to prevent copper-mediated oxidative DNA and protein damage by inhibiting copper redox cycling (*33*).

Selenoneine, the selenium-containing analog of ergothioneine, has not been as extensively studied due to its very recent discovery in the blood of bluefin tuna (*4*). Since then, selenoneine has been identified in humans in very low concentrations (5-10 nM (*34*)); similar to ergothioneine, selenoneine is taken up from the diet. It is thought that selenoneine may be metabolized through selenium methylation, and this metabolic product was identified from human urine by mass spectrometry (*35*). Selenoneine's biological function is not known, but Yamishita, *et al.* reported that selenoneine is a more potent radical scavenger than ergothioneine, as determined by *in vitro* DPPH radical scavenging studies, and hence may play an important role in the redox cycle in animals (*34*).

Many of the drugs used to treat hyperthyroidism have thione functional groups, the most common of which is methimazole (Figure 1) (*36, 37*). In addition to its anti-thyroid activity, methimazole prevents oxidative DNA damage by hydrogen peroxide scavenging (*38, 39*). Other compounds that fall into this thione-containing class of hyperthyroid inhibitors are the thiourea drugs 6-*n*-propyl-2-thiouracil (PTU) and 6-methyl-2-thiouracil (MTU; Figure 1), but these drugs have not been studied as potential antioxidants. Although these compounds are the most commonly employed drugs for treating hyperthyroidism, their detailed mechanism of action is still unproven (*26*). It has been proposed that these drugs may block thyroid hormone synthesis by coordinating to the iron center of thyroid peroxidase (*37*).

Imidazole Thione Compounds

The heterocyclic *N,N'*-dimethylimidazole thione (dmit) and (*N,N'*-dimethylimidazole selone (dmise) compounds (Figure 3) synthesized by Roy, *et al.* (*37*) resemble methimazole and are structurally similar to ergothioneine (*24*) and selenoneine (*4*), respectively. Similar compounds 1,1'-(1,2-ethanediyl)-bis(3-methyl-imidazole-2-thione) (ebit) and 1,1'-(1,2-ethanediyl)bis(3-methyl-imidazole-2-selone) (ebis; Figure 3) also have been synthesized by Jia, *et al.* (*40*) to provide bidentate thione and selone ligands for metal coordination. These heterocyclic chalcogenones are strong σ- and π-donors, and similar compounds, such as the class of imidazoline-2-thione ligands (Figure 3), display a diversity of metal binding modes (*41*).

Trithiocyanuric acid (2,4,6-trimercaptotriazine, ttcH$_3$; Figure 3) and its trisodium salt have a wide range of applications: they can be used to precipitate heavy metals (*42–44*), and the acid shows anti-toxoplasmal activity more effective than the presently used drugs 5-fluorouracil and emimicin (*45*). The protonation state and solubility of trithiocyanuric acid depends strongly on pH. At pH values below 5, only the water-insoluble ttcH$_3$ exists, whereas increasingly soluble ttcH$_2^-$, ttcH^{2-}, and ttc^{3-} anions are formed as the pH increases. Unsurprisingly, the different protonation states of this compound greatly affect its metal binding abilities (*46*). Depending on the extent of ttcH$_3$ deprotonation, different metal coordination modes are possible, and transition metal complexes with various nuclearities can be formed (*46*).

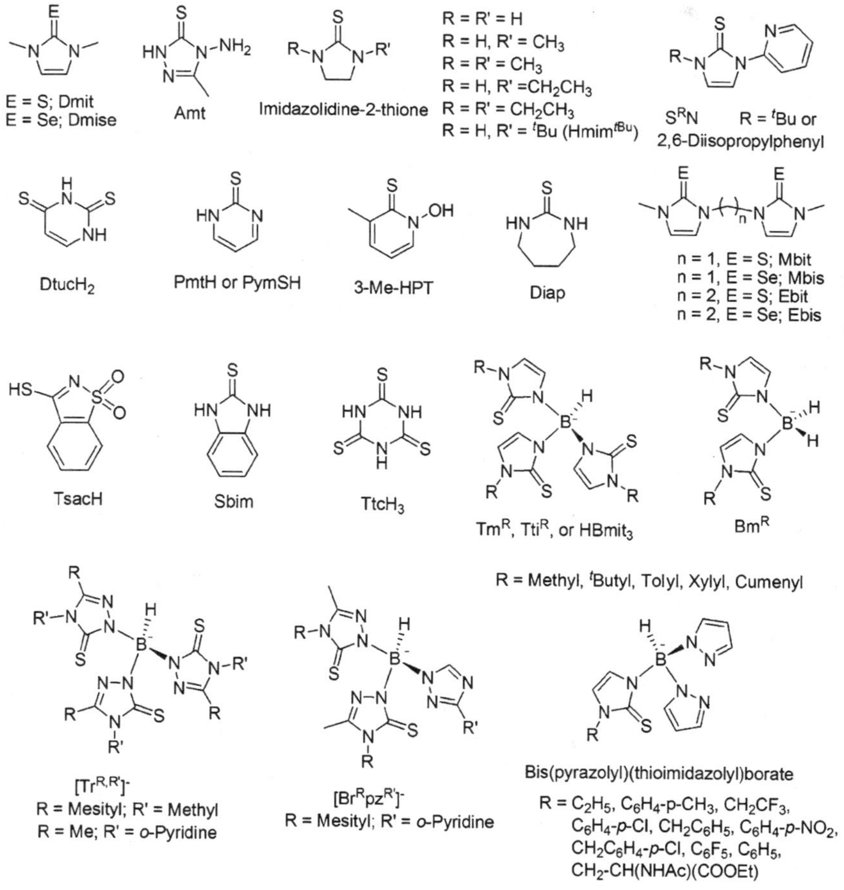

Figure 3. Thione- and selone-containing ligands discussed in this review.

Scorpionate-Thione Compounds

The so-called scorpionate compounds have become a hugely popular class of ligands in coordination chemistry since they mimic biological coordination sites (*17, 47–53*). This perceived analogy, coupled with their ease of synthesis and their remarkable versatility in transition-metal complexation, rapidly established them as a widely used class of ligands. One of the first examples, tris(pyrazolyl)borate, was described by Trofimenko in 1966 (*54*). These are tripodal, anionic ligands with an N3 donor set that caps one face of a coordination polyhedron. The ability to synthesize a wide range of substituted pyrazolyl ligands facilitates tuning of their steric, and to some degree, electronic properties (*18, 55*).

The thione-containing hydrotris(methimazolyl)borate (Tm^R), hydrobis(methimazolyl)borate (Bm^R), and hydrotris(thioxotriazolyl)borate (Tr$^{R,R'}$) ligands (Figure 3), like the pyrazolyl borates, have great potential for modification. The *N*-methyl group can be readily replaced by a range of alkyl (ethyl, *tert*-butyl, benzyl, and cyclohexyl) or aryl (phenyl, *p*-tolyl, *o*-tolyl, mesityl, cumenyl, 2-biphenyl, 2,6-xylyl, and 2,6-diisopropylphenyl) groups. In keeping with Trofimenko's synthesis of the pyrazolyl borate ligands, Spicer and Armstrong (*17, 56, 57*) synthesized Tm^R and Bm^R (*54*) by reaction of methimazole with sodium borohydride in a solvent-free system. Hydrogen evolves as the methimazole melts, forming sodium hydrotris(methimazolyl)-borate in 73-88% yields. Both lithium (*58*) and potassium (*59–61*) borohydrides have also been used successfully in these syntheses. Coordination complexes of these soft scorpionate ligands have been used as structural enzyme models for nickel-containing hydrogenases (*62*) and zinc-containing alcohol dehydrogenases (*63*).

Thione Compounds as Ligands

Thiones and selones readily bind transition metals such as Fe(II), Co(II/III), Ni(II), Cu(I/II), and Zn(II). Recently, there has been increased interest in the chemistry of soft Lewis base donors such as thiolates, thioamides, selenolates, and selenoamides for use in catalysis (*64*) and in bioinorganic chemistry for the study of copper metallothioneins, metallochaperones, and methinobactin (*65–67*). Metal-thione complexes also have been studied for a broad range of biological activities, including applications as antitumor, antibacterial, antifungal, and antiviral agents (*68–71*).

Many publications describe metal-thione complexes and, in a few cases, the structures and properties of these complexes have been compared to analogous metal-selone complexes. Unsurprisingly given the polarizability of the sulfur donor atoms, thione complexes have been largely reported only for the late first-row d-block metal ions, typically only in lower oxidation states.

Iron-Thione Complexes

Iron-thione complexes have not been extensively studied, likely because Fe(II) is unstable to oxidation and, thus, these complexes must be synthesized using air-free conditions. Iron-thione complexes are of interest as model complexes to study sulfur-rich metal centers in proteins and to determine potential biological behavior of ergothioneine and thione-containing drugs. Sanina, *et al.* reported the synthesis of [Fe(1,2,4-triazole-3-thione)$_2$(NO)$_2$]·0.5H$_2$O (Figure 4A) by combining 3-mercapto-1,2,4-triazole

(Figure 3) with [Na$_2$Fe$_2$(S$_2$O$_3$)$_2$(NO)$_4$]·2H$_2$O and Na$_2$S$_2$O$_3$·5H$_2$O in a 10:1:2 molar ratio in aqueous alkaline solution (*72*). The heterocyclic thione ligands coordinate iron through the sulfur atom, rather than through the available nitrogen atoms, with Fe–S bond distances of 2.298(1) and 2.318(1) Å (*72*).

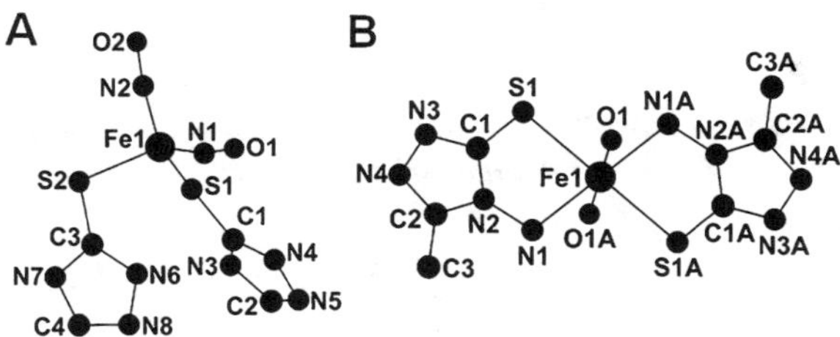

Figure 4. Crystal structures of A) Fe(1,2,4-triazole-3-thione)$_2$(NO)$_2$]·0.5H$_2$O (72), and B) Fe(amt)$_2$(H$_2$O)$_2$][ClO$_4$]$_2$ (68). Hydrogen atoms, solvent, and counterions are omitted for clarity.

The infrared spectrum of Fe(1,2,4-triazole-3-thione)$_2$(NO)$_2$·0.5H$_2$O has a single band at 702 cm^{-1} due to the iron-bound C=S that shifts to lower energy compared to the unbound 1,2,4-triazole-3-thione (750 or 745 cm^{-1}) (*72*). This shift upon complex formation indicates significant weakening of the C=S bond, and iron backbonding to the thione sulfur. Sanina, *et al.* reported that one of the sulfur atoms, S(1), is negatively charged and covalently binds to iron, whereas the other sulfur atom, S(2), is formally neutral and forms a donor–acceptor bond with iron (*72*). This description of the bonding is not supported by the single infrared C=S band and the very similar Fe-S bond lengths reported for this complex.

A similar complex was synthesized using 4-amino-3-methyl-1,2,4-triazole-5-thione (amt; Figure 3) and iron(II)perchlorate to yield [Fe(amt)$_2$(H$_2$O)$_2$][ClO$_4$]$_2$ (Figure 4B) (*72*). This iron compound decomposed during X-ray structural analysis as demonstrated by a ~30% decrease in the intensity of the standard reflections over the course of data collection. As expected, the symmetrically equivalent Fe-S bond distances of 2.474(1) Å (*72*) in this octahedral iron complex are significantly longer than the Fe-S bonds in the tetrahedral Fe(1,2,4-triazole-3-thione)$_2$(NO)$_2$·0.5H$_2$O. Crystals of [Fe(amt)$_2$(H$_2$O)$_2$][ClO$_4$]$_2$, initially pale green and transparent, slowly become yellowish brown and cloudy within a few days of exposure to air, characteristic of Fe(II) to Fe(III) oxidation (*68*).

The monodentate iron-thione complexes Fe(dmit)$_2$Cl$_2$ (Figure 5A), [Fe(dmit)$_4$][BF$_4$]$_2$, and [Fe(dmit)$_4$][OTf]$_2$ (Figure 5B), were synthesized by addition of dmit (Figure 3) to FeCl$_2$·H$_2$O, Fe(BF$_4$)$_2$, or Fe(OTf)$_2$, respectively (*73*). These distorted tetrahedral Fe(II) complexes bind though the sulfur atom of the monodentate thione ligand, with bond distances ranging from 2.3395(9) to 2.395(2) Å.

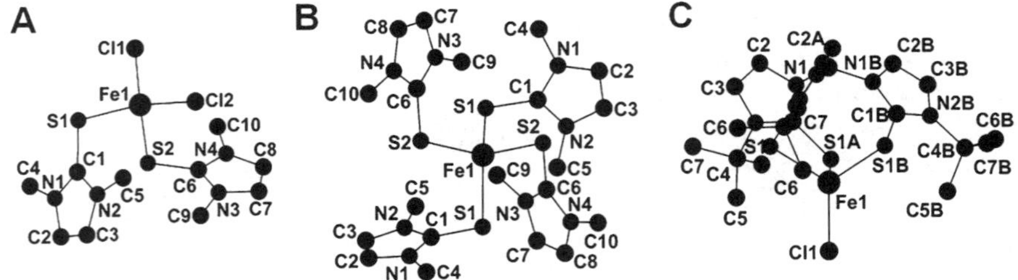

Figure 5. Crystal structures of A) Fe(dmit)$_2$Cl$_2$ (73), B) [Fe(dmit)$_4$][BF$_4$]$_2$ (73), and C) Tm^{tBu}FeCl (74). Hydrogen atoms and counterions are omitted for clarity.

Table 1. Infrared bands for C=S and C=Se bond vibrations in thione and selone compounds

Metal Complex	*C=S/Se Infrared Band (cm^{-1})*	*Change From Unbound Ligand (cm^{-1})*	*Ref.*
Fe-Thiones			
Fe(dmit)$_2$Cl$_2$	1176	-5	(73)
[Fe(dmit)$_4$][BF$_4$]$_2$	1178	-3	(73)
[Fe(dmit)$_4$][OTf]$_2$	1177	-4	(73)
Fe-Selones			
Fe(dmise)$_2$Cl$_2$	1149	1	(73)
[Fe(dmise)$_4$][BF$_4$]$_2$	1155	7	(73)
[Fe(dmise)$_4$][OTf]$_2$	1151	3	(73)
Co-Thiones			
Tm^{tBu}$_2$Co$_2$(μ-Tm^{tBu})$_2$Cl$_2$	1087	-30	(77)
Co-Selones			
[Co(mbis)Cl$_2$]$_n$	1096	-32	(81)
[Co(ebis)Cl$_2$]$_n$	1135	9	(81)
Ni-Thiones			
Ni(mbit)Br$_2$	1153	15	(81)
Ni(ebit)Br$_2$	1146	10	(81)
Ni(S^{R}N)Br$_2$	1156	2	(82)
Ni-Selones			
Ni(mbis)Br$_2$	1133	5	(81)

Continued on next page.

Table 1. (Continued). Infrared bands for C=S and C=Se bond vibrations in thione and selone compounds

Metal Complex	C=S/Se Infrared Band (cm⁻¹)	Change From Unbound Ligand (cm⁻¹)	Ref.
Ni(ebis)Br$_2$	1128	2	(81)
Cu-Thiones			
trans-Cu(dmit)$_2$Cl	1173	-8	(19)
trans-Cu(dmit)$_2$Br	1173	-8	(19)
trans-Cu(dmit)$_2$I	1171	-10	(19)
cis-Cu(dmit)$_2$I	1171	-10	(19)
[TpmCu(dmit)][BF$_4$]	1174	-7	(14)
[Tpm*Cu(dmit)][BF$_4$]	1171	-10	(14)
[Tpm^{iPr}Cu(dmit)][BF$_4$]	1180	-1	(14)
Tp*Cu(dmit)	1175	-6	(14)
[Cu(dmit)$_3$][OTf]	1147	-34	(20)
[Cu(dtucH$_2$)(PPh$_3$)Cl]$_n$	1090	-45	(102)
[Cu(dtucH$_2$)(PPh$_3$)Br]$_n$	1105	-30	(102)
[Cu(dtucH$_2$)(PPh$_3$)I]$_n$	1125	-10	(102)
Cu-Selones			
trans-Cu(dmise)$_2$Cl	1149	1	(19)
cis-Cu(dmise)$_2$Br	1150	2	(19)
cis-Cu(dmise)$_2$I	1148	0	(19)
Cu$_4$(μ-dmise)$_4$(μ-Br)$_2$Br$_2$	1150	2	(19)
Cu$_4$(μ-dmise)$_4$(μ-I)$_2$I$_2$	1144	-4	(19)
[TpmCu(dmise)][BF$_4$]	1150	2	(14)
[Tpm*Cu(dmise)][BF$_4$]	1150	2	(14)
[Tpm^{iPr}Cu(dmise)][BF$_4$]	1150	2	(14)
Tp*Cu(dmise)	1146	-2	(14)
[Cu(dmise)$_4$][OTf]	1146	-2	(20)

When dmit is bound to iron(II), the infrared stretches for the C=S bonds shift very slightly to lower energies (3 to 5 cm⁻¹) compared to unbound dmit (1181 cm⁻¹; Table 1), indicating little iron-thione backbonding. The ¹H NMR spectra of these three complexes show a downfield shift in the imidazole proton resonances by at least δ 0.82 compared to unbound dmit, suggesting increased aromaticity of the metal-bound imidazole ring. This increased aromaticity favors the resonance structure with increased negative charge on the thione sulfur atom (Figure 2), promoting metal coordination. This effect has been observed previously with copper- and platinum-thione complexes (*13, 14, 19*). The electrochemistry of the dmit ligand also changes when bound to iron, with the Ep_a shifting by 99 mV compared to unbound dmit (455 mV), indicating that the Fe(II)-bound thione becomes more readily oxidizable (*73*). This increase in oxidation potential suggests that the bound dmit may protect Fe(II) from oxidation.

Scorpionate-Thione Complexes with Iron

Treatment of $FeCl_2$ and LiTm^R at room temperature over four days yields the scorpionate-thione complexes Tm^{R}FeCl (R = Me, ^{t}Bu, Ph, and 2,6-^{i}PrC$_6$H$_3$; Figure 3); the structure of the *t*-butyl derivative, Tm^{tBu}FeCl, is shown in Figure 4C (*74*). These Fe(II) complexes adopt distorted tetrahedral geometry with Fe-S bond distances ranging from 2.366(1) to 2.3881(1) Å (*74*). As observed for the monodentate thione complexes, these chelating thione complexes are moderately air- and moisture-stable in the solid state but unstable in solution.

The reported iron–thione complexes are only found as Fe(II) and adopt tetrahedral or octahedral geometry around the Fe(II) metal center. They are easily synthesized, binding through the sulfur atom rather than an available nitrogen atom, as observed for Fe(1,2,4-triazole-3-thione)$_2$(NO)$_2$]·0.5H$_2$O (*72*). Fe(II)-thione complexes oxidize when exposed to oxygen, but these oxidized products have not been characterized. Because of iron's importance in generating ROS such as hydroxyl radical, and the demonstrated ability of biological thiones to prevent oxidative damage from these species, further studies of the relationship between iron-thione binding and reactivity of these complexes with ROS will help elucidate the mechanisms for thione antioxidant activity.

Cobalt-Thione Complexes

In the only report of a monodentate thione complex with cobalt, Castro, *et al.* (*75*) synthesized Co(C$_6$H$_{12}$N$_2$S)$_2$Cl$_2$ with the heterocyclic thione 1-propylimidazolidine-2-thione-κS (Figure 6A). Similar to the Fe(dmit)$_2$Cl$_2$ complex, the Co(II) center adopts distorted tetrahedral geometry with the two thione ligands bound through the S atoms and Co-S bond distances of 2.341(2) and 2.330(2) Å (*75*).

Scorpionate-Thione Complexes with Cobalt

The scorpionate-thione ligand hydrotris(1-methylimidazol-2-ylthio)borate (HBmit$_3$, Figure 3) was treated with CoF$_3$ to afford the cobalt complexes [Co(III)(HBmit$_3$)$_2$][HBmit$_3$] (Figure 6B) and [Co(II)(HBmit$_3$)$_2$]·4H$_2$O (*76*). X-ray structural analysis indicated that these complexes have octahedral and tetrahedral geometries, respectively, with the cobalt bound to all three thione sulfur atoms. In [Co(III)(HBmit$_3$)$_2$][HBmit$_3$], the Co-S distances range from

2.293(6) to 2.321(5) Å, and in [Co(II)(HBmit$_3$)$_2$]·4H$_2$O, the Co-S bond distances are somewhat longer, ranging from 2.34(2) to 2.391(4) Å (*76*). As expected with the change in oxidation state, the geometry of the cobalt metal center also changes: Co(III) adopts the expected octahedral geometry, whereas Co(II) adopts tetrahedral geometry. Interestingly, coordination of the HBmit ligand does not favor a single cobalt oxidation state, the only example of a thione ligand to do so.

Similar to the synthesis of the Tm^R Fe(II) complexes, treatment of CoCl$_2$ with LiTm^R (Figure 3) at room temperature over four days yielded the Tm^{R}CoCl complexes (R = Me, ^{t}Bu, Ph, and 2,6-^{i}PrC$_6$H$_3$). The tetrahedral Tm^{tBu}CoCl complex (Figure 6C) has Co-S bond distances of 2.312(1) to 2.331(1) Å (*74*). The reaction of cobalt(II) chloride hexahydrate with *N-tert*-butyl-2-thioimidazole (Tm^{tBu}) yields both [κ^2-(Tm^{tBu})$_2$]CoCl$_2$ and Tm^{tBu}$_2$Co$_2$(μ-Tm^{tBu})$_2$Cl$_2$·CH$_3$CN (Figure 6D) depending on the reaction time. In both complexes, each cobalt(II) center is four-coordinate with distorted tetrahedral geometry (*77*). Unexpectedly, two of the sulfur atoms of the Tm^{tBu} ligand in [κ^2-(Tm^{tBu})$_2$]CoCl$_2$ do not coordinate cobalt, but instead form a S–S bond with a typical bond distance of 2.050(4), while the available nitrogen atoms on the imidazole coordinate Co(II). The infrared spectrum of this complex shows a weak absorption band at 472 cm^{-1}, confirming disulfide bond formation, with no observable band at 1117 cm^{-1} for the C=S stretch of free thioimidazole.

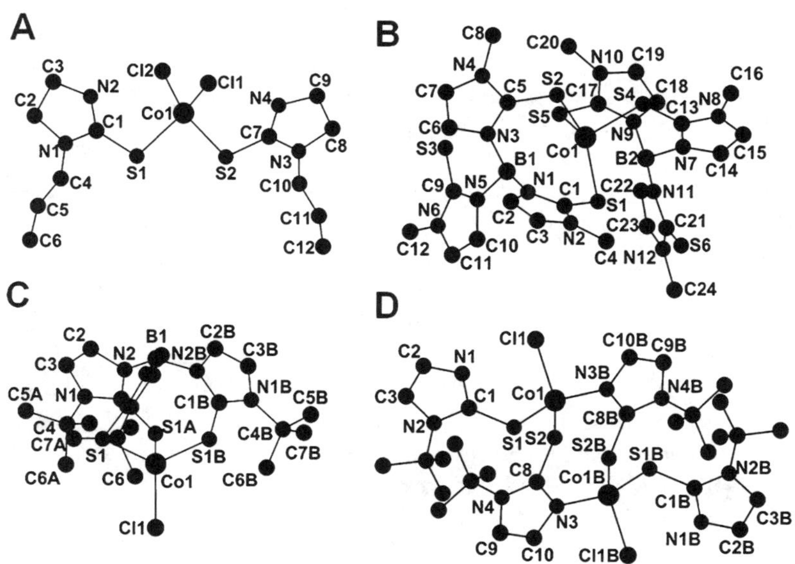

Figure 6. Crystal structures of A) Co(C$_6$H$_{12}$N$_2$S)$_2$Cl$_2$ (75), B) [Co(III)(HBmit$_3$)$_2$][HBmit$_3$] (76), C) Tm^{tBu}CoCl (74), and D) Tm^{tBu}$_2$Co$_2$(μ-Tm^{tBu})$_2$Cl$_2$ (77). Hydrogen atoms and counterions are omitted for clarity.

The Tm$^{tBu}_2$Co$_2$(μ-Tm^{tBu})$_2$Cl$_2$ complex is a sulfur-bridged dimer where four Tm^{tBu} ligands are coordinated to both Co(II) centers. Two of the tridentate thione ligands bind in a bidentate fashion through nitrogen as well as the sulfur atoms, with the other two thione ligands coordinated only through the sulfur atoms. The infrared spectrum of Tm$^{tBu}_2$Co$_2$(μ-Tm^{tBu})$_2$Cl$_2$·CH$_3$CN exhibits a lower energy C=S stretching band at 1087 cm^{-1} compared to the unbound ligand (1117 cm^{-1}), indicative of cobalt backbonding to the sulfur atoms (*77*).

The majority of cobalt-thione complexes form with Co(II), in tetrahedral structures similar to Fe(II)-thione complexes, but thione ligands bind both Co(II) and Co(III) oxidation states, a feature not observed for iron complexes. In addition, the Tm^{tBu} ligand in [κ^2-(Tm^{tBu})]$_2$CoCl$_2$ does not bind through the sulfur atoms, but through the available nitrogen donors of the ligand, with concomitant formation of a diselenide bond from the thione sulfur atoms. No reactivity studies have been reported for these cobalt-thione complexes.

Nickel-Thione Complexes

Ni-thiolate/thione chemistry is of significant interest because of its relevance to hydrogenase chemistry (*78–80*). In addition, since ergothioneine coordination chemistry is not well established, a recent study utilized a simpler version of this thione to investigate its nickel binding (*78*). Mononuclear Ni(II) dithione complexes with the bidentate ligands mbit or ebit (Figure 3) were synthesized by thione addition to Ni(PPh$_3$)Br$_2$ under nitrogen for 10 h to yield Ni(mbit)Br$_2$ or Ni(ebit)Br$_2$ (Figure 7A) (*81*). Both Ni(II)-thione complexes are air and moisture stable and adopt distorted tetrahedral geometry with symmetrically equivalent Ni-S bond distances of 2.283(1) and 2.286(2) Å, respectively. Ni(mbit)Br$_2$ and Ni(ebit)Br$_2$ infrared spectra show C=S stretches at 1153 and 1146 cm^{-1}, respectively, stretches that are shifted to higher energy compared to the unbound ligands (1138 and 1136 cm^{-1}, respectively), indicating primarily donor bond formation (*81*).

The tetrakis(1-*tert*-butyl-imidazole-2-thione) nickel complex [Ni(Hmim^{tBu})$_4$][I]$_2$ (Figure 7B) is readily synthesized by treating Hmim^{tBu} (Figure 3) with NiI$_2$ in acetonitrile (*82*). The Ni(II) adopts square planar geometry, coordinated to four thione ligands with symmetrically equivalent Ni-S bond distances of 2.220(2) and 2.236(3) Å. Two iodide ions bridge the deprotonated nitrogen atoms from separate Hmim^{tBu} ligands. When the coordinated Hmim^{tBu} ligands are deprotonated by KH in the presence of a second equivalent of NiI$_2$, the dinuclear complex Ni$_2$(mim^{tBu})$_4$ is formed. Each of the four mim^{tBu} ligands bridge the two Ni centers through N and S atoms, forming N2S2 Ni(II) coordination with symmetrically equivalent Ni-S bond distances of 2.218(1) Å and 2.238(1) Å (*78*), similar to Ni-S distances from the monomeric [Ni(Hmim^{tBu})$_4$][I]$_2$.

Mixed thione/nitrogen and thione/oxygen donor ligands have also been used to synthesize Ni(II) complexes. Reaction of Ni(DME)Br$_2$ (DME = dimethoxyethane) with 1-pyridyl-(3-*tert*-butylimidazole-2-thione) or 1-pyridyl-(3-(2,6-diisopropylphenylimidazole)-2-thione) yields Ni(S^{R}N)Br$_2$ complexes (*82*). The structure of Ni(S^{R}N)Br$_2$ (S^{R}N = 1-pyridyl-(3-(2,6-

diisopropylphenylimidazole)-2-thione) is shown in Figure 7C. Both complexes adopt tetrahedral geometry, with a Ni-S bond distance of 2.282(3) Å. The infrared C=S stretching frequencies for both complexes (1156 cm^{-1}) show little change compared to the unbound ligand (1154 cm^{-1}), indicating primarily donor Ni-S interactions (*82*).

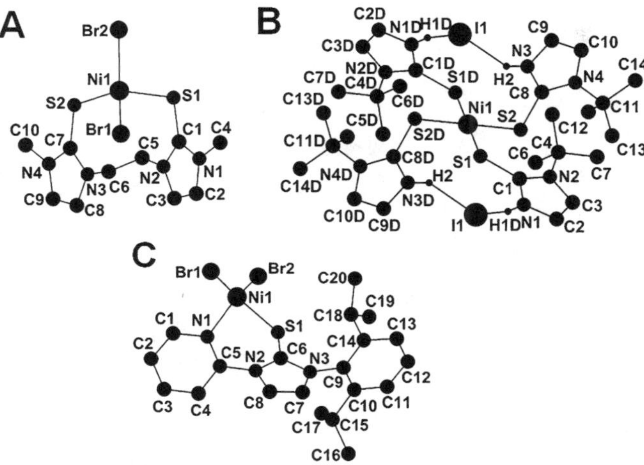

Figure 7. Crystal structures of A) Ni(ebit)Br$_2$ (81), B) [Ni(Hmim^{tBu})$_4$][I]$_2$ (78), and C) Ni(S^{R}N)Br$_2$ (S^{R}N = 1-pyridyl-(3-(2,6-diisopropylphenylimidazole)-2-thione)) (82). Hydrogen atoms and counterions are omitted for clarity.

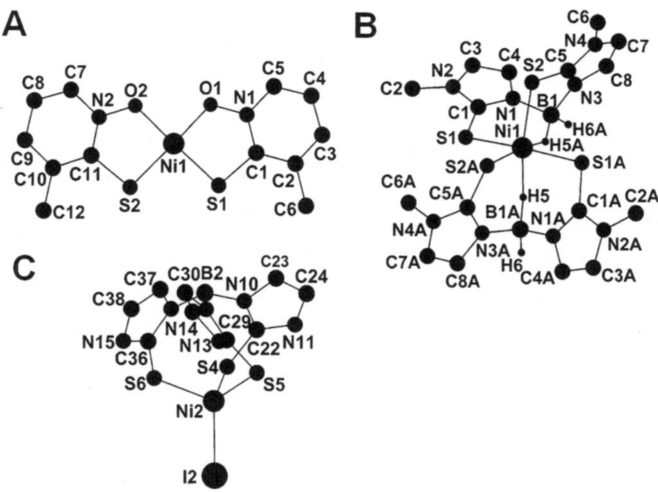

Figure 8. Crystal structures of A) Ni(3-Me-HPT)$_2$ (85), B) Ni(Bm^{Me})$_2$ (62), and C) Tm^{tBu}NiI (74). Hydrogen atoms and t-butyl groups are omitted for clarity.

A separate crystallographic study used the 1-hydroxy-3-methylpyridine-2($1H$)-thione ligand (3-Me-HPT, Figure 3), a compound widely investigated for its antifungal and antibacterial properties (*83*) as well as its potential as a ligand for metalloenzyme modeling (*84*). Addition of 3-Me-HPT to $NiCl_2 \cdot 6H_2O$ affords $Ni(3\text{-Me-HPT})_2$ (Figure 8A), a complex with distorted square-planar geometry, O2S2 coordination, and an Ni-S bond distance of 2.131(1) Å for the symmetrically equivalent sulfur atoms (*85*).

Scorpionate-Thione Complexes with Nickel

Seeking to investigate new structural and functional model compounds for nickel hydrogenases, the scorpionate-thione ligand bis(2-mercapto-1-methylimidazolyl)borate (Bm^{Me}, Figure 3) was treated with $NiCl_2 \cdot 6H_2O$ to afford $Ni(Bm^{Me})_2$ (Figure 8B). This air-stable complex adopts octahedral geometry around Ni(II) with Ni-S bond distances of 2.362(2) and 2.389(1) (*62*). The S4H2 coordination around the Ni(II) center is unique among nickel complexes and provides a structural model for the active form of NiFe hydrogenase enzymes.

Scorpionate thione complexes of Ni(II) also have been synthesized from NiX_2 (X = Cl, Br, or I) and $LiTm^R$ to give a series of Tm^RNiX (R = Me, ^tBu, Ph, and 2,6-$^iPrC_6H_3$) complexes. The structure of $Tm^{tBu}NiI$ is provided in Figure 8C. Each complex adopts distorted tetrahedral geometry with Ni-S bond distances ranging from 2.278 to 2.300 Å, shorter than the M-S bond distances in the analogous Fe(II) (2.366-2.388 Å) and Co(II) (2.298-2.333 Å) complexes (*74*).

As observed for Fe(II)- and Co(II)-thione complexes, the few reported Ni(II)-thione complexes adopt primarily tetrahedral geometry, with similar Ni-S bond distances ranging from 2.130 to 2.300 Å. All of these Ni-thione complexes were studied only for their structural characteristics, so additional reactivity and electrochemical studies to understand properties of these Ni(II)-thione complexes still remain to be performed.

Copper-Thione Complexes

Coordination of heterocyclic thiones to copper results in diverse architectures ranging from mononuclear complexes to polynuclear networks with monodentate or bidentate coordination of the sulfur atoms. Mono-alkylated heterocyclic thiones can also bind metals *via* the non-alkylated nitrogen atom. Many different types of copper-thione complexes have been synthesized and characterized to study the sulfur-rich environments of copper coordination in proteins (*17*).

The biological coordination chemistry of methimazole (Figure 1) is of particular interest, due to its use as an anti-thyroid drug. One structural study reported addition of CuI, methimazole, and PPh_3 to form $Cu(methimazole)(PPh_3)_2I$ (Figure 9A). In this complex, Cu(I) is tetrahedrally coordinated with a Cu-S bond distance of 2.369(1) Å (*86*), consistent with similar copper(I) thione complexes (*87*, *88*).

Heterocyclic thioamides are also of interest for their potential luminescence properties (*89*) as well as their medicinal purposes (*90, 91*), and complexes of these ligands have been studied with numerous metals, including copper, mercury, silver, and palladium (*92*). One example of this diversity of heterocyclic thioamide complexes was reported when pyrimidine-2-thione (pymSH, Figure 3) was treated with CuX (X = I, Br, or Cl) and PPh_3, resulting in synthesis of the unusual dimer $Cu_2(pymSH)(PPh_3)_2I_2$, where the two copper ions are bridged by N and S atoms of pymSH and two iodide ions, as well the monomers $Cu(pymSH)(PPh_3)_2Br$ (Figure 9B) and $Cu(pymSH)(PPh_3)_2Cl$ (*92*). Changing the halide from I to Cl or Br results in formation of monomeric rather than dimeric complexes; however, all three complexes have the expected tetrahedral Cu(I) coordination and similar Cu-S bond distances (2.3530(3), 2.3340(6), and 2.3720(3) Å, respectively) (*92*).

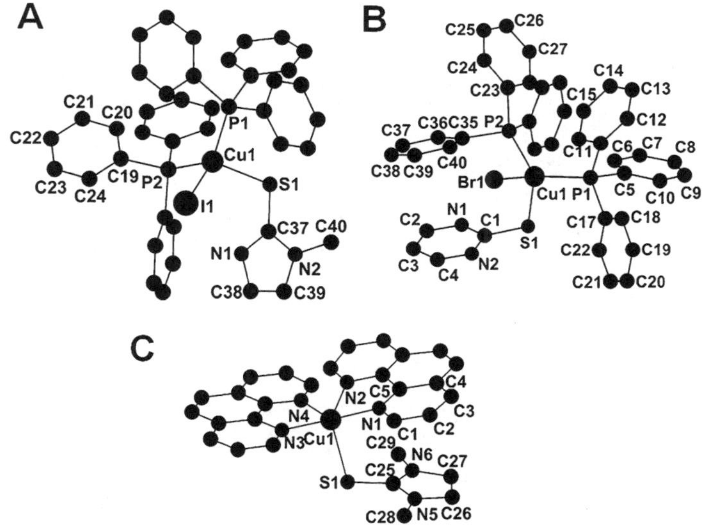

Figure 9. Crystal structures of A) Cu(methimazole)(PPh₃)₂I (86), B) Cu(pymSH)(PPh₃)₂Br (92), and [Cu(phen)₂(dmit)][ClO₄]₂ (93). Hydrogen atoms and counterions are omitted for clarity.

Cu(II) phenanthroline (phen) complexes have been widely examined *in vivo* and *in vitro* for their cytotoxic (*94*), genotoxic (*95*), and antitumor effects (*96*). Pivetta, *et al.* (*93*) synthesized and characterized four penta-coordinate copper(II) complexes $[Cu(phen)_2(L)][ClO_4]_2$ (L = disubstituted imidazolidine-2-thione ligands with mixed H, Me, and Et substituents; Figure 3) to study their formation constants, cytotoxicity, and antiprion activity. In all four complexes, copper(II) adopts distorted trigonal–bipyramidal geometry, as seen for $[Cu(phen)_2(dmit)][ClO_4]_2$ (Figure 9C), with similar but fairly long Cu-S bond distances ranging from 2.385(2) to 2.427(1) Å. Although at least one imidazole nitrogen is unsubstituted in three of the four compounds, Cu(I) binds exclusively through the thione sulfur atoms. In the first study to quantify thione ligand binding, formation constants (log K values) for these copper-thione complexes in acetonitrile were measured between 1.75 and 3.20, following the same trend as the Cu-S bond lengths (*93*). From these data, Pivetta, *et al.* concluded that alkyl substitution of the thione ligand, even if increasing thione nucleophilicity, did not improve the donor ability of the thione ligand.

No antiprion activity was observed when these [Cu(phen)$_2$(L)][ClO$_4$]$_2$ complexes were tested in scrapie-prion-protein-infected mouse neuroblastoma cells at non-cytotoxic levels; however, these compounds exhibited significant cytotoxicity, with cytostatic concentrations (CC$_{50}$ values) between 0.41 and 0.54 μM (*93*). These effects were consistent with the cytotoxicity of other Cu-phen complexes and proposed to occur through oxidative damage to DNA and other cellular targets.

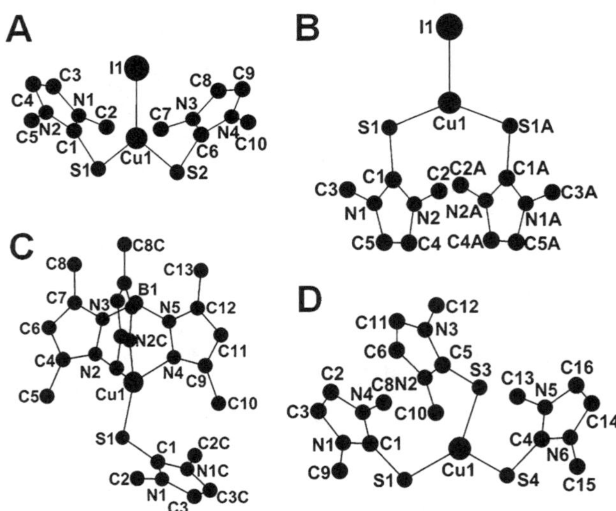

*Figure 10. Crystal structures of A) trans-Cu(dmit)$_2$I (19), B) cis-Cu(dmit)$_2$I (19), C) Tp*Cu(dmit) (14), and D) [Cu(dmit)$_3$][OTf] (20). Hydrogen atoms and counterions are omitted for clarity.*

In addition to phenanthroline-thione copper complexes, complexes of the formulae Cu(dmit)$_2$X (X = I, Br, and Cl) were synthesized and characterized by Kimani, *et al.* (*19*) Changes in the halide ligand have little effect on Cu–S bond distances: 2.2376(6) Å for *trans*-Cu(dmit)$_2$Cl, 2.2298(9) Å for *trans*-Cu(dmit)$_2$Br, 2.234(1) Å for *trans*-Cu(dmit)$_2$I (Figure 10A), and 2.240(1) Å for *cis*-Cu(dmit)$_2$I (Figure 10B). All of these Cu(I) complexes are unstable to copper oxidation in air. Only the iodide complex has both *trans* and *cis* structural isomers, with pi-stacking of the dmit imidazolium rings in the *cis* conformer observed in the solid state and confirmed by density functional theory calculations (*19*). Infrared spectra of these complexes show C=S stretching frequencies 8-10 cm^{-1} lower than unbound dmit (1181 cm^{-1}), indicating weak Cu-S backbonding. The ^{1}H NMR resonance for the olefinic protons of the copper- bound dmit ligand (δ 7.27-7.32) shift downfield relative to unbound dmit (δ 6.64),and the ^{13}C{^{1}H} resonance of the C=S carbon also shifts downfield by δ 3 to 8 upon copper complexation, characteristic increasing the zwitterionic form of the thione ligand (Figure 2) upon metal binding (*19*).

The same thione ligand was also used to synthesize the copper-thione complexes [TpmCu(dmit)][BF$_4$], [Tpm*Cu(dmit)][BF$_4$], and Tp*Cu(dmit) (Tpm = tris(pyrazolyl)methane, Tpm* = tris(3,5-dimethylpyrazolyl)methane, and Tp* = hydrotris(3,5-dimethylpyrazolyl)borate; Figure 10D) with trinitrogen donor scorpionate ligands to determine how thione coordination alters the redox potentials of the copper center (*14*). In these complexes, the Cu(I) ion is in a distorted tetrahedral environment with Cu-S bond lengths ranging from 2.191(8) to 2.202(7) Å (*14*). Thione C=S infrared stretching frequencies in these complexes range from 1172 to 1178 cm^{-1} (Table 1), and the olefinic ^{1}H NMR and C=S ^{13}C{^{1}H} NMR resonances shift downfield by δ 0.18-0.40 and δ 2.3-7.8, respectively, relative to unbound dmit (*14*), similar to coordination effects observed for the Cu(dmit)$_2$X complexes (*19*).

Electrochemical studies of these tris(pyrazolyl) copper-thione complexes in acetonitrile showed Cu(II/I) potentials between 70 and -160 mV, lowered by 374 to 617 mV compared to the non-thione-coordinated [TpmCu(NCCH$_3$)]$^+$ complex (*14*). In addition, increasing steric bulk on the tris(pyrazolyl) ligand stabilizes Cu(II) relative to Cu(I). Although this represents a significant shift in this copper potential upon thione binding, the resulting potentials are well above that of the cellular reductant NADH (-324 mV (*97*)), indicating that similar Cu(I) compounds *in vivo* still may be capable of redox cycling and generation of damaging reactive oxygen species.

Adding four equivalents of dmit to Cu(OTf)$_2$ yields the reduced [Cu(dmit)$_3$][OTf] complex (Figure 10D) with concomitant oxidation of the thione to the imidazolium disulfide (*20*), the first imidazolium disulfide synthesized by Cu(II) reduction. The [Cu(dmit)$_3$]$^+$ cation adopts distorted trigonal-planar geometry with Cu-S bond distances ranging from 2.226(1) to 2.260(1) Å. Kinetic studies were performed in acetonitrile, demonstrating that dmit reduces Cu(II) to Cu(I) at a rate of 0.18 s^{-1} (*20*), and establishing that these thione ligands are capable of promoting copper redox cycling under these conditions.

Only a few structurally characterized complexes of 2,4-dithiouracil (dtucH$_2$; Figure 3) coordinated to Cu(I), Ti(III), and Ru(III) have been reported (*98–101*). Cu(I) halides combined with this ligand form [Cu(dtucH$_2$)(PPh$_3$)X]$_n$ (X = Cl, Br, and I; Figure 11A with X = Br), with tetrahedral Cu(I) geometry and an average Cu–S bond length of 2.332 Å (*102*). The two sulfur atoms of the 2,4-dithiouracil ligands are coordinated to separate copper ions, forming a polymeric chain. The infrared spectra of the [Cu(dtucH$_2$)(PPh$_3$)X]$_n$ complexes display a γ(C=S) band between 1090 and 1125 cm^{-1}, significantly shifted from that of the unbound ligand at 1135 cm^{-1}, and consistent with copper coordination through the sulfur rather than through the nitrogen atoms as well as backbonding of copper to the thione (*102*).

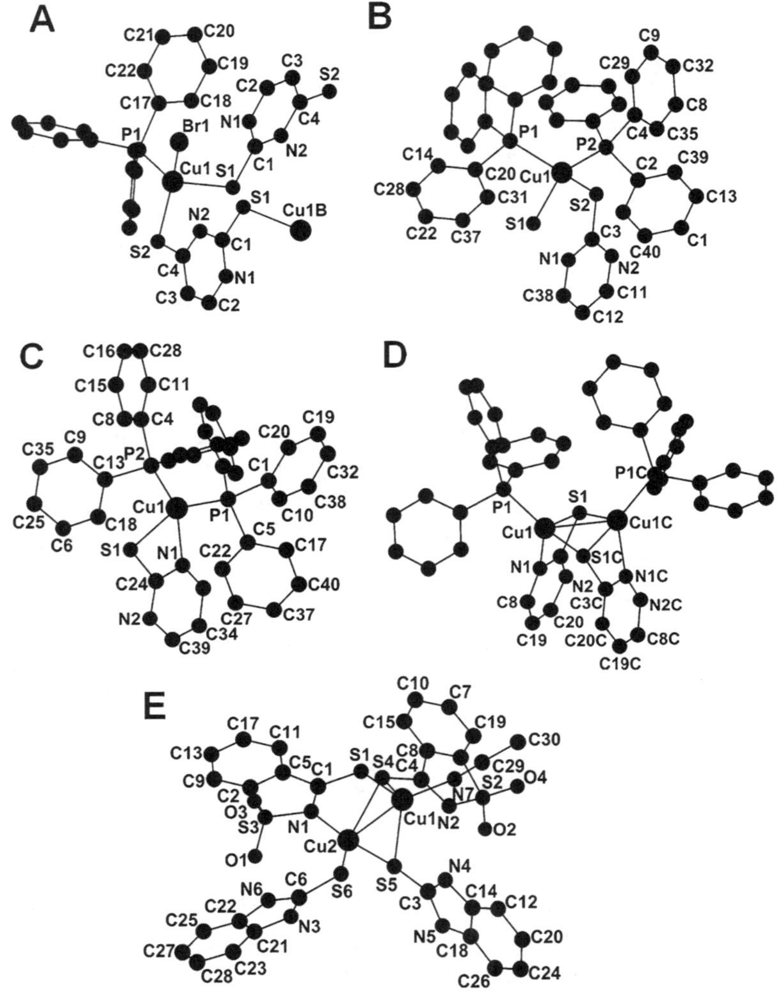

Figure 11. Crystal structures of A) [Cu(l-S,S-dtucH₂)(PPh₃)Br]ₙ (102), B) Cu(PPh₃)₂(pmtH)(SH) (103), C) Cu(PPh₃)₂(pmt) (103), D) Cu₂(PPh₃)₂(pmt)₂·0.5CH₃OH (103), and E) [Cu₂(tsac)₂(sbim)₂(CH₃CN)] (104). Hydrogen atoms are omitted for clarity.

The structural and antioxidant properties of Cu(PPh₃)₂(pmtH)Cl, Cu(PPh₃)₂(pmtH)(SH) (Figure 11B), Cu(PPh₃)₂(pmt) (Figure 11C), and Cu₂(PPh₃)₂(pmt)₂·0.5CH₃OH (Figure 11D; pmtH = 2-mercaptopyrimidine; Figure 3) have also been reported (*103*). 2-Mercaptopyrimidine readily binds copper(I) chloride in the presence of PPh₃ to form the monomeric complex Cu(PPh₃)₂(pmtH)Cl. The acidic nature of the H(N) proton on the pmtH ligand in Cu(PPh$_3$)$_2$(pmtH)Cl was established by treating this complex with sodium hydroxide to form the dimeric Cu(PPh$_3$)(pmt)$_2$·0.5CH$_3$OH (Figure 11 D) (*103*). In this complex of the deprotonated ligand, pmt binds each tetrahedral Cu(I) through the nitrogen atom and a bridging sulfur atom from the thione (Cu-S bond distances of 2.67(1) and 2.31(1) Å. A terminal PPh$_3$ ligand completes the tetrahedral copper coordination. The thione-bridged copper ions are close enough (2.699(2) Å) to form a Cu-Cu bond. In contrast, treatment of pmtH with Cu(CH$_3$COO) and PPh$_3$ in ethanol results in the formation of the monomeric Cu(PPh$_3$)$_2$(pmt) (Figure 11C), with the tetrahedral Cu(I) terminally bound through the nitrogen and sulfur atoms of the deprotonated pmt ligand (*103*). As expected, the Cu-N distance in this complex (2.144(1) Å) is similar to the Cu-N distance in the dimeric complex (2.052(1) Å), but the terminal Cu-S bond (2.400(1) Å) is significantly shorter.

Reaction of copper(II) sulfate or nitrate with pmtH and PPh$_3$ in refluxing methanol/acetonitrile solution forms the unusual Cu(PPh$_3$)$_2$(pmtH)(SH) complex (Figure 11B) with two PPh$_3$ ligands and the sulfur atom of the pmtH ligand (Cu-S 2.3652(6) Å) coordinated to Cu(I) (*103*). Tetrahedral geometry around Cu(I) is completed by a terminal thiol group, with a Cu-S bond distance of 2.3718(5) Å.

These Cu(I) pmtH and pmt complexes were tested for their ability to inhibit linoleic acid peroxidation by lipogenase (LOX), an enzyme that controls cancer cell proliferation. Although all four complexes prevent LOX activity with 50% inhibitory concentrations of 7 to 47 µM, the dimeric Cu$_2$(PPh$_3$)$_2$(pmt)$_2$ was the most effective (*103*). Direct LOX binding by these complexes was observed using ^{1}H NMR spectroscopy, and computational studies suggested that the most effective inhibitors interact close to the LOX active site but not in a single binding pocket (*103*).

Dennehy, *et al.* report the synthesis of copper complexes of the ternary thiosaccharin (tsac) and 1*H*-benzimidazole-2(3*H*)-thione (sbim; Figure 3) to form Cu$_2$(tsac)$_2$(sbim)$_2$(NCCH$_3$) (Figure 11E) and Cu$_2$(tsac)$_2$(sbim)$_3$ (*104*). In both these complexes, the tsac and the sbim ligands have monodentate and bidentate coordination. In Cu$_2$(tsac)$_2$(sbim)$_2$(NCCH$_3$), the two copper ions are bridged by the N atom and thione S of one tsac ligand as well as through the sulfur of the second tsac ligand. The two sbim ligands are bound only through the sulfur atoms, in monodentate and bidentate fashion, to give a NS3 core around one Cu(I) and an S3 core around the second Cu(I). Distorted tetrahedral geometry around the second Cu(I) is completed by coordinated acetonitrile. The Cu-S bond distance for the terminal sbim is 2.209(1) Å, somewhat shorter than for the bridging sbim ligand (2.238 Å). The Cu-S bond distances for the tsac ligand vary more significantly, with a µ-N,S-Cu distance of 2.2799(9) compared to the average Cu-S distance of 2.523 Å for the tsac bridging through only the sulfur atom (*104*).

Monodentate thione ligands bind to Cu(I) in trigonal planar or tetrahedral geometries and the Cu(II) in trigonal planar geometry. The majority of the monodentate thione-copper complexes contain Cu(I), as expected for polarizable sulfur donor ligands, but electrochemistry of the tris(pyrazolyl) Cu(I) thione complexes indicate that thione coordination destabilizes Cu(I) relative to acetonitrile coordination (*14*). In all cases, the thiones predominantly bind through the sulfur atom rather than other available nitrogen atoms as observed with the pmt, a characteristic also observed for both iron and cobalt complexes.

Although these complexes are relatively easy to synthesize, the reactions to synthesize Cu(I) complexes typically require air-free techniques and the products oxidize over time. Similar to iron-thione complexes, copper-bound thione ligands favor the zwitterionic resonance structure (Figure 2), as determined by NMR spectroscopy showing increased aromaticity within the ring upon metal binding.

Scorpionate-Thione Complexes with Copper

Many copper coordination studies have used tridentate thione-containing scorpionate ligands that permit tetrahedral Cu(I) coordination with a substitutable binding site. One example of this type of complex is $Tr^{R,R'}Cu(PPh_3)$ ($Tr^{R,R'}$ = tris(mercaptoimidazoyl)borate; Figure 12A), where the Cu(I) is bound to the three thiones of the $Tr^{R,R'}$ ligand (Figure 3), with average Cu-S bond distances of 2.367 Å, and a PPh_3 ligand (*105*). Further characterization, including infrared and electrochemical studies, would be useful to determine details of the thione-copper interactions and how coordination of these chelating ligands affect copper redox behavior.

A similar study was conducted by Rabinovich, *et al.* (*106*) to study the effects on bonding and structure of the sterically bulkier thione chelating ligand Tm^{tBu} (Figure 3). $Tm^{tBu}Cu(PPh_3)$ was prepared by combining equimolar amounts of $NaTm^{tBu}$, PPh_3, and CuCl. In contrast to Cu(I) complexes of monodentate thione ligands, the $Tm^{tBu}Cu(PPh_3)$ complex is air and light stable, a feature typical of these thione-scorpionate complexes. The symmetry equivalent Cu–S bond distance of 2.3435(7) Å in $Tm^{tBu}Cu(PPh_3)$ (Figure 12B) is similar to those in $Tm^{Me}Cu(PAr_3)$ (2.357(2) and 2.332(1) Å for Ar = *m*-tolyl and *p*-tolyl groups, respectively), suggesting that the steric bulk of the secondary ligand does not affect the binding modes of the scorpionate Tm^R ligand (*106*).

In some cases thione-scorpionate ligands bridge multiple metal centers. The dimeric Cu(I) complexes $Tr^{Mes,Me}_2Cu_2$ and $Tr^{Mes,o\text{-}Py}_2Cu_2$ (Tr^R = hydrotris(thioxotriazolyl)borate; Figure 12C), are air stable and exhibit similar coordination geometries in the solid state (*107*), where every copper ion is surrounded by three thione groups from two Tr^R ligands in a trigonal arrangement. The presence of a [B-H-Cu] three-center-two-electron interaction in both compounds causes the overall coordination to become tetrahedrally distorted (with S3H coordination for each copper center). In $Tr^{Mes,Me}_2Cu_2$, the symmetrically equivalent non-bridging Cu-S distances are 2.265(1) and 2.355(1) Å, and the bridging Cu-S distance is 2.253(2) Å. Similarly, the $Tr^{Me,o\text{-}Py}_2Cu_2$ complex has symmetrically equivalent, non-bridging Cu-S bond distances of 2.285(1) and 2.305(1) Å and bridging Cu-S bond lengths of 2.252(1) Å (*107*). The tridentate ligand hydro[bis(thioxotriazolyl)-3-(2-pyridyl)pyrazolyl]borate ($Br^{Mes}pz^{o\text{-}Py}$; Figure 3) with only two thione groups also forms a dimeric complex, ($Br^{Mes}pz^{o\text{-}Py})_2Cu_2$, with trigonal copper(I) coordination from a bridging pyrazolyl nitrogen atom and two terminal thione sulfur atoms (*107*). The symmetrically equivalent Cu-S distances in ($Br^{Mes}pz^{o\text{-}Py})_2Cu_2$ are 2.245(2) and 2.268(3) Å, significantly shorter than the terminal Cu-S bonds in $Tr^{Me,o\text{-}Py}2Cu_2$ with the tridentate thione ligand $Tr^{Me,o\text{-}Py}$.

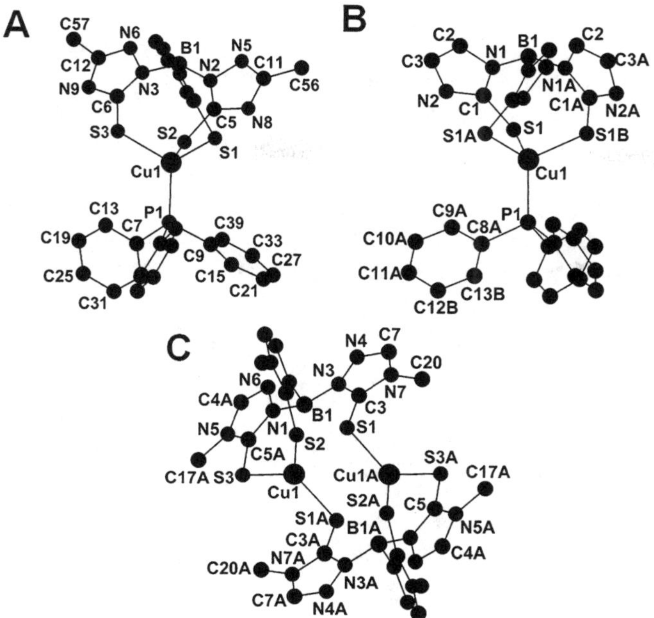

Figure 12. Crystal structures of A) Tr$^{R,R\,spm}$Cu(PPh$_3$) (105), B) Tm^{tBu}Cu(PPh$_3$) (106), and C) Tr$^{Me,o-Py}_2$Cu$_2$ (107). Hydrogen atoms, t-butyl groups, o-pyridyl groups, and phenyl groups are omitted for clarity.

To examine the stability of these dinuclear complexes and to determine relative formation constants of monodentate and scorpionate ligands bound to Cu(I), NMR titrations of Tr$^{Mes,Me}_2$Cu$_2$, Tr$^{Mes,o-Py}_2$Cu$_2$, and (Br^{Mes}pz^{o-Py})$_2$Cu$_2$ with PPh$_3$, thiourea, and pyridine were performed (*105*). Addition of two equivalents of PPh$_3$ ligand to each of the dinuclear complexes was sufficient to give the corresponding monomeric Cu(I) PPh$_3$ complexes, whereas thiourea titrations reached completion only at a much higher ligand to complex ratios. Pyridine titrations showed no reactivity with any dimeric complexes even upon addition of 40-64 equivalents, indicating that these Cu(I) complexes have a much greater affinity for phosphorus- and sulfur-donor ligands compared to nitrogen-donor ligands.

The Tr^R, Tm^R, and Br^{R}pz^R ligands bind Cu(I) in two coordination environments: mononuclear complexes with tetrahedral geometry and dinuclear complexes with trigonal planar geometry. In contrast to the majority of mononuclear Cu(I) complexes, Cu(I) complexes with these tridentate thione ligands are substantially more air stable. Since most reports of copper-thione complexes are entirely structural, there is a great need for characterizing the reactivity of these complexes to determine how thione coordination affects copper oxidation and redox cycling, a major factor in ergothioneine antioxidant behavior, and how well these copper-thione complexes model sulfur-rich copper sites in metalloproteins.

Zinc-Thione Complexes

A large number of zinc enzymes perform thiolate alkylations for various purposes (*50*), the most prominent of which are cobalamin-independent methionine synthase (*108*) and the Ada DNA repair protein (*109*). In these enzymes, zinc is bound to histidine and cysteine residues in N2S, NS2, or S3 donor environments, and it is generally believed that thiolate alkylation is facilitated by zinc binding (*110*). Although thiols and thiones do not behave identically, synthesis of zinc-thione complexes allows the study of sulfur-rich zinc sites in these metalloproteins (*110*).

The [Zn(diap)$_2$(OAc)$_2$]·H$_2$O complex (Figure 13A) was synthesized with 1,3-diazepane-2-thione (Figure 3), a ligand similar to thiourea but with sufficient steric bulk to restrict the coordination number. The distorted tetrahedral coordination around the zinc ion in [Zn(diap)$_2$(OAc)$_2$] consists of two terminal S-donating thione ligands and one oxygen atom from each of two acetate anions. The Zn–S bond distances (2.322(1) and 2.340(1) Å) are very similar to Zn–S bond distances in other reported pseudo-tetrahedral zinc–thiolate complexes (*111*).

A similar coordination study of *N,N'*-bis(3-aminopropyl)ethylene-diamine (bapen) was performed by the synthesis of the mononuclear [Zn(bapen)(ttcH)]·CH$_3$CH$_2$OH complex (Figure 13B) (ttcH$_3$ = trithiocyanurate dianion, Figure 3), where ttcH is bound to the zinc as a bidentate ligand in the *cis* configuration through N and S donors. The Zn(II) ion in this complex adopts a substantially distorted octahedral geometry, coordinated to the four N atoms of the bapen ligand (Zn-N = 2.104(2) Å) and the two S and N atoms of the ttcH^{2-} anion (Zn-S = 2.5700(7) Å and Zn-N = 2.313(2) Å) in the [cis] configuration (*46*).

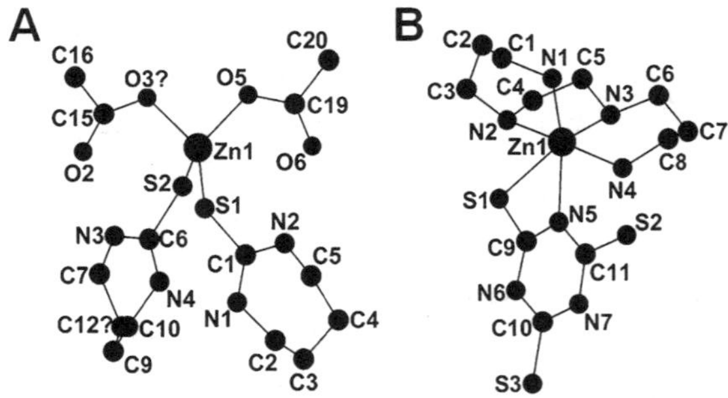

Figure 13. Crystal structures of A) Zn(diap)$_2$(OAc)$_2$ (111) and B) Zn(bapen)(ttcH) (46). Hydrogen atoms are omitted for clarity.

Scorpionate Complexes with Zinc

The tripodal sulfur ligand hydrotris(*t*-butylthioimidazolyl)borate (Tti[tBu]) was also reported to coordinate zinc to form Tti[tBu]Zn(OClO$_3$). Recrystallization of Tti[tBu]Zn(OClO$_3$) from methanol in the presence of 2-butylmercaptan forms the cationic zinc tetramer [Tti^{tBu}Zn]$_4$[ClO$_4$]$_4$ (Figure 14A) as determined by X-ray structural analysis. In this tetramer, each Tti^{tBu} ligand is coordinated through one of its sulfur atoms to each of two neighboring zinc ions, with the third sulfur atom bridging them. The bridging Zn–S bonds (2.402(3) - 2.479(2) Å) are approximately 0.15 Å longer than those to the terminal sulfur atoms (2.291(5) - 2.342 (2) Å), and the intraring S–Zn–S angles of 91.90(6) to 105.16(7)° are among the smallest of all S–Zn–S angles in the molecule (*112*).

The tris(thioimidazolyl)borate scorpionate ligands (Tti^R; Figure 3) provide a S3 donor set and were used to synthesize 11 new complexes of the formula Tti^{R}Zn(SR′) (R = xylene, *t*-butyl, tolyl, cumenyl; SR′ = Et, ^{i}Pr, tolyl, Ph, C$_6$H$_4$(*p*-NO$_2$), C$_6$H$_3$(*o*-OCH$_3$)$_2$, C$_6$F$_5$, and others) to understand why unusual sulfur-rich donor environments are common for zinc-containing enzymes (*110*). To synthesize these zinc-thione-thiolate complexes, KTti^R was combined with zinc nitrate or zinc perchlorate to generate Tti^{R}Zn(ONO$_2$) or Tti^{R}Zn(OClO$_3$) *in situ*, and then the desired sodium thiolate was added. The structure of an example thione-thiolate complex, Tti^{Xyl}Zn(SEt), is shown in Figure 14B. Tti^R ligand conformation in this series of Zn(II) complexes varies little, with the Zn-S distances of 2.33(1) to 2.39(1) Å. The terminal Zn-S thiolate bond lengths range from 2.252(6) to 2.305(3) Å in these complexes, with a slight tendency for the aliphatic thiolates to have shorter Zn-S bonds than the aromatic thiolates (*110*).

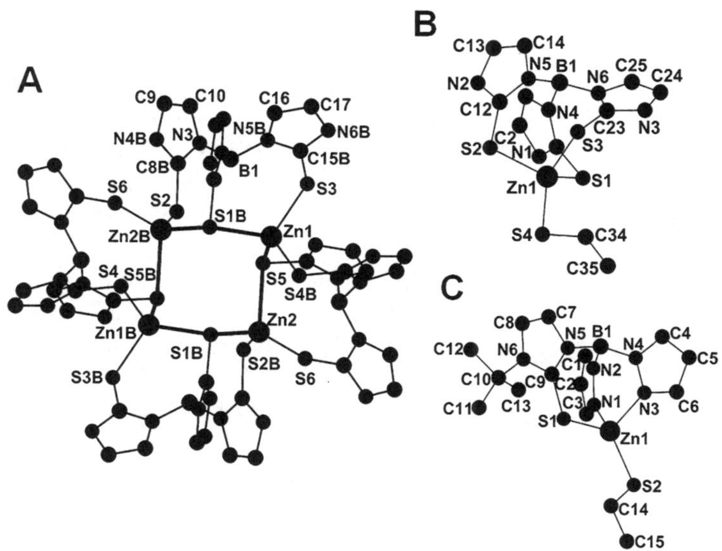

Figure 14. Crystal structures of A) [Tti^{tBu}Zn]$_4$[ClO$_4$] (112), B) Tti^{Xyl}Zn(SEt) (110), and C) L^{1}Zn(SEt) (113). Xylyl groups, t-butyl groups, and hydrogen atoms are omitted for clarity.

The tripodal thione ligands bis(pyrazolyl)(3-*tert*-butyl-2-thioimidazol-1-yl)hydroborate) (**L¹**) and bis(pyrazolyl)(3-isopropyl-2-thioimidazol-1-yl)hydroborate (**L²**, Figure 3) are similar to the tris(thioimidazolyl)borates, but provide N2S coordination. Combined with zinc nitrate or zinc chloride and the corresponding thiolate, these ligands yielded a total of 17 (**L¹**)Zn(SR) and (**L²**)Zn(SR) complexes (R = Et, CH_2CF_3, tolyl, Ph, $C_6H_4(p\text{-}NO_2)$, $C_6H_3(o\text{-}OCH_3)_2$, C_6F_5, and others) (*113*). X-ray structural analysis of (**L¹**)Zn(SEt) (Figure 14C) confirmed tetrahedral N2S2 coordination of zinc, with Zn-thione and Zn-SEt bond distances of 2.349(3) and 2.228(3) Å, respectively (*113*). There is a remarkable similarity between all 17 structures; the comparable bond lengths vary not more than 0.02 Å, and the Zn-S thiolate bonds are characteristically shorter than the Zn-S thione bonds by 0.132 Å.

To understand why sulfur-rich coordination environments are necessary for zinc-catalyzed thiolate alkylations, methylation reactions were performed with Zn(II)-scorpionate complexes. Effects of systematic variation on the zinc coordination environment were determined using zinc-thiolate complexes of tris(3,5-dimethylpyrazolyl)borate (Tp*) ligands as N3 donors (*114*), **L¹** and **L²** ligands as N2S donors (*113*), and the Tti^R ligands as S3 donors (*110*). Methyl iodide in chloroform solution was used as the methylating agent instead of the biological methyltetrahydrofolate, a methylammonium species. Methylammonium salts and trimethyl phosphate were not reactive enough for efficient alkylation of the model zinc-thiolate complexes (*114*). Methylation reactions of the Tti^RZn(SR′) complexes to yield Tti^RZnI and MeSR′, were the fastest of the tested model complexes, several orders of magnitude faster (10 min to several hours for complete methylation) (*110*) than methylation of the Tp*Zn-thiolate complexes (*114*), and significantly faster than the **L¹** and **L²** copper complexes (one day to two weeks) (*113*). These methylation studies indicate that zinc coordination by sulfur donors is the most effective way to increase zinc-thiolate electron density and nucleophilicity, and support the zinc-bound thiolate mechanism of alkylation, but methylation rates for these model complexes are far lower than those observed in enzymes (*110, 113*). Thus, these methylation studies complexes have established mechanistic details for this reaction, but researchers are still far from understanding the factors controlling alkylation efficiency in zinc enzymes.

Unsurprisingly, thione ligands typically bind zinc(II) though the sulfur atom in a tetrahedral geometry; however the six-coordinate Zn(bapen)(ttcH) complex (*46*) demonstrates that thione ligands may also bind through available nitrogen atoms, similar to a few reported Fe(II), Ni(II), and Cu(I) thione complexes (*68, 103*). Synthesis of Zn(II) complexes with scorpionate thione ligands to study thiolate methylation reactions has resulted in a multitude of structurally characterized, tetrahedral Zn(II) thione/thiolate complexes. Zinc-thione complexes with the Tti^R ligands are the most promising models for zinc-containing thiolate alkylation enzymes, but their slow alkylation rates demonstrate the need to adjust the electronic properties of the thione scorpionate ligands to better model these reactions.

Selone Compounds as Ligands

Reports of metal-selone complexes are rare and primarily limited to structural studies, likely because biological selone compounds were not identified until recently (*4, 12, 34, 115*). Selone compounds are expected to be more nucleophilic than their thione analogs, resulting in stronger coordination bonds (*116–118*), although formation constants of these ligands have never been measured for comparison. Metal-selone complexes are more readily oxidizable than their thione counterparts, resulting in increased air sensitivity, but their synthesis is no more difficult than that of the air-sensitive thione complexes.

Iron-Selone Complexes

Very few iron-selone complexes have been reported, possibly due to the facile oxidation of Fe(II) or the selone ligand, and the need for air-free synthesis. The only report of iron-selone complexes is the synthesis and characterization of the mononuclear complexes Fe(dmise)$_2$Cl$_2$, [Fe(dmise)$_4$][OTf]$_2$, and [Fe(dmise)$_4$][BF$_4$]$_2$ (dmise = 1,3-*N*,*N*-dimethylimidazole selone; Figure 3) (*73*). These selone compounds are synthesized by addition of dmise (Figure 3) to FeCl$_2$·H$_2$O, Fe(OTf)$_2$, and Fe(BF$_4$)$_2$, respectively (*73*), and are analogous to the previously discussed iron-thione compounds. The selone complexes Fe(dmise)$_2$Cl$_2$ (Figure 15A) and [Fe(dmise)$_4$][BF$_4$]$_2$ (Figure 15B) crystallize in distorted tetrahedral geometry with Fe-Se bond distances ranging from 2.455(2) to 2.510(1) Å. When dmise coordinates Fe(II), the infrared stretches for the Se=C bonds shift very slightly to higher energies (1149-1155 cm^{-1}) compared to unbound dmise (1148 cm^{-1}), indicating primarily donor bond formation. This strengthening of the Se=C bond in Fe(II) selone complexes upon dmise binding is in contrast to the slight Fe-thione backbonding observed for the analogous thione complexes Fe(dmise)$_2$Cl$_2$, [Fe(dmise)$_4$][OTf]$_2$, and [Fe(dmise)$_4$][BF$_4$]$_2$ (*73*).

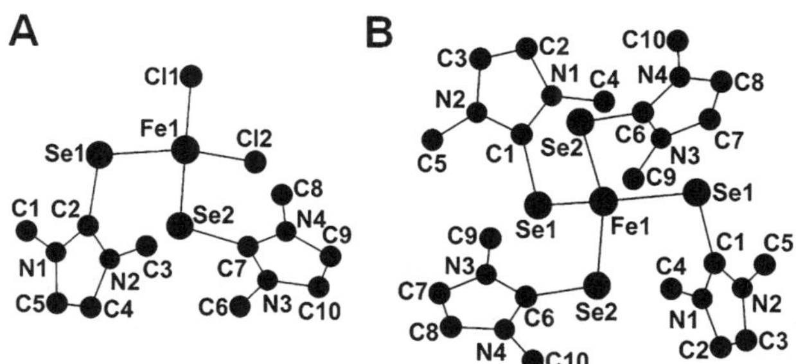

Figure 15. Crystal structures of A) Fe(dmise)$_2$Cl$_2$ and B) [Fe(dmise)$_4$][BF$_4$]$_2$ (73). Hydrogen atoms and counterions are omitted for clarity.

The ^{1}H NMR spectra of these three Fe(II)-selone complexes show imidazole proton resonances shifting downfield by δ 0.4-0.5 compared to the unbound ligand (δ 6.77), indicating increased aromaticity of the selone imidazole ring (Figure 2) upon coordination, similar to analogous Fe-thione complexes (*73*). As observed for the analogous thione complexes, the electrochemistry of these iron-bound selone ligands also change when bound to iron, with the Ep_a of dmise shifting by 40 mV compared to unbound dmise (372 mV), indicating that iron-bound selones are more readily oxidizable (*73*).

Cobalt-Selone Complexes

In 1997, Williams, *et al.* synthesized Co(dmise)₂Cl₂ (Figure 16A), the first metal halide complex to be prepared with the dmise ligand, by combining CoCl₂ and dmise in methanol and heating. Upon cooling, blue-green crystals of Co(dmise)₂Cl₂ (Figure 16A) precipitated (*119*). In this complex, the Co(II) center adopts distorted tetrahedral geometry with Co-Se distances of 2.277(3) and 2.456(3) Å (*120*). No further characterization was reported; thus, additional infrared and electrochemical studies are required to understand cobalt-selone binding and for comparison with selone complexes of other metal ions.

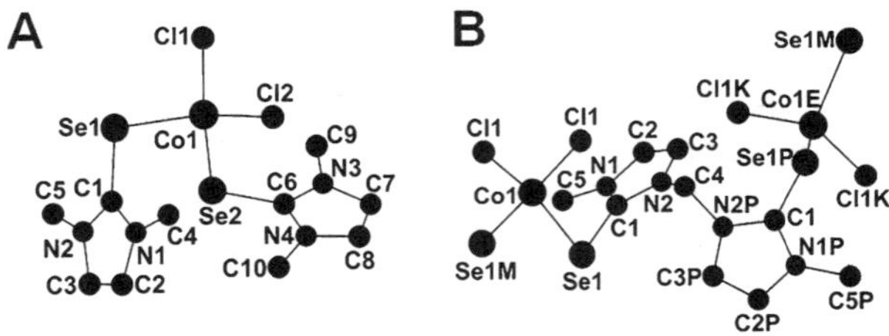

Figure 16. Crystal structures of A) Co(dmise)₂Cl₂ (119) and B) [Co(mbis)Cl₂]ₙ (81). Hydrogen atoms are omitted for clarity.

Similar reactions were also carried out with the ethyl- and methyl-bridged selone ligands, ebis and mbis (Figure 3). Treating CoCl₂ with mbis or ebis (Figure 3) in tetrahydrofuran forms the polymeric [Co(mbis)Cl₂]ₙ (Figure 16B) or [Co(ebis)Cl₂]ₙ complexes (*81*). In both complexes, the Co(II) metal center adopts distorted tetrahedral geometry, bound to two terminal chloride ligands and two Se atoms from different bidentate selone ligands to form diselone-Co(II) chains. In these mbis and ebis complexes, the symmetrically equivalent Co–Se bond distances are 2.447(4) and 2.423(5) Å, respectively. The most unusual structural feature in [Co(mbis)Cl₂]ₙ is the formation of infinite helical chains that are of different chiralities (right and left handed), despite the fact that they contain the same components and linking sequence (*81*). In contrast, the [Co(ebis)Cl₂] molecules form zigzag chains. The infrared spectrum of [Co(mbis)Cl₂]ₙ shows a shift of the C=Se stretching frequency to lower energy (1096 cm⁻¹) compared to unbound mbit (1128 cm⁻¹), indicating weak backbonding of Co(II) to mbis. In [Co(ebis)Cl₂]ₙ, however, the C=Se stretching frequency is shifted to higher energy (1135 cm⁻¹) relative to unbound ebis (1126 cm⁻¹; Table 1), indicating donor bond formation.

As expected, the few structurally characterized cobalt-selone complexes demonstrate that selone ligands bind primarily to Co(II) through the selenium atom with distorted tetrahedral geometry. Similar to their thione analogues mbit and ebit, the bidentate ligands mbis and ebis form polymeric complexes with Co(II). The shifts in C=Se stretching frequencies observed for [Co(mbis)Cl$_2$]$_n$ and [Co(ebis)Cl$_2$]$_n$ indicate that there is no single binding mode of these multidentate ligands to Co(II). Further synthesis and structural characterization of Co(II)-selone complexes as well as infrared and electrochemical studies are required for a comprehensive understanding of cobalt-selone binding.

Nickel-Selone Complexes

Reports of Ni-selone complexes are limited to structural studies. A single paper investigates the substitution reaction of [NiL(MeCN)][BF$_4$]$_2$ (L = 2,5,8-trithia(2,9)-1,10-phenanthrolinophane) with dmise to yield [NiL(dmise)][BF$_4$]$_2$ (Figure 17A) (*121*). This Ni(II) complex has octahedral geometry, with a Ni-Se bond distance of 2.592(1) Å.

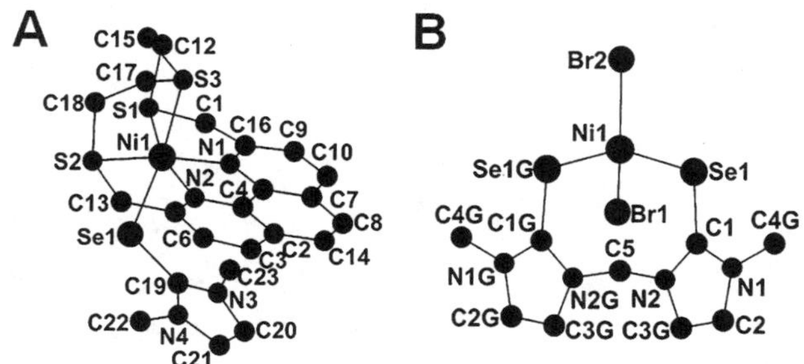

Figure 17. Crystal structures of A) [NiL(dmise)][BF$_4$]$_2$ (L = 2,5,8-trithia(2,9)-1,10-phenanthrolinophane) (121) and Ni(mbis)Br$_2$ (81). Hydrogen atoms are omitted for clarity.

Mononuclear Ni(II) diselone complexes were also synthesized by addition of mbis or ebis to Ni(PPh$_3$)$_2$Br$_2$ under nitrogen to afford Ni(mbis)Br$_2$ (Figure 17B) and Ni(ebis)Br$_2$ (*81*). Unlike the analogous thione complexes, these selone complexes are very sensitive to air and moisture. Ni(mbis)Br$_2$ has distorted tetrahedral geometry around the Ni(II), with symmetry equivalent Ni-Se bond distances of 2.394(1)Å. Infrared C=Se stretches for nickel-bound mbis and ebis are at 1133 and 1128 cm^{-1}, respectively, very slightly shifted to higher energies compared to the unbound ligands (Table 1), indicating no Ni(II)-selone backbonding (*81*).

Nickel-selone complexes have not been widely investigated; however, the few complexes that have been synthesized demonstrate both octahedral and distorted tetrahedral geometry around Ni(II). In contrast to their polymeric complexes with Co(II), the bidentate selone ligands mbis and ebis form monodentate complexes with Ni(II). The infrared C=Se stretching frequencies for Ni(mbis)Br$_2$ and Ni(ebis)Br$_2$ are similar to that of [Co(ebis)Cl$_2$]$_n$, Fe(dmise)$_2$Cl$_2$, and [Fe(dmise)$_4$]$^{2+}$ indicating primarily donor bond formation in these complexes. In contrast, weak backbonding is observed for the Co(II)-selone bond in [Co(mbis)Cl$_2$]$_n$. Since so little is known, any additional synthetic, computational modeling, or reactivity studies would greatly elucidate metal-selone bonding in these complexes.

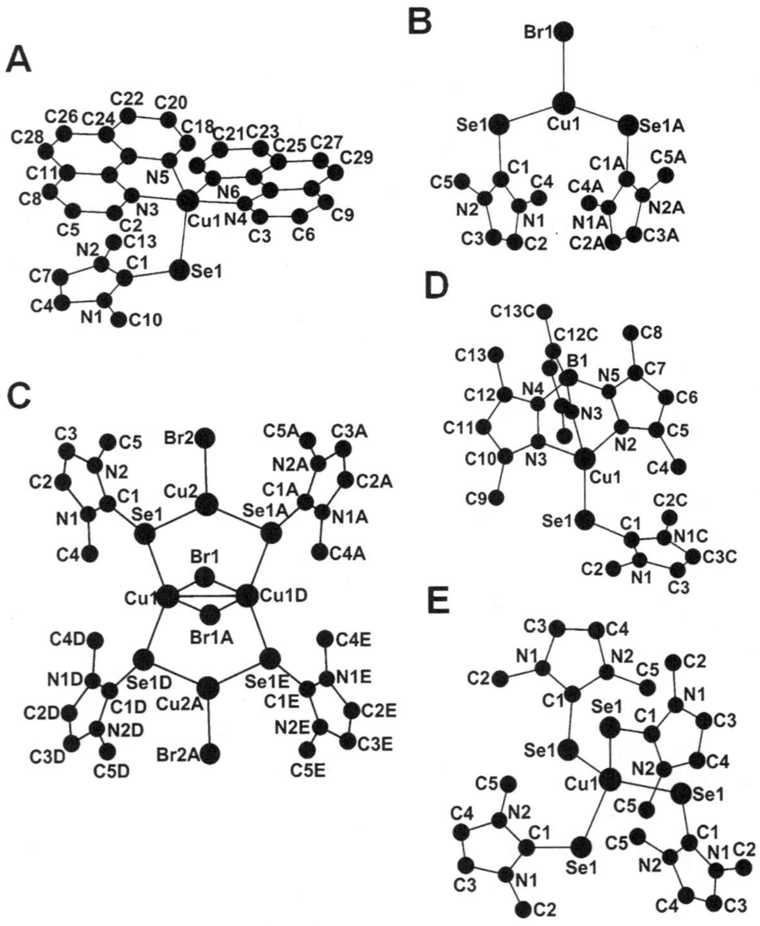

*Figure 18. Crystal structures of A) [Cu(phen)$_2$(dmise)](ClO$_4$)$_2$ (122), B) cis-Cu(dmise)$_2$Br (19), C) Cu(μ-dmise)$_4$(μ-Br)$_2$Br$_2$ (19), D) Tp*Cu(dmise) (14), and E) [Cu(dmise)$_4$](OTf)$_2$ (20). Hydrogen atoms and counterions are omitted for clarity.*

Copper-Selone Complexes

In the only reported example of a Cu(II) selone complex, the distorted trigonal-bipyramidal Cu(II) center in $[Cu(phen)_2(dmise)][ClO_4]_2$ (phen = 1,10-phenanthroline; Figure 18A) is coordinated by four nitrogen donors from the two phenanthroline ligands and by the selenium atom from the dmise ligand. In this complex, the Cu-Se distance of 2.491(3) Å is slightly longer than expected for a Cu(II)-selone complex, likely due to the strong π-conjugation properties of the phenanthroline ligands (*122*). Trigonal-bipyramidal coordination geometry is not common for Cu(II) complexes, particularly when they contain ligands with reducing properties (*123*).

Using the same dmise ligand, the distorted-trigonal-planar copper-selone complexes *trans*-Cu(dmise)$_2$Cl, *cis*-Cu(dmise)$_2$Br (Figure 18B), and *cis*-Cu(dmise)$_2$I were synthesized by combining the respective Cu(I) halides with two equivalents of dmise (*19*). Similarly to their thione counterparts, these copper selone complexes have identical Cu–Se bond distances of 2.34 Å, but unlike the thione analogs, *cis* geometry is favored upon selone coordination, likely due to the greater polarizability of the selenium atom in addition to the polarizable halides. When CuI and CuBr are treated with only one equivalent of dmise, the tetrameric $Cu_4(\mu\text{-dmise})_4(\mu\text{-Br})_2Br_2 \cdot 0.5CH_3CN$ (Figure 18C) and $Cu_4(\mu\text{-dmise})_4(\mu\text{-I})_2I_2 \cdot 1.5CH_3CN$ complexes are formed (*19*). The structures of both these complexes are similar, and notable for the two different coordination geometries around the copper(I) ions. In $Cu_4(\mu\text{-dmise})_4(\mu\text{-Br})_2Br_2$, two selenium atoms and one bromide ion each coordinate the Cu(2) and Cu(2A) ions in a distorted trigonal planar geometry, with average Cu–Se bond distances of 2.41 Å. The two additional copper centers (Cu(1) and Cu(1A)) adopt distorted tetrahedral geometries, with a $Cu_2(\mu\text{-Br})_2$ core, two bridging selenium atoms from the dmise ligand, and an average Cu–Cu distance of 2.631(2) Å (*19*).

In contrast to the backbonding observed for their thione analogs, the infrared spectra of *trans*-Cu(dmise)$_2$Cl, *cis*-Cu(dmise)$_2$Br (Figure 18B), and *cis*-Cu(dmise)$_2$I, $Cu_4(\mu\text{-dmise})_4(\mu\text{-Br})_2Br_2 \cdot 0.5CH_3CN$, and $Cu_4(\mu\text{-dmise})_4(\mu\text{-I})_2I_2 \cdot 1.5CH_3CN$ show copper-bound C=Se bond stretching frequencies (1149-1163 cm^{-1}) shifting to higher energies relative to unbound dmise (1148 cm^{-1}), indicating Cu-Se donor bonding (*19*). $^{77}Se\{^1H\}$ NMR spectra of these complexes show significant upfield shifts for the copper-bound dmise resonance, ranging from δ -13.4 to -75.0 compared to unbound dmise (δ 21.8). Cu(II/I) reduction potentials for these Cu(I) selone complexes range from -340 to -360 mV, significantly lower than their thione counterparts (-177 to -284 mV) (*19*).

In an effort to understand the role of copper-selenium coordination in metal-mediated DNA damage prevention, Kimani, *et al.* synthesized biologically relevant Cu(I) dmise complexes with trinitrogen donor ligands (*14*). For each of the complexes, [TpmCu(dmise)][BF$_4$], [Tpm*Cu(dmise)][BF$_4$], [Tpm^{iPr}Cu(dmise)][BF$_4$], and Tp*Cu(dmise) (Tpm = tris(pyrazolyl)methane, Tpm* = tris(3,5-dimethylpyrazolyl)methane, Tpm^{iPr} = tris(3,5-isopropylpyrazolyl)methane, Tp* = hydrotris(3,5-dimethylpyrazolyl)borate) (*14*), each copper ion adopts distorted tetrahedral coordination with similar Cu-Se bond lengths ranging from 2.294(6) to 2.33(8) Å. The short Cu-Se bond distances for all four complexes suggest strong donor interactions between the soft selenium ligand and the soft copper metal ion, consistent with slight shifts to higher C=S bond stretching energies in the infrared spectra (2-3 cm^{-1}; Table 1) (*14*), but only a limited number of non-bridging copper selone complexes are available for structural comparison. As is the case with Fe(II) selone complexes, ^{1}H NMR resonances of the imidazole protons and ^{13}C{^{1}H} NMR resonances of the C=S carbon atom shift downfield by δ 0.20-0.41 and δ 2.4-8.0 compared to unbound dmise, comparable to NMR shifts observed for the analogous Cu(I)-thione complexes (*14*). Thus, both copper-bound dmit and dmise favor the zwitterionic resonance form of the chalcogenone ligand (Figure 2) in these complexes, despite the slight backbonding observed for Cu-S interactions and primarily donor interactions for the Cu-Se bonding as determined by infrared spectroscopy.

Electrochemical studies of [TpmCu(dmise)][BF$_4$], [Tpm*Cu(dmise)][BF$_4$], [Tpm^{iPr}Cu(dmise)][BF$_4$], and Tp*Cu(dmise) in acetonitrile showed very low Cu(II/I) redox potentials for these complexes (-283 to 390 V), lower by 0.635 to 0.874 V relative to the analogous complexes with acetonitrile ligands in place of dmise, and an average of 0.224 V lower than their thione analogs (*17*). Thus, selone binding to Cu(I) stabilizes Cu(II) in these complexes much more effectively than analogous thione binding. In fact, the very low potentials for these copper-selone complexes are near or below the potential of NADH (-324 mV (*97*)), suggesting that similar Cu(I)-selone binding *in vivo* may prevent copper redox cycling and generation of damaging reactive oxygen species.

Additional studies to determine the ability of selone coordination to reduce Cu(II) to Cu(I) were carried out by the addition of 5.5 equiv dmise to Cu(OTf)$_2$ to afford the reduced [Cu(dmise)$_4$][OTf] complex (Figure 18E) and the oxidized [Cu(dmise)$_3$][OTf)$_2$] (*20*). This reaction represents the first synthesis of an imidazolium triselenide by Cu(II) reduction. The Cu(I) center of the [Cu(dmise)$_4$]$^+$ cation has tetrahedral geometry with symmetrically equivalent Cu-Se bond distances of 2.454(1) Å (*20*). A similar reaction with the analogous thione dmit results in formation of the trigonal planar complex [Cu(dmit)$_3$][OTf], illustrating the significant differences in binding and coordination chemistry that can exist between similar ligands. This difference is also observed in the infrared spectra of the two complexes, with the C=Se stretching vibration shifting much less to lower energy than the C=S stretching vibration (2 cm^{-1} vs. 34 cm^{-1}) upon chalcogenone binding, indicating significantly more backbonding interactions in the Cu-S bond.

Dmise was also added to Cu(OTf)$_2$ in acetonitrile, and the increasing absorbance of [Cu(NCCH$_3$)$_4$]$^+$ was monitored to determine the kinetics of copper reduction by this selone. Initial rates indicated that dmise reduces Cu(II) to Cu(I) three times faster than dmit (0.50 vs. 0.18 s^{-1}) in a first-order reaction with respect to ligand concentration (*20*). Under oxygen-free conditions, this reduction was four orders of magnitude slower, indicating the involvement of O$_2$ in this reaction. Thus, biological selone coordination to Cu(II) in the presence of oxygen may promote reduction to ROS-generating Cu(I) faster than thione binding.

Only a few copper-selone complexes have been reported, but these complexes display a wide variety of coordination environments, ranging from tetrahedral and trigonal geometry with Cu(I) to trigonal bipyramidal geometry with Cu(II). Similar to the thiones, selone ligands can be terminal and bridging ligands, although so few complexes have been reported that it is difficult to compare the prevalence of selone binding modes with their thione analogs. Copper-selone complexes behave similarly to other late-transition-metal-selone complexes in that infrared spectroscopy indicates little or no backbonding in the metal-selone interaction, in contrast to their thione analogs that can show significant weakening of the C=S bond due to Cu-S backbonding (Table 1). In the few electrochemical studies conducted, copper-selone binding significantly lowers the Cu(II/I) redox potential compared to thione binding, a result that may indicate increased antioxidant ability of selones relative to analogous thione compounds.

Zinc-Selone Complexes

Only a single zinc-selone complex has been reported, $Zn(dmise)_2Cl_2$ (Figure 19), prepared by combining $ZnCl_2$ and dmise in boiling acetonitrile. X-ray crystallographic studies of this complex show the expected tetrahedral geometry about the zinc(II) center with Zn-Se bond distances of 2.4691(4) Å and 2.4873(4) Å (*124*). Although this is the only example of a zinc-selone complex, this and future complexes may be of interest in the development of precursors for zinc selenide-based materials used in optical electronics, laser technology, semiconductor research, and blue-green light emitting diode materials (*125–127*).

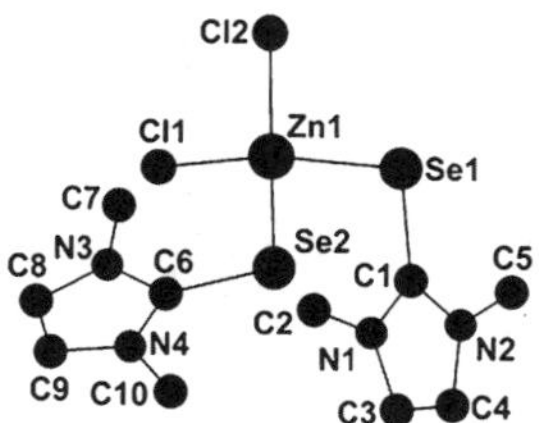

Figure 19. Crystal structure of Zn(dmise)₂Cl₂ (124). Hydrogen atoms are omitted for clarity.

Metal-selone complexes are rare; however, they have become more investigated in the past few years due to the recent discovery of selenoneine and other biological selones (*11, 128*). The few reported metal-selone complexes are easily synthesized with a variety of late-transition-metal ions, including Fe(II), Co(II), Ni(II), Cu(I), Cu(II), and Zn(II). Typical metal thione interactions involve significant backbonding from the metal center, but most metal-selone complexes interact primarily through donor bonds with little or no evidence of metal backbonding. Selone complexes are often similar to analogous thione complexes, but seemingly minor M-S and M-Se bonding differences can result in different coordination numbers and binding modes. Drawing wide-ranging conclusions about selone coordination or comparisons to analogous thione complexes is difficult, however, due to the paucity of well-characterized complexes and lack of reactivity studies.

Conclusions

The metal coordination properties of thione- and selone-containing ligands have been investigated for multiple applications, including antitumor and antioxidant properties, protein active-site models, and optical materials. The wide structural diversity exhibited by thione and selone donor ligands upon binding to first-row-transition metals explains the considerable recent interest in their coordination chemistry. Many studies of thione and selone complexes with these metal ions demonstrate a rich coordination chemistry with varying coordination numbers and geometries, reinforcing the fact that subtle differences in both metal and ligand properties can have significant consequences for resulting structure and reactivity.

Although electrochemical studies of metal-thione and -selone complexes are still uncommon, they indicate that both ligand and metal redox properties are altered upon coordination. Copper-selone complexes have lower Cu(II/I) reduction potentials compared to their thione analogs, and this difference might be at least partly responsible for the greater antioxidant activity observed for selones. The thione-scorpionate class of ligands have provided a facile system for modeling sulfur-rich metalloenzyme active sites, including nickel-containing hydrogenases and zinc-containing thiolate alkylation enzymes. Despite the utility of thione-scorpionate ligands, no similar selone-scorpionate metal complexes have been reported. Reactivity studies of thione- and selone-metal complexes are also scarce, and such studies related to antioxidant activity and functional metalloenzyme modeling would be particularly relevant.

Sulfur and selenium should behave similarly due to their near-identical electronegativities; however, their different atomic radii (1.04 and 1.17 Å, respectively) (*129*) result in increased bond distances and polarizability for selenium compared to sulfur. Selenium's longer bond distances and greater polarizability may reduce backbonding and result in the observed differences in analogous thione- and selone-metal coordination, but current studies of these ligands do not provide decisive conclusions as to the specific properties that account for differences in their coordination chemistry. Irrespective of the specific cause, these differences between sulfur and selenium as well as differences in the properties of thione and selone ligands result in varied coordination complexes. This is not surprising, since organisms go to significant lengths to include selenium in place of sulfur in amino acids. In large part, it remains to be determined how much of the rich coordination chemistry demonstrated by both thiones and selones impacts the biological functions of naturally occurring thione and selone compounds.

Acknowledgments

We thank the National Science Foundation (CHE 1213912) for financial support, and B. S. S. thanks the Clemson University Chemistry Department for a graduate fellowship.

Abbreviation list

DNA = deoxyribonucleic acid
ROS = reactive oxygen species
NMR = nuclear magnetic resonance
GPx = glutathione peroxidase
DPPH = 2,2-diphenyl-1-picrylhydrazyl
PTU = 6-*n*-propyl-2-thiouracil
MTU = 6-methyl-2-thiouracil
amt = 4-amino-3-methyl-1,2,4-triazole-5-thione
dmit = *N,N'*-dimethylimidazole thione
dmise = *N,N'*-dimethylimidazole selone
mbit = 1,1'-methylenebis(1,3-dihydro-3-methyl-2*H*-imidazole-2-thione)
mbis = 1,1'-methylenebis(1,3-dihydro-3-methyl-2*H*-imidazole-2-selone)
ebit = 1,1'-(1,2-ethanediyl)bis(3-methyl-imidazole-2-thione)
ebis = 1,1'-(1,2-ethanediyl)bis(3-methyl-imidazole-2-selone)
ttcH$_3$ = 2,4,6-trimercaptotriazine
sbim = 1*H*-benzimidazole-2(3*H*)-thione
tsacH = thiosaccharinate
diap = 1,3-diazepane-2-thione
3-Me-HPT = 1-hydroxy-3-methylpyridine-2(1H)-thione
Tp^R = tris(pyrazolyl)borate
Tm^R = hydrotris(methimazolyl)borate
Bm^R = hydrobis(methimazolyl)borate
Tr^R = hydrotris(thioxotriazolyl)borate
Br^R = hydrobis(thioxobiazolyl)borate
HBmit$_3$ = hydrotris(1-methylimidazol-2-ylthio)borate
Ep_a = anodic (oxidation) potential
LOX = lipogenase

References

1. Battin, E. E.; Zimmerman, M. T.; Ramoutar, R. R.; Quarles, C. E.; Brumaghim, J. L. *Metallomics* **2010**, *3*, 503–512.
2. Ramoutar, R. R.; Brumaghim, J. L. *Cell Biochem. Biophys.* **2010**, *58*, 1–23.
3. Ramoutar, R. R.; Brumaghim, J. L. *J. Inorg. Biochem.* **2007**, *101*, 1028–1035.
4. Yamashita, Y.; Yamashita, M. *J. Biol. Chem.* **2010**, *285*, 18134–18138.
5. Franzoni, F.; Colognato, R.; Galetta, F.; Laurenza, I.; Barsotti, M.; Di Stefano, R.; Bocchetti, R.; Regoli, F.; Carpi, A.; Balbarini, A.; Migliore, L.; Santoro, G. *Biomed. Pharmacother.* **2006**, *60*, 453–457.
6. Aruoma, O. I.; Whiteman, M.; England, T. G.; Halliwell, B. *Biochem. Biophys. Res. Commun.* **1997**, *231*, 389–391.
7. Battin, E. E.; Brumaghim, J. L. *Cell Biochem. Biophys.* **2009**, *55*, 1–23.
8. Brown, K. M.; Arthur, J. R. *Public Health Nutr.* **2001**, *4*, 593–599.
9. Schrauzer, G. N. *J. Am. Coll. Nutr.* **2001**, *20*, 1–4.

10. Papp, L. V.; Holmgren, A.; Khanna, K. K. *Antioxid. Redox Signaling* **2010**, *12*, 793–795.

11. Papp, L. V.; Lu, J.; Holmgren, A.; Khanna, K. K. *Antioxid. Redox Signaling* **2007**, *9*, 775–806.

12. Lu, J.; Holmgren, A. *J. Biol. Chem.* **2009**, *284*, 723–727.

13. Silvestri, A.; Ruisi, G.; Girasolo, M. A. *J. Inorg. Biochem.* **2002**, *92*, 171–176.

14. Kimani, M. M.; Brumaghim, J. L.; VanDerveer, D. *Inorg. Chem.* **2010**, *49*, 9200–9211.

15. Raper, E. S. *Coord. Chem. Rev.* **1985**, *61*, 115–184.

16. Akrivos, P. D. *Coord. Chem. Rev.* **2001**, *213*, 181–210.

17. Spicer, M. D.; Reglinski, J. *Eur. J. Inorg. Chem.* **2009**, 1553–1574.

18. Pettinari, C. In *Scorpionates II: Chelating Borate Ligands*; Imperial College Press: London, 2008; pp 381–415.

19. Kimani, M. M.; Bayse, C. A.; Brumaghim, J. L. *Dalton Trans.* **2011**, *40*, 3711–3723.

20. Kimani, M. M.; Wang, H. C.; Brumaghim, J. L. *Dalton Trans.* **2012**, *41*, 5248–5259.

21. Battin, E. E.; Perron, N. R.; Brumaghim, J. L. *Inorg. Chem.* **2006**, *45*, 499–501.

22. Briggs, I. J. *Neurochemistry* **1972**, 27–35.

23. Epand, R. M.; Epand, R. F.; Wong, S. C. *J. Clin. Chem. Clin. Biochem.* **1988**, *26*, 623–626.

24. Ey, J.; Schomig, E.; Taubert, D. J. *J. Agric. Food Chem.* **2007**, *55*, 6466–6474.

25. Short, R. D. J.; Baskin, S. I. Antioxidative activity of ergothioneine and ovothiol. In *Oxidants, Antioxidants and Free Radicals*; Baskin, S. I., Salem, H., Eds.; Taylor&Francis: Washington, DC, 1997; pp 203–206.

26. Aruoma, O. I.; Spencer, J. P. E.; Mahmood, N. *Food Chem. Toxicol.* **1999**, *37*, 1043–1053.

27. Deiana, M.; Rosa, A.; Casu, V.; Piga, R.; Assunta, D. M.; Aruoma, O. I. *Clin. Nutr.* **2004**, *23*, 183–193.

28. Jang, J.-H.; Aruoma, O. I.; Jen, L.-S.; Chung, H. Y.; Surh, Y.-J. *Free Radicals. Biol. Med.* **2004**, *36*, 288–299.

29. Colognato, R.; Laurenza, I.; Fontana, I.; Coppede, F.; Siciliano, G.; Coecke, S.; Aruoma, O. I.; Benzi, L.; Migliore, L. *Clin. Nutr.* **2006**, *25*, 135–145.

30. Markova, N. G.; Karaman-Jurukovska, N.; Dong, K. K.; Damaghi, N.; Smiles, K. A.; Yarosh, D. B. *Free Radicals Biol. Med.* **2009**, *46*, 1168–1176.

31. Paul, B. D.; Snyder, S. H. *Cell Death Differ.* **2010**, *17*, 1134–1140.

32. De Luna, P.; Bushnell, E. A.; Gauld, J. W. *J. Phys. Chem.* **2013**, *A117*, 4057–4065.

33. Zhu, B. Z.; Mao, L.; Fan, R. M.; Zhu, J. G.; Zhang, Y. N.; Wang, J.; Kalyanaraman, B.; Frei, B. *Chem. Res. Toxicol.* **2011**, *24*, 30–34.

34. Yamashita, Y.; Yabu, T.; Yamashita, M. *World. J. Biol. Chem.* **2010**, *1*, 144–150.

35. Klein, M.; Ouerdane, L.; Bueno, M.; Pannier, F. *Metallomics* **2011**, *3*, 513–520.
36. Cooper, D. S. *N. Engl. J. Med.* **2005**, *352*, 905–917.
37. Roy, G.; Das, D.; Mugesh, G. *Inorg. Chim. Acta* **2007**, *360*, 303–316.
38. Roy, G.; Mugesh, G. *Chem. Biodiversity* **2008**, *5*, 414–439.
39. Ferreira, A. C.; de Carvalho Cardoso, L.; Rosenthal, D.; de Carvalho, D. P. *Eur. J. Biochem.* **2003**, *270*, 2363–2368.
40. Jia, W.; Huang, Y.; Lin, Y.; Jin, G. *Dalton Trans.* **2008**, 5612–5620.
41. Lobana, T. S.; Sharma, R.; Butcher, R. J. *Z. Anorg. Allg. Chem.* **2008**, *634*, 1785–1790.
42. Henke, K. R.; Robertson, D.; Krepps, M. K.; Atwood, D. A. *Water Res.* **2000**, *34*, 3005–3013.
43. Matlock, M. M.; Henke, K. R.; Atwood, D. A. *J. Hazard. Mater.* **2002**, *B92*, 129–142.
44. Matlock, M. M.; Henke, K. R.; Atwood, D. A.; Robertson, D. *Water Res.* **2001**, *35*, 3349–3655.
45. Iltzsch, M. H.; Tankersley, K. O. *Biochem. Pharmacol.* **1994**, *48*, 781–792.
46. Marek, J.; Kopel, P.; Travnicek, Z. *Acta Crystallogr.* **2003**, *C59*, m558–m560.
47. Han, R.; Looney, A.; Mcneill, K.; Parkin, G.; Rheingold, A. L.; Haggerty, B. S. *J. Inorg. Biochem.* **1993**, *49*, 105–121.
48. Vahrenkamp, H. *Acc. Chem. Res.* **1999**, *32*, 589–596.
49. Parkin, G. *Chem. Commun.* **2000**, 1971–1985.
50. Parkin, G. *Chem. Rev.* **2004**, *104*, 699–767.
51. Parkin, G. *New J. Chem.* **2007**, *31*, 1996–2014.
52. Kitajima, N.; Moro-oka, Y. *Chem. Rev.* **1994**, *94*, 737–757.
53. Tolman, W. B. *J. Biol. Inorg. Chem.* **2006**, *11*, 261–271.
54. Trofimekno, S. *J. Am. Chem. Soc.* **1966**, *88*, 1842–1844.
55. Trofimekno, S. *The Coordination Chemistry of Polypyrazolylborate Ligands*; Imperial College Press: London, 1999.
56. Garner, M.; Reglinski, J.; Cassidy, I.; Spicer, M. D.; Kennedy, A. R. *Chem. Commun.* **1996**, 1975–1976.
57. Reglinski, J.; Garner, M.; Cassidy, I. D. S., P. A.; Spicer, M. D.; Armstrong, D. R. *Dalton Trans.* **1999**, 2119–2126.
58. Kimblin, C.; Bridgewater, B. M.; Chruchhill, D. G.; Parkin, G. *Chem. Commun.* **1999**, 2301–2302.
59. Santini, C.; Lobbia, G. G.; Pettinari, C.; Pellei, M.; G., V.; Calogero, S. *Inorg. Chem.* **1998**, *37*, 2301–2302.
60. Soares, L. F.; Silva, R. M. *Inorg. Synth.* **2002**, *33*, 199–202.
61. Soares, L. F.; Silva, R. M.; Doriguetto, A. C.; Ellena, J.; Mascarenas, Y. P.; Castellano, E. E. *J. Braz. Chem. Soc.* **2004**, *15*, 695–700.
62. Alvarez, H. M.; Krawiec, M.; Donovan-Merkert, B. T.; Fouzi, M.; Rabinovich, D. *Inorg. Chem.* **2001**, *40*, 5736–5737.
63. Cassidy, I.; Garner, M.; Kennedy, A. R.; Potts, G. B. S.; Reglinski, J.; Slavin, P. A.; Spicer, M. D. *Eur. J. Inorg. Chem.* **2002**, 1235–1239.
64. Kim, H. R.; Jung, I. G.; Yoo, K.; Jang, K.; Lee, E. S.; Yun, J.; Son, S. U. *Chem. Commun.* **2010**, *46*, 758–760.

65. Kim, H. J.; Graham, D. W.; DiSpirito, A. A.; Alterman, M. A.; Galeva, N.; Larive, C. K.; Asunskis, D.; Sherwood, P. N. A. *Science* **2004**, *305*, 1612–1615.

66. Davis, A. V.; O'Halloran, T. V. *Nat. Chem. Biol.* **2008**, *4*, 148–151.

67. Holm, R. H.; Kennepohl, P.; Solomon, E. I. *Chem. Rev.* **1996**, *96*, 2239–2314.

68. Clark, R. W.; Squattrito, P. J.; Sen, A. K.; Dubey, S. N. *Inorg. Chim. Acta* **1999**, *293*, 61–69.

69. Eweiss, N.; Bahajaj, A.; Elsherbini, E. *Heterocycl. Chem.* **1986**, *23*, 1451–1459.

70. Awad, I.; Abdel-Rahman, A.; Bakite, E. *J. Chem. Technol. Biotechnol.* **1991**, *51*, 483–488.

71. Tiekink, E. R. T.; Cookson, P. D.; Linahan, B. M.; Webster, L. K. *Met.-Based Drugs* **1994**, 299–304.

72. Sanina, N. A.; Rakova, O. A.; Aldoshin, S. M.; Shilov, G. V.; Shulga, Y. M.; Kulikov, A. V.; Ovanesyan, N. S. *Mendeleev Commun.* **2004**, *14*, 7–8.

73. Stadelman, B. S.; Kimani, M. M.; Brumaghim, J. L. manuscript in preparation.

74. Senda, S.; Ohki, Y.; Hirayama, T.; Toda, D.; Chen, J.; Matsumoto, T.; Kawaguchi, H.; Tatsumi, K. *Inorg. Chem.* **2006**, *45*, 9914–9925.

75. Castro, J.; Lourido, P. P.; Sousa-Pedrares, A.; Labisal, E.; Piso, J.; Garcia-Vazquez, J. A. *Acta Crystallogr.* **2002**, *C58*, m319–m322.

76. Baba, H.; Nakano, M. *Polyhedron* **2011**, *30*, 3182–3185.

77. Aggarwal, V.; Kumar, V. R.; Singh, U. P. *J. Chem. Crystallogr.* **2011**, *41*, 121–126.

78. Pang, K.; Figueroa, J. S.; Tonks, I. A.; Sattler, W.; Parkin, G. *Inorg. Chim. Acta* **2009**, *362*, 4609–4615.

79. Bouwman, E.; Reedijk, J. *Coord. Chem. Rev.* **2005**, *249*, 1555–1581.

80. Gloaguen, F.; Rauchfuss, T. B. *Chem. Soc. Rev.* **2009**, *38*, 100–108.

81. Jia, W.; Huang, Y.; Lin, Y.; Wang, G.; Jin, G. *Eur. J. Inorg. Chem.* **2008**, 4063–4073.

82. Huang, Y.; Jia, W.; Jin, G. *J. Organmet. Chem.* **2009**, *694*, 86–90.

83. West, D. X.; Liberlon, A. E.; Rajendran, K. G.; Hall, I. H. *Anti-Cancer Drugs* **1993**, *4*, 241–249.

84. Xiong, R. G.; Song, B. L.; You, X. Z.; Mak, T. C. W.; Zhou, Z. Y. *Polyhedron* **1996**, *15*, 991–996.

85. Raj, S. S. S.; Fun, H.; Niu, D.; Lu, Z.; Sun, B.; Song, B.; You, X. *Acta Crystallogr.* **1999**, *C55*, 1410–1411.

86. Li, D.; Luo, Y. F.; Wu, T.; Ng, S. W. *Acta. Crystallogr.* **2004**, *E60*, m726–m727.

87. Aslanidis, P.; Cox, P. J.; Karagiannidis, P.; Hadjikakou, S. K.; Antoniadis, C. D. *Eur. J. Inorg. Chem.* **2002**, 2216–2222.

88. Aslanidis, P.; Hadjikakou, S. K.; Karagiannidis, P.; Gdaniec, M.; Kosturkiwicz, Z. *Polyhedron* **1993**, *12*, 2221–2226.

89. Li, R. Z.; Li, D.; Huang, X. C.; Qi, Z. Y.; Chen, X. M. *Inorg. Chem. Commun.* **2003**, *6*, 1017–1019.

90. Katiyar, D.; Tiwari, V. K.; Tripathi, R. P.; Srivastava, A.; Chaturvedi, V.; Srivastava, R.; Srivastava, B. S. *Bioorg. Med. Chem.* **2003**, *11*, 4369–4375.

91. Scheit, K. H.; Gartner, E. *Biochim. Biophys. Acta* **1969**, *183*, 10–18.

92. Lobana, T. S.; Kaur, P.; Castineiras, A.; Turner, P.; Failes, T. W. *Struct. Chem.* **2008**, *19*, 727–733.

93. Pivetta, T.; Cannas, M. D.; Demartin, F.; Castellano, C.; Vascellari, S.; Verani, G.; Isaia, F. *J. Inorg. Biochem.* **2011**, *105*, 329–338.

94. Gracia-Mora, L.; Ruiz-Ramirez, L.; Gomez-Ruiz, C.; Tinoco-Mendez, M.; Marquez-Quinones, A.; Romero-De Lira, L.; Martin-Hernandez, A.; Macias-Rosales, L.; Bravo-Gomez, M. E. *Met.-Based Drugs* **2001**, *8*, 19–28.

95. Alemon-Medina, R.; Brena-Valle, M.; Munoz-Sanchez, J. L.; Gracia-Mora, M. I.; Ruiz-Azuara, L. *Cancer Chemother. Pharmacol.* **2007**, *60*, 219–228.

96. Carvallo-Chaigneau, F.; Trejo-Solis, C.; Gomez-Ruiz, C.; Rodriguez-Aguilera, E.; Macias-Rosales, L.; Cortes-Barberena, E.; Cedillo-Pelaez, C.; Gracia-Mora, I.; Ruiz-Azuara, L.; Madrid-Marina, V.; Constantino-Casas, F. *BioMetals* **2008**, *21*, 17–28.

97. Pierre, J. L.; Fontecave, M. *BioMetals* **1999**, *12*, 195–199.

98. Yamanari, K.; Kushi, Y.; Fuyuhiro, A.; Kaizaki, S. *Dalton Trans.* **1993**, 403–408.

99. Corbin, D. R.; Francesconi, L. C.; Hendrickson, D. N.; Stucky, G. *Chem. Commun.* **1979**, 248–249.

100. Landgrafe, C.; Sheldrick, W. S. *Dalton Trans.* **1996**, 989–998.

101. Aslanidis, P.; Cox, P. J.; Kaltzoglou, A.; Tsipis, A. C. *Eur. J. Inorg. Chem.* **2006**, 334–344.

102. Sultana, R.; Lobana, T. S.; Sharma, R.; Castineiras, A.; Akitsu, T.; Yahagi, K.; Aritake, Y. *Inorg. Chim. Acta* **2010**, *363*, 3432–3441.

103. Batsala, G. K.; Dokorou, V.; Kourkoumelis, N.; Manos, M. J.; Tasipoulos, A. J.; Mavromoustakos, T.; Simcic, M.; Golic-Grdadolnik, S.; Hadjikakou, S. K. *Inorg. Chim. Acta* **2012**, *382*, 146–157.

104. Dennehy, M.; Quinzani, O. V.; Faccio, R.; Freire, E.; Mombru, A. W. *Acta Crystallogr.* **2012**, *C68*, m12–m16.

105. Gennari, M.; Lanfranchi, M.; Marchio, L.; Pellinghelli, M. A.; Tegoni, M.; Cammi, R. *Inorg. Chem.* **2006**, *45*, 3456–3466.

106. Patel, D. V.; Mihalcik, D. J.; Kreisel, K. A.; Yap, G. P. A.; Zakharov, L. N.; Kassel, W. S.; Rheingold, A. L.; Rabinovich, D. *Dalton Trans.* **2005**, 2410–2416.

107. Cammi, R.; Gennari, M.; Giannetto, M.; Lanfranchi, M.; Marchio, L.; Mori, G.; Paiola, C.; Pellinghelli, M. A. *Inorg. Chem.* **2005**, *44*, 4333–4345.

108. Matthews, R. G. *Acc. Chem. Res.* **2001**, *34*, 681–689.

109. Myers, L. C.; Jackow, F.; Verdine, G. L. *J. Biol. Chem.* **1995**, *270*, 6664–6670.

110. Ibrahim, M. M.; Seebacher, J.; Steinfield, G.; Vahrenkamp, H. *Inorg. Chem.* **2005**, *44*, 8531–8538.

111. Beheshti, A.; Clegg, W.; Dale, S. H.; Hyvadi, R. *Inorg. Chim. Acta* **2007**, *260*, 2967–2972.

112. Seebacher, J.; Vahrenkamp, H. *J. Mol. Struct.* **2003**, *656*, 177–181.

113. Ji, M.; Benkmil, B.; Vahrenkamp, H. *Inorg. Chem.* **2005**, *44*, 3518–3523.

114. Brand, U.; Rombach, M.; Seebacher, J.; Vahrenkamp, H. *Inorg. Chem.* **2001**, *40*, 6151–6157.

115. Yamashita, Y.; Amlund, H.; Suzuki, T.; Hara, T.; Hossain, M. A.; Yabu, T.; Touhata, K.; Yamashita, M. *Fish. Sci.*, *77*, 679–686.

116. Minoura, M.; Landry, V. K.; Melnick, J. G.; Pang, K. L.; Marchio, L.; Parkin, G. *Chem. Commun.* **2006**, *38*, 3990–3992.

117. Landry, V. K.; Buccella, D.; Pang, K. L.; Parkin, G. *Dalton Trans.* **2007**, 866–870.

118. Landry, V. K.; Parkin, G. *Polyhedron* **2007**, *26*, 4751–4757.

119. Williams, D. J.; Fawcett-Brown, M. R.; Raye, R. R.; VanDerveer, D.; Tang, Y. P.; Jones, R. L. *Heteroatom Chem.* **1993**, *4*, 409–414.

120. Williams, D. J.; Jones, T. A.; Rice, E. D.; Davis, K. J.; Ritchie, J. A.; Pennington, W. T.; Schimek, G. L. *Acta Crystallogr.* **1997**, *C53*, 837–838.

121. Blake, A. J.; Casabo, J.; Devillanova, F. A.; Escriche, L.; Garau, A.; Isaia, F.; Lippolis, V.; Kivekas, R.; Muns, V.; Schroder, M.; Sillanpaa, R.; Verani, G. *Dalton Trans.* **1999**, 1085–1092.

122. Blake, A. J.; Lippolis, V.; Pivetta, T.; Verani, G. *Acta Crystallogr.* **2007**, *C63*, m364–m367.

123. Barclay, G. R.; Hoskins, B. F.; Kennard, C. H. L. *J. Chem. Soc.* **1963**, 5691–5699.

124. Williams, S. J.; White, K. M.; VanDerveer, D.; Wilkinson, A. P. *Inorg. Chem. Commun.* **2002**, *5*, 124–126.

125. Tournie, E.; Morhain, C.; Ongaretto, C.; Bousquet, V.; Brunet, P.; Neu, G.; Faurie, J. P.; Triboulet, R.; Ndap, J. O. *Mater. Sci. Eng.* **1997**, *B43*, 21–28.

126. Lee, H.; Kim, T. S.; Jeong, T. S.; An, H. G.; Kim, J. Y.; Youn, C. J.; Yu, P. Y.; Hong, K. J.; Lee, H. J.; Shin, Y. J. *J. Cryst. Growth* **1998**, *191*, 59–64.

127. Wright, A. C. *J. Cryst. Growth* **1999**, *203*, 309–316.

128. Papp, L. V.; Holmgren, A.; Khanna, K. K. *Antioxid. Redox Signaling* **2010**, *12*, 793–795.

129. Miessler, G. L.; Tarr, D. A. *Inorganic Chemistry*, 3rd ed.; Pearson Prentice Hall: Upper Saddle River, New Jersey, 2004; p 45.

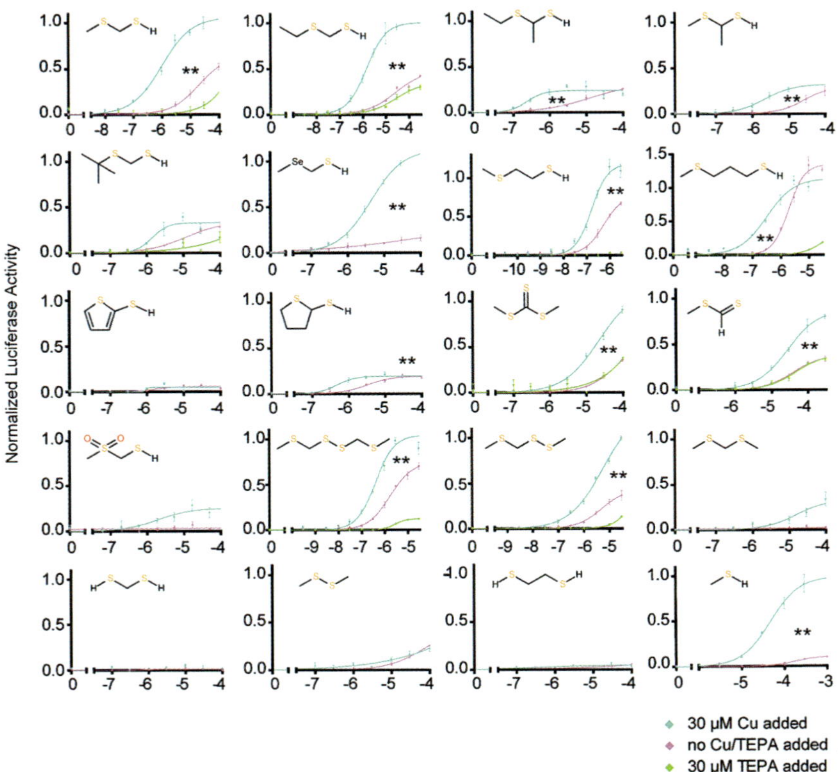

Figure 6. Dose-response curves of MOR244-3 to representative sulfur-containing compounds with and without 30 µM exogenous copper ion. The horizontal scale shows the exponent of the odorant molar concentration (e.g., $^{-6} = 10^{-6}$ M), and the vertical axis shows the normalized luciferase activity, an indirect measure of the response of the receptor to substrates. For odors with a significant response in the absence of exogenous copper ion, as defined arbitrarily by a top value greater than 0.32, dose–response curves with 30 µM of TEPA are also shown. F-tests were used to compare the pairs of dose–response curves with or without copper ion; asterisks shown represent significance of p-values after Bonferroni corrections. Adapted with permission from (29). Copyright 2012, PNAS.
(Chapter 1)

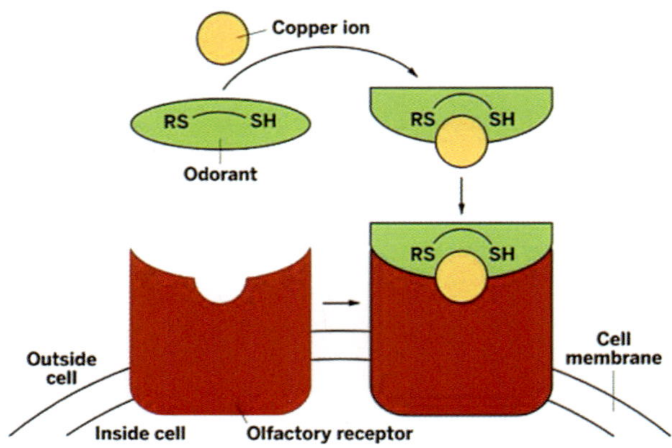

Figure 7. Schematic showing docking of copper-coordinated odorants with odorant receptor, for example, MOR244-3. The copper ion binds to the ligand followed by binding of the copper ion-odor complex to the OR. Adapted with permission from Chemical & Engineering News, 90 (7), p. 9, February 13, 2012. Copyright 2012, American Chemical Society. (Chapter 1)

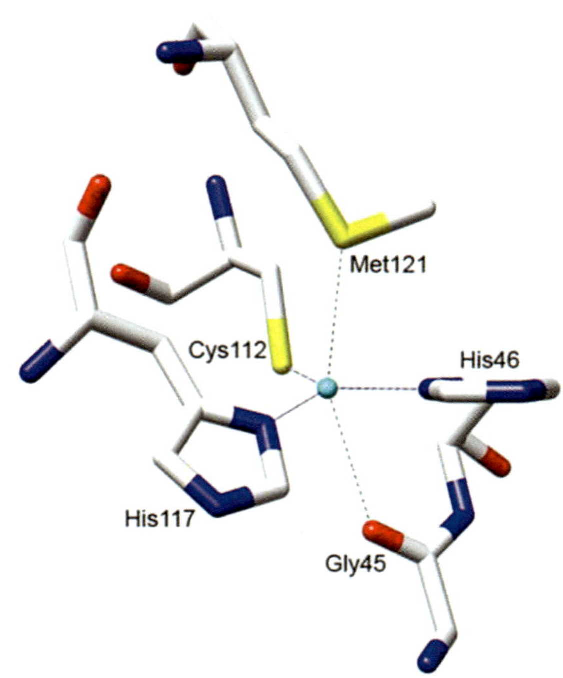

Figure 8. Type 1 copper site in Pseudomonas aeruginosa azurin. Copyright 2010, American Chemical Society (41). (Chapter 1)

Color Insert - 2

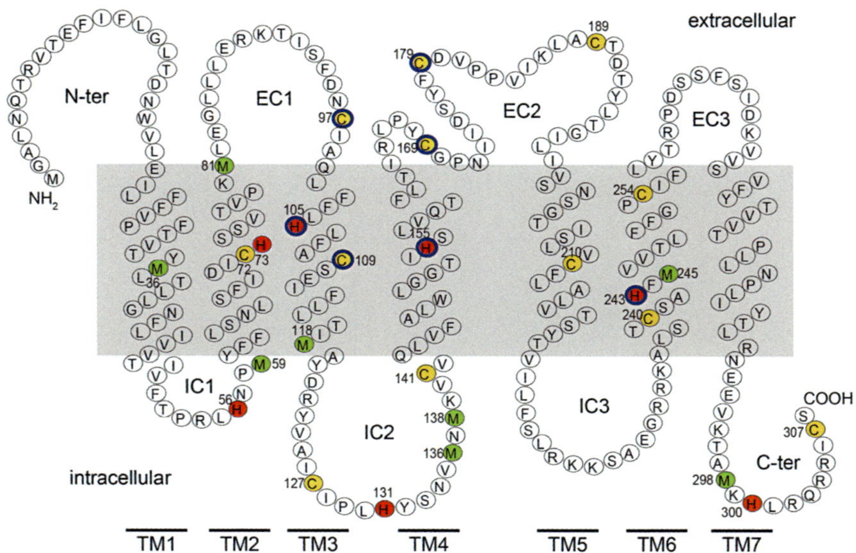

Figure 9. A serpentine model of the MOR244-3 receptor color-coded for the methionine (green), histidine (red), and cysteine (yellow) residues that were subjected to mutagenesis analysis. Residues circled in blue are those that exhibited complete loss-of-function phenotypes in the luciferase assay. Transmembrane domains, as predicted by the TMHMM server, are indicated by "TM". Adapted with permission from (29). Copyright 2012, PNAS. (Chapter 1)

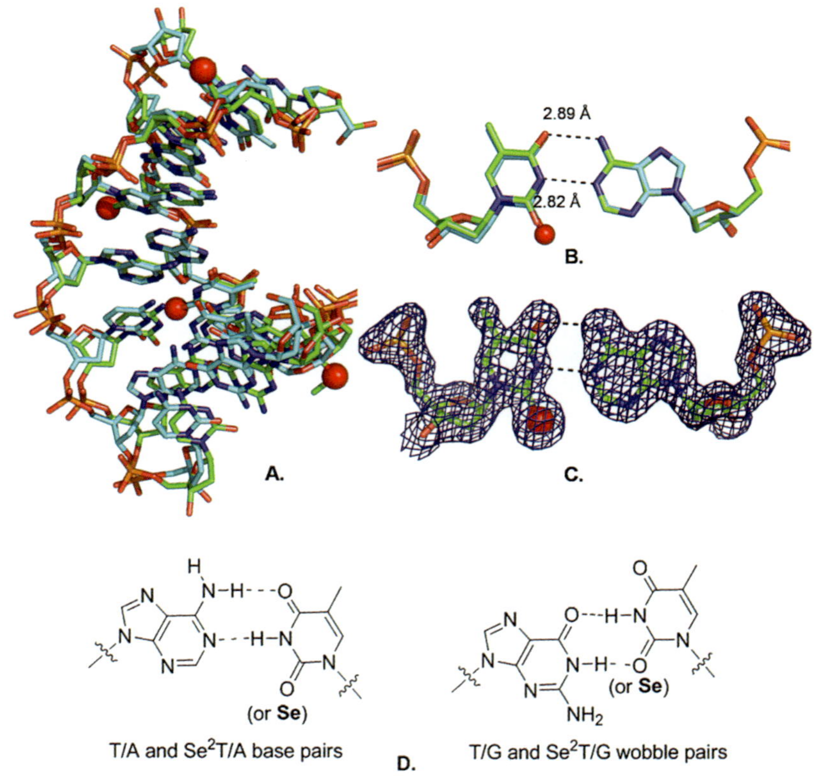

*Figure 4. Global and local structures of the Se²T-DNA [(5'-GdU₂'-ₛₑ-G-ˢᵉT-ACAC-3')₂], with a resolution of 1.58 Å. (**A**) The superimposed comparison of the Se-DNA duplex (3HGD, in green) with its native counterpart (1D78, in cyan). The red balls represent the selenium atoms. (**B**) The superimposed comparison of the local ˢᵉT4/A5 (in green) and native T4/A5 (in cyan) base pairs. (**C**) The experimental electron density map of the ˢᵉT4/A5 base pair with σ = 1.0. (**D**) Native and Se-modified T/A base pair and T/G wobble pairs. (Chapter 5)*

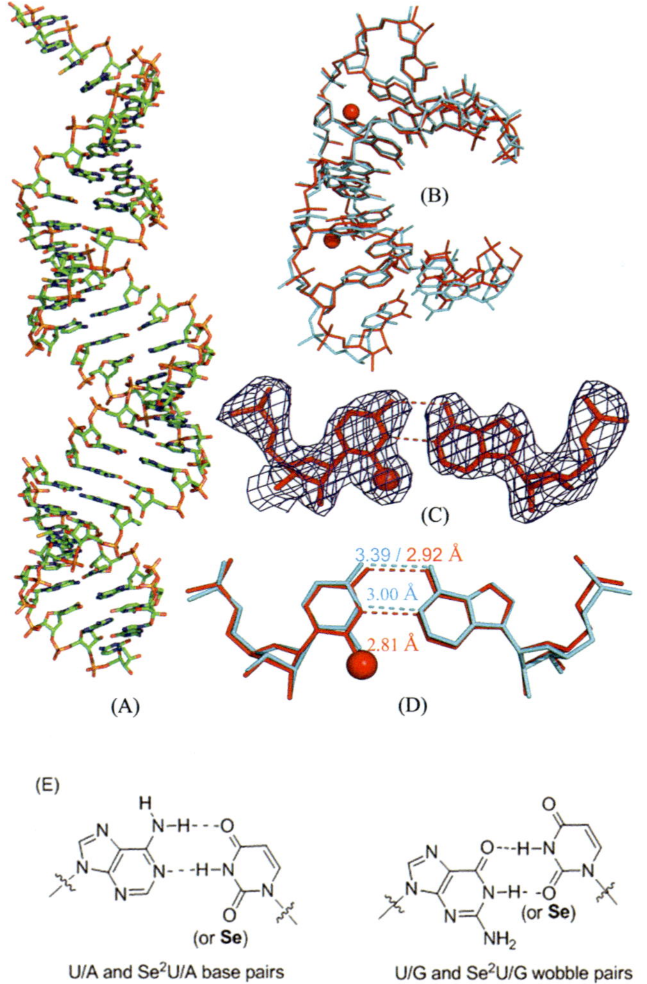

*Figure 5. Global and local structures of the Se²U-containing RNA r[(5'-GUAUA(ᔆᵉU)AC-3']₂ with a resolution of 2.3 Å. (**A**) The overall structure of the duplex. (**B**)The superimposed comparison of one ᔆᵉU-RNA duplex (red; PDB ID: 3S49) with its native counterpart [5'-r(GUAUAUA)-dC-3']₂ (cyan; PDB ID: 246D) with a RMSD value 0.55. The two red balls represent the selenium atoms. (**C**) The experimental electron density of ᔆᵉU6/A11 basepair with σ=1.0. (**D**) The superimposed comparison of the local base pair ᔆᵉU6/A11 (red) and the native U6/A11 (cyan). The numbers indicate distances between the corresponding atoms. (**E**) Native and Se-modified U/A base pair and U/G wobble pairs. (Chapter 5)*

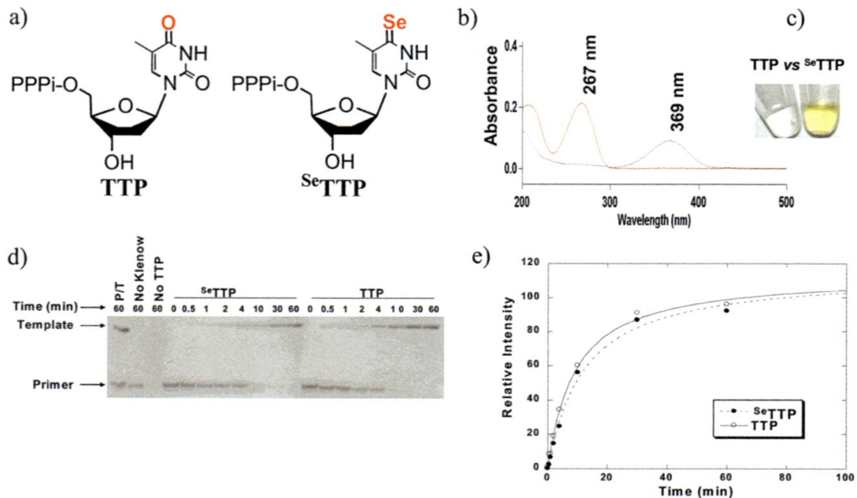

*Figure 8. The color of ^{Se}TTP and its Se-DNA. **a**) TTP and ^{Se}TTP chemical structures; **b**) UV profiles of TTP and ^{Se}TTP; **c**) colorless TTP and yellow ^{Se}TTP; **d**) Gel analysis of DNA polymerization with TTP and ^{Se}TTP; **e**) Plots of the incorporation of TTP and ^{Se}TTP into DNA. (Chapter 5)*

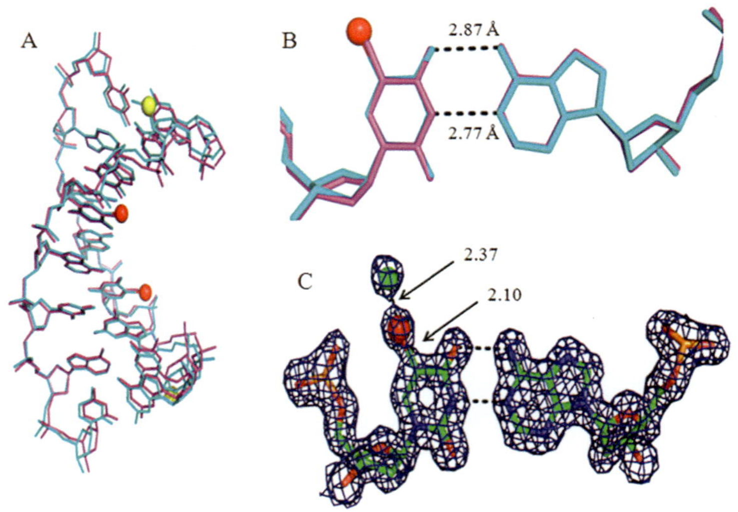

*Figure 10. The global and local structures of the Te-DNA duplex [5'-G(2'-SeMe-dU)G($^{5-Ph-Te}$T)ACAC-3']$_2$. The red and yellow balls represent Te and Se atoms, respectively. (**A**) Te-dsDNA structure (in cyan; 1.50 Å resolution; PDB ID: 3FA1) superimposed over the native structure (in magenta; PDB ID: 1DNS). (**B**) Local structure of the $^{5-Ph-Te}$T /A base pair (in cyan) superimposed over the native T/A base pair (in magenta). (**C**) The experimental electron density (2|F$_o$|_|F$_c$| map) of the $^{5-Ph-Te}$T /A base pair (σ =1.0). The green ball represents approximately one-sixth of a phenyl group (one carbon). (Chapter 5)*

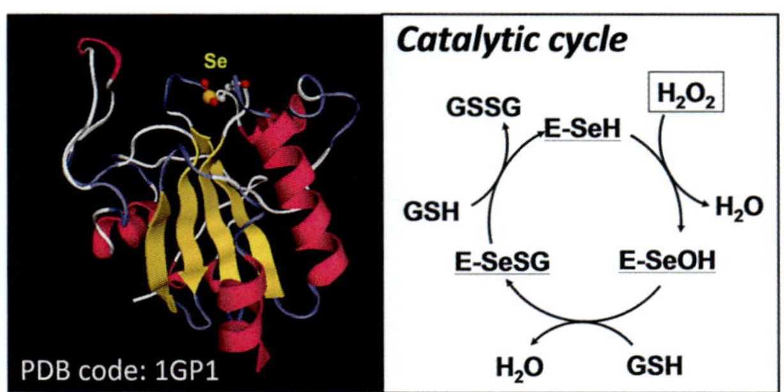

Figure 1. Structure of glutathione peroxidase (GPx) and the catalytic cycle. (Chapter 8)

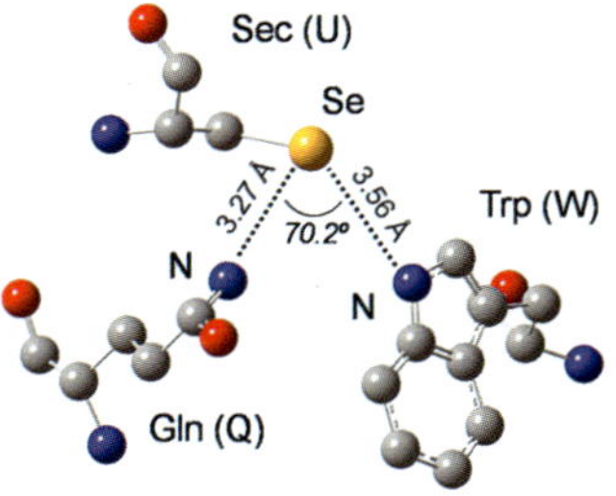

Figure 2. The active site of GPx revealed by X-ray analysis. The two oxygen atoms attached to the Se atom are omitted. (Chapter 8)

Chapter 4

Sulfur Oxygenation Enhances Ligand Exchange in Nitrile-Hydratase-Inspired Ruthenium(II) Complexes

César A. Masitas, Mark S. Mashuta, and Craig A. Grapperhaus*

Department of Chemistry, University of Louisville,
Louisville, Kentucky 40292
*E-mail: grapperhaus@louisville.edu.

Ligand-exchange reactions on a series of sulfur-oxygenated derivatives of the $N_2S_3RuPPh_3$ complex (bmmp-TASN)Ru-PPh$_3$ (bmmp-TASN = 4,7-bis(2′-methyl-2′-mercaptopropyl)-1-thia-4,7-diazacyclononane) reveal enhanced ligand substitution with increased sulfur oxidation. Phosphine exchange occurs readily in methanol at 50°C for the thiolate/sulfinate (RS⁻/RSO$_2$⁻) and sulfenate/sulfinate (RSO⁻/RSO$_2$⁻) derivatives yielding (bmmp-O$_2$-TASN)Ru-PPh$_2$CH$_3$ and (bmmp-O$_3$-TASN)Ru-PPh$_2$CH$_3$, which were characterized by ^{31}P NMR, +ESI-MS, and single crystal X-ray diffraction. The parent bis-thiolate (RS⁻/RS⁻) requires higher reaction temperatures or extended reaction times for ligand exchange. Single crystals of the phosphine-free complex [(bmmp-TASN)Ru]$_2$ were obtained upon prolong standing in methanol.

The enzyme nitrile hydratase (NHase) catalyzes the hydration of nitriles to amides at a metal-containing active site (Figure 1). Recent reviews have focused on structural and functional models of NHase (*1–5*). The active site contains either a mononuclear low spin non-heme iron(III), which favors the hydration of aliphatic nitriles, or a non-corrin cobalt(III) that preferentially hydrates aromatic nitriles in a highly conserved -C-S-L-C-S-C- metal binding domain with an N_2S_3 primary coordination sphere (*1–3, 5*). In iron-containing NHase, the nitrogen donors are deprotonated carboxamido nitrogens from ser-113 and cys-114 with cysteine sulfur donors from cys-110, cys-113, and cys-115. While cys-109 is in the typical thiolate oxidation state (RS-; ox. state = -2), cys-114 has been oxidized to a sulfenate (RSO-; ox. state = 0) and cys-112 is present as a sulfinate (RSO$_2$-; ox. state = +2) (6). The oxidized sulfur donors are co-facial with the substrate binding site, while the thiolate sits trans to the sixth coordination site that is presumably occupied by water or nitrile during catalytic turnover. A similar donor arrangement, including the oxidation states of the cysteine donors, is found in Co-NHase (7).

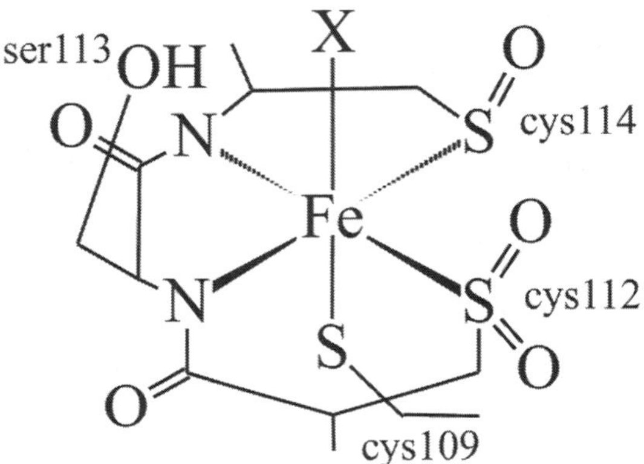

Figure 1. Fe-containing nitrile hydratase active site (X = substrate binding site).

The post-translational sulfur oxygenation at the NHase active site is required for hydratase activity (8). Proposed roles for the S-oxygenation include modulation of Lewis acidity (9), facilitation of reversible NO bonding (10, 11), activation of H$_2$O as a proximal base (12–14), and alteration of ligand binding (15). Density functional theory calculations by Solomon suggest sulfur-oxygenation strengthens axial binding between Fe-NHase and nitrile due to the loss of π-bonding with the S-donors, whereas binding to Co-NHase is relatively unaffected due to the lack of Co-S π-bonding (15). In the two model complex systems reported to date with variable sulfur-oxygenation levels and exchangeable ligands (11, 16–18), sulfur-oxygenation enhances ligand lability.

Asymmetric Sulfur Oxygenation

Controlled Sulfur Oxygenation

The selective oxidation of the three active site cysteine residues at the NHase active site yielding a mixed sulfenate/sulfinate/thiolate donor set is critical for catalytic activity. "Over–oxidation" mutants in which both Cys-112 and Cys-114 are oxygenated to sulfinates (RSO$_2$-) are inactive (19). NHase deactivation upon prolonged oxidation has also been noted by Song et al (20).

The isolation and characterization of synthetic model complexes with asymmetric sulfur oxygenation has proven difficult. To date, only five structurally characterized, mono-nuclear complexes with a mixed sulfenate/sulfinate donor set are known, Figure 2 (*16, 21–24*).   These include Ni(II) (*22*) and Pt(II) (*24*) complexes with stable 4-coordintate, d^8-metal ions with no affinity for additional ligands.   The Co(III) sulfenate/sulfinate prepared by Kung et al. contains an η^2-sulfenate that blocks the sixth-coordination site (*23*), whereas the Ru(II) thiolate/sulfenate/sulfinate reported by Dilworth is coordinatively saturated with three bidentate chelates (*21*).   Only our Ru(II) complex based on the pentadentate N_2S_3 ligand (bmmp-TASN)$^{2-}$ (bmmp-TASN = 4,7-bis(2'-methyl-2'-mercaptopropyl)-1-thia-4,7-diazacyclononane)   combines the mixed sulfenate/sulfinate donor set with a variable ligand binding site (*16*). Co-containing peptide-based models with asymmetric sulfur oxygenation have also been reported, but the structures were not crystallographically determined (*25, 26*).

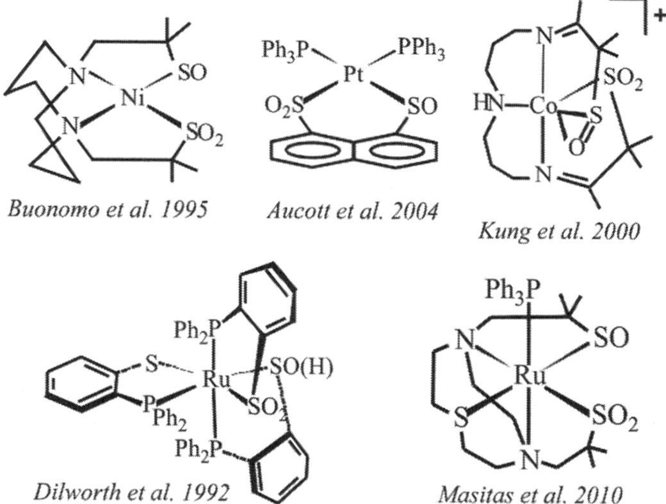

Figure 2. Structurally characterized mononuclear sulfenate/sulfinate complexes.

We previously employed the bmmp-TASN ligand to prepare iron-containing mimics of NHase.   The facial coordination of the TASN backbone positions the potentially reactive thiolate donors cis to one another and co-facial with the variable ligand binding site, Figure 3.   Iron(III) complexes of bmmp-TASN bind a variety of ligands in the sixth coordination site (*4, 27–29*).   The spin-state of iron is dependent on the nature of the sixth ligand with weak-field ligands giving high-spin complexes and strong-field donors yielding low-spin complexes (*28*). Notably, the spin-state of the iron influences the final outcome of the oxidation product. Whereas the high-spin complexes (bmmp-TASN)FeCl reacts with O_2 to yield ligand disulfide and a structurally characterized iron–oxo cluster degradation product (*30*), the low-spin (bmmp-TASN)FeCN yields an insoluble O–bound bis-sulfonate (RSO$_3^-$/RSO$_3^-$), Scheme 1 (*29*). Although the observed product is over-oxidized with respect to NHase, these results demonstrate the importance of spin-state in directing S-oxygenation as supported by DFT calculations (*29*).

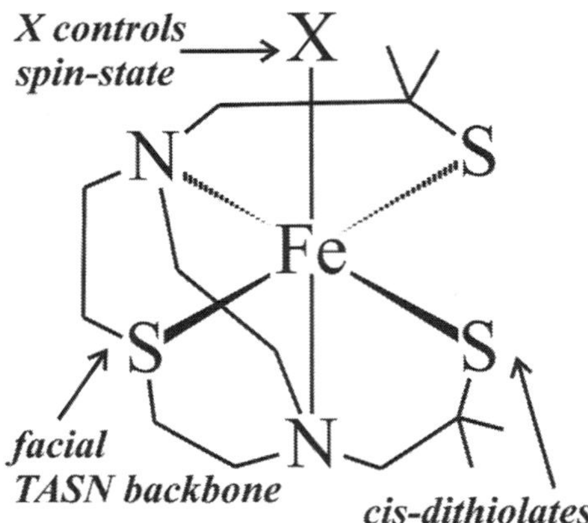

Figure 3. Features of (bmmp-TASN)FeX model complexes.

Ruthenium Sulfur Oxygenates

The tendency of low-spin iron(III) complexes to favor sulfur oxygenation inspired us to investigate the oxygen sensitivity of related low-spin complexes. Our initial efforts focused on the ruthenium(II) complex (bmmp-TASN)RuPPh$_3$ (**1**). The complex is readily prepared from H$_2$(bmmp-TASN), base, and RuCl$_2$(PPh$_3$)$_3$ (*16*). Oxygen exposure of (bmmp-TASN)RuPPh$_3$ proceeds through a series of stepwise reactions that are kinetically well-separated, Scheme 2 (*16, 17*). Reaction times of several minutes selectively yields (bmmp-O$_2$-TASN)RuPPh$_3$ (**2**) in which one thiolate donor is oxidized to a sulfinate. The synthesis of the mixed sulfenate/sulfinate derivative (bmmp-O$_3$-TASN)RuPPh$_3$ (**3**) requires several hours several hours of oxygen exposure, whereas the bis-sulfinate (bmmp-O$_4$-TASN)RuPPh$_3$ (**4**) is only observed along with uncharacterized decomposition products after several days.

Analysis of the structural parameters of **1** – **3** from single crystal X-ray diffraction studies, Table 1, combined with optimized DFT structures of **1** – **4** reveals a systematic increase in the Ru-P bond distances as a function of S-oxygenation. This trend opposes expectations of tighter Ru-P binding as a function of decreased sulfur donor ability upon oxygenation. The structural results are supported experimentally by multi-edge X-ray absorption spectroscopy (*31*). The XAS studies include N K-edge spectroscopy, which to our knowledge, has not been previously utilized to extract covalency parameters in metal complexes. The results demonstrate dramatic reduction in metal-ligand covalency with each sequential oxygenation, with significant hardening of the Ru(II) center leading to enhanced lability of the soft PPh$_3$ donor.

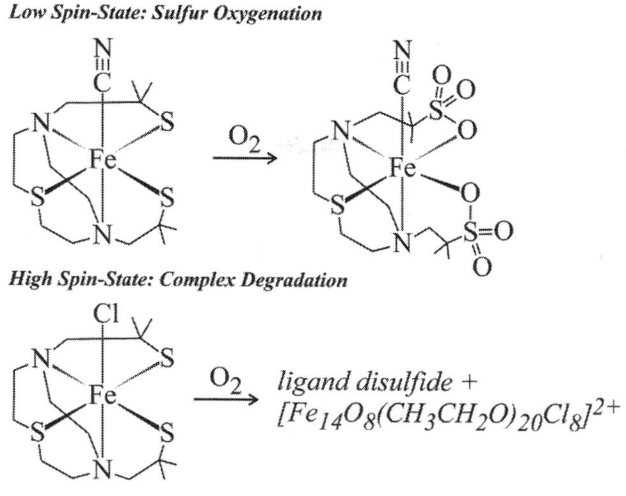

Scheme 1. Spin state of iron influences Fe-SR reactivity with O_2

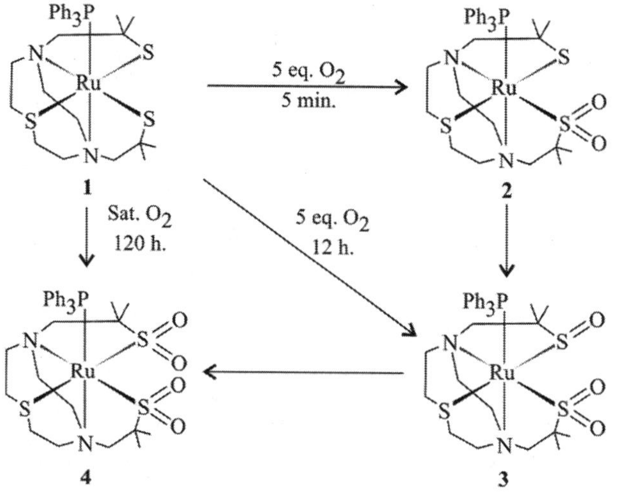

*Scheme 2. Sulfur-oxygenation pathway of **1**. (Reproduced with permission from reference (17). Copyright 2010 American Chemical Society.)*

Table 1. Selected Experimental Bond Distances (Å) and Angles (°) for 1 – 3

	1	2	3
Ru-S1	2.2900(10)	2.3102(6)	2.3622(9)
Ru-S2	2.4057(9)	2.2473(6)	2.2548(9)
Ru-S3	2.3754(10)	2.3943(6)	2.3493(9)
Ru-P1	2.2911(10)	2.3519(6)	2.3790(9)
Ru-N1	2.198(2)	2.1927(19)	2.178(3)
Ru-N2	2.178(2)	2.200(2)	2.192(3)
S2-O1		1.4906(17)	1.489(3)
S2-O2		1.4658(18)	1.471(3)
S3-O3			1.556(3)
P1-Ru-S2	95.44(2)	93.16(2)	90.72(3)
P1-Ru-S3	92.42(3)	93.99(2)	91.59(3)
S2-Ru-S3	94.93(3)	97.09(2)	94.13(3)

Enhanced Ligand Lability

Catalytic Nitrile Hydration

The increased Ru-P bond distances as a function of sulfur-oxygenation in the experimental and computational structures of **1 – 4** suggest enhanced lability of the coordinated PPh$_3$. Phosphine exchange reactions are known for select Ru-thiolates reported by Sellmann (*32, 33*). The inclusion of an exchangeable donor in a NHase mimic with variable S-oxygenation levels offers an unparalleled opportunity to investigate the influence of sulfur modification on substrate binding, hydration, and product release.

Preliminary studies of catalytic benzonitrile hydration trials for **1 – 3** were performed using biphasic benzonitrile/water mixtures at elevated temperature, Table 2 (*18*). Under these conditions, all three complexes readily dissociate PPh$_3$ as described below. Using 1:3 benzonitrile:water mixtures, the parent bis-thiolate **1** supports 66 ± 1 turnovers after 18 h, whereas up to 188 ± 1 turnovers are found when the substrate ratio is changed to 3:1. In contrast, the sulfur-oxygenate derivatives **2** and **3** are most active at the lower nitrile:water ratio with 238 ± 23 and 242 ± 23 turnovers, respectively. This is attributed to decreased product inhibition in **2** and **3** as the benzamide product and catalysts selectively partition in the organic layer (*18*). Control reactions in the absence of catalyst yield 8 ± 2 turnovers over the same time period (*18*). Detailed kinetic analyses in homogeneous mixtures are currently underway to further refine the influence of sulfur-oxygenation on hydration rates and binding affinities.

Phosphine Exchange Reactions

To further confirm PPh3 dissociation, a series of phosphine exchange products were examined. Complexes **1 – 3** readily react with methyldiphenylphosphine (PPh$_2$CH$_3$) at elevated temperatures in solution to yield the phosphine exchange products **5 – 7**, Scheme 3. For the sulfur-oxygenated derivatives **2** and **3**, six equivalents of PPh$_2$CH$_3$ were added to 5 mL of a 3 mM solution of the metal precursor. The reaction was heated to reflux (65 °C) for at least 5 hours. Reaction products were isolated by crystallization upon slow evaporation of methanol to yield **6** or **7** in 92% and 94% yield, respectively. Similar attempts with **1** failed to yield the phosphine exchange product **5**, although this complex could be readily prepared by extended heating (36 hours) of 1.4 mM chlorobenzene solution of **1** with 5 equivalents PPh$_2$CH$_3$ at 100 °C.

Table 2. Benzonitrile hydration turnover numbers (TONs) for catalysts 1 – 3 at 124 °C in neat substrate mixtures at neutral pH (*18*)

Catalyst	Mole %	V_{PhCN}/V_{H2O}	TON
1	0.0055	0.33	66 ± 1
1	0.0027	1.00	91 ± 3
1	0.0018	3.00	188 ± 32
1	0.0017	4.00	176 ± 26
1	0.0015	7.00	28 ± 4
2	0.0055	0.14	238 ± 23
2	0.0055	0.33	88 ± 9
2	0.0027	1.00	33 ± 8
3	0.0055	0.14	242 ± 23
3	0.0055	0.33	86 ± 11
3	0.0027	1.00	38 ± 10
3	0.0018	3.00	23 ± 7

The exchange reaction is proposed to proceed via a dissociative mechanism. This is consistent with the partial dissociation of PPh$_3$ observed upon heating solutions of **1 – 3**. The ^{31}P NMR spectra of crystalline samples of **1 – 3** display resonance peaks for coordinated and free PPh$_3$ triphenylphosphine and the corresponding Ru complex. Similar experiments over a temperature range of 50 °C – 100 °C in chlorobenzene indicate increased phosphine dissociation at elevated temperature with nearly complete dissociation at 100 °C. Free PPh$_3$ is observed at -6.23 ppm, whereas coordinated PPh$_3$ resonance frequencies are seen at 43.37, 34.34, and 33.68 ppm for **1 – 3**, respectively. As shown in Table 3, the ^{31}P resonances shift by -8.7 to -11.1 ppm to 34.64, 25.54, and 22.54 ppm for the PPh$_2$CH$_3$ derivatives **5 – 7**, respectively.

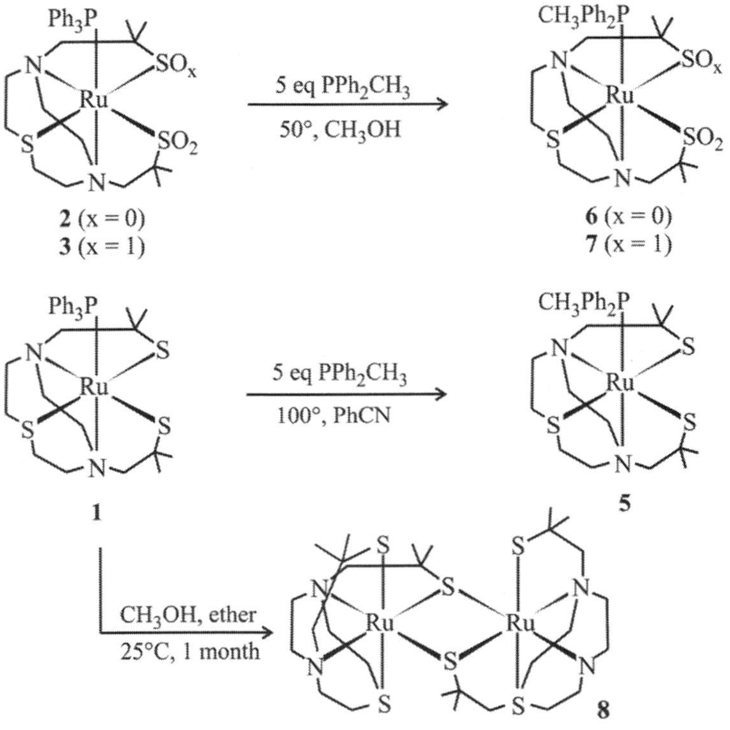

Scheme 3. Phosphine exchange reaction pathways

While **1 – 3** readily dissociate PPh$_3$ at elevated temperatures, the products are stable with respect to phosphine or solvent exchange at 25 °C. Room temperature cyclic voltammetry and square wave voltammetry reveal a single oxidation-reduction event in compounds for **1 – 3** associated with the PPh$_3$ bound complex, Figure 4. Each complex displays a single, quasi-reversible event in the cyclic voltammogram (*17*) indicative of a single PPh$_3$ coordinated complex in solution. The squarewave voltammogram, Figure 4, clearly shows a progressive increase in the Ru$^{III/II}$ reduction potential as a function of sulfur-oxygenation. The Ru$^{III/II}$ shifts by +458 mV from -851 mV (versus ferrocenuim/ferrocene) to -393 mV upon oxidation of **1** to **2**. This is similar to the shift observed upon oxidation of other ruthenium-thiolates to the corresponding sulfinate (*34*). Further oxidation of **2** to **3** results in an additional +130 mV shift to -263 mV, consistent with reports by Darensbourg that addition of a single O-atom to sulfur shifts nickel and palladium reduction potentials by ~100 mV (*35*).

Table 3. ^{31}P NMR resonance frequencies in chlorobenzene for 1 – 3 and 5 – 7

	^{31}P resonance (ppm)
PPh$_3$	-6.23
1	43.37
2	34.34
3	33.67
PPh$_2$CH$_3$	-27.83
5	34.64
6	25.49
7	22.54

The Ru-phosphine complexes also retain the coordinated phosphine upon ionization. The +ESI-MS for each complex corresponds with the parent peak matching the expected *m/z* value. Compound **6** displays a parent peak at *m/z* = 677.1140, consistent with a theoretical value of 677.1009 *m/z* for [**6** + Na]$^+$. The isotopic distribution shows an accuracy of 0.0131 *m/z*. Compound **7** displays a parent peak at *m/z* = 693.1192, also consistent with the theoretical value of its + Na$^+$ peak, *m/z* 693.0958 for [**7** + Na]$^+$. No significant fragmentation of the phosphine dissociated metal-core is observed.

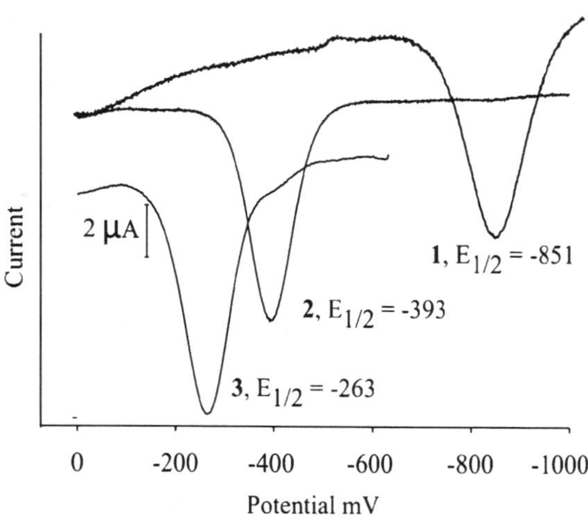

Figure 4. Square voltammograms of 1 – 3 in acetonitrile with 0.1 TBAHFP as supporting electrolyte. Potentials referenced to Fc$^+$/Fc. (Reproduced with permission from reference (17). Copyright 2010 American Chemical Society.)

X-ray Crystal Structure Analyses

The solid-state structures of the phosphine exchange products **6** and **7** were elucidated by single crystal X-ray diffraction. Each complex displays Ru(II) in a pseudo-octahedral N_2S_3P donor environment, Figure 5. Additionally, the phosphine bimetallic complex [(bmmp-TASN)Ru]₂ (**8**) was structurally characterized, Figure 6. Selected bond distances and angles for **6 – 8** are provided in Table 4. X-ray structural data in CIF format have been deposited with the Cambridge Crystallographic Data Centre (CCDC# 940595 - 940597).

As in the precursor complexes **2** and **3**, both **6** and **7** contain a facially coordinated TASN ring (N1, N2, and S1) with two pendant sulfur donors S2 and S3 positioned trans to N2 and S1, respectively. The S2 sulfur is in the sulfinato oxidation state in both complexes, whereas S3 is a thiolate in **6** and a sulfenate in **7**. The phosphorus (P1) of the PPh_2CH_3 donor sits trans to N1. The two oxygen atoms, O1 and O2, of the sulfinate donor (S2) in **6** are directed roughly along the S1-Ru-S3 bond axis with torsion angles of -11.74(13)° and +40.42(12)° for O1-S2-Ru1-S1 and O2-S2-Ru1-S3, respectively. A similar arrangement with torsion angles of -12.31(6)° and +35.82(6)° for O1-S2-Ru1-S1 and O2-S2-Ru1-S3, respectively, are observed in **7**. The sulfenate oxygen (O3) of **7** is oriented anti to the phosphine donor with an O3-S3-Ru1-N1 torsion angle of -16.46(8)°.

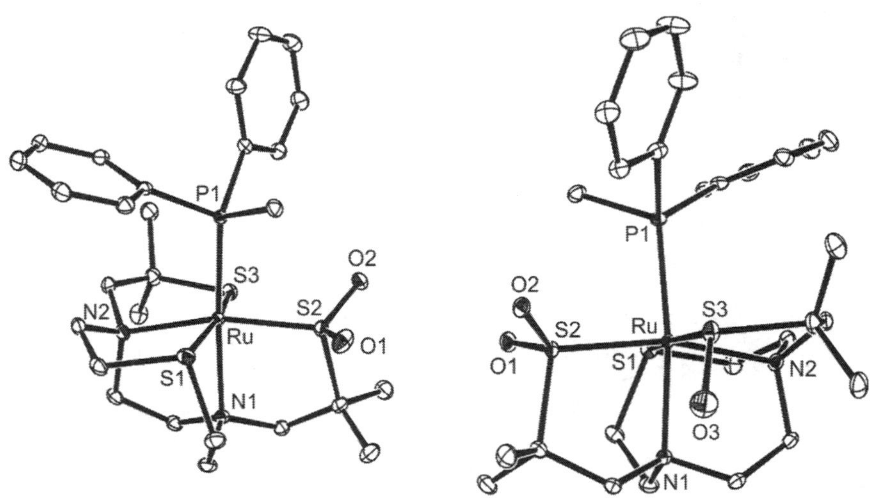

*Figure 5. ORTEP (36) representations of **6** (left) and **7** (right).*

The Ru-P1 bond distance of **6**, 2.3045(8) Å, is 0.0231(9) Å shorter than in **7**, 2.3276(4) Å. This is similar to the 0.027(1) Å difference between the related sulfur-oxygenates **2** to **3**. The Ru-P1 bond distance to PPh_2CH_3 in **6** and **7** is approximately 0.05 Å shorter than in the corresponding PPh_3 derivatives **2** and **3**. This is consistent with the increased donor ability of PPh_2CH_3 relative to PPh_3 as noted in the voltammetry experiments and the smaller cone angle of PPh$_2$CH$_3$. No significant changes are observed in Ru ligand bond distances to the (bmmp-TASN) chelate upon exchange of PPh$_3$ with PPh$_2$CH$_3$. As illustrated for **3** and **7**, the Ru-N1 (2.178(3) and 2.1746(13) Å), Ru-S2 (2.2548(9) and 2.2562(4) Å), and Ru-S3 (2.3493(9) and 2.3532(4) Å) are statistically equivalent in the structures of **3** and **7**, respectively.

Complex **8** was obtained upon slow crystallization of **1** by liquid-vapor diffusion of methanol-ether under air-free conditions over a period of one month. Repeated attempts to reproduce **8** were unsuccessful. The coordinated PPh$_3$ in **1** has been replaced by a bridging thiolate (S3′) in **8**. The Ru-S3′ bond distance to the bridging thiolate is 2.2899(4) Å with a Ru1-Ru1′ interatomic distance of 2.8350(4) Å. Similar dinuclear Ru$_2$S$_2$ cores have been observed for other Ru(II) complexes with pentadentate chelates (*37, 38*).

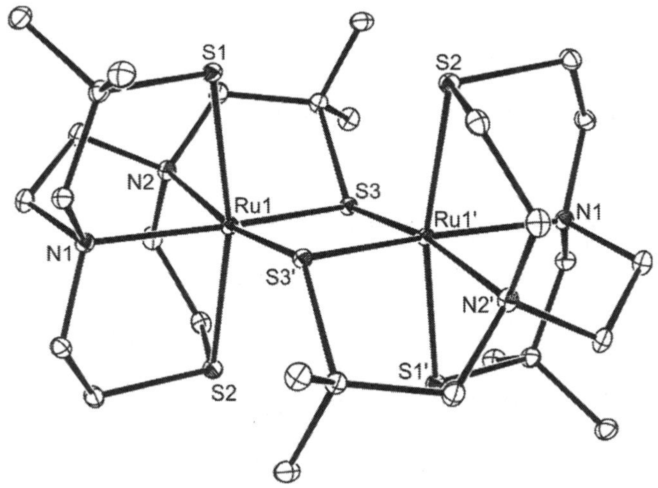

*Figure 6. ORTEP (36) representation of **8**.*

Relevance to NHase

The active site of NHase requires post-translational sulfur oxygenation for catalytic activity. Synthetic models of the asymmetric sulfur-oxygenate environment (sulfenate/sulfinate) in small models are rare. Complexes **1 – 4** represent the only series of S-oxygenates with a variable ligand binding site. In this manuscript, we demonstrate the enhanced lability of the coordinated PPh$_3$ in **1 – 3** as a function of S-oxygenation through a series of phosphine exchange reactions. The structurally characterized sulfur-oxygenates **5** and **6** represent, to our knowledge, the first sulfur-oxygenate models of NHase prepared via ligand exchange. The accessibility of such facile preparation of S-oxygenate complexes provides a new route to the synthesis of related S-oxygenate derivatives. The accessibility of the ligand binding site correlates with our previously reported catalytic nitrile hydration studies. Ongoing studies in our laboratories to address the influence of sulfur oxygenation on substrate/product binding affinities and hydration rate constants will shed further light on the role of S-oxygenation at the NHase active site.

Table 4. Selected Experimental Bond Distances (Å) and Angles (°) for 6 – 8

	6	*7*	*8*
Ru-S1	2.3163(8)	2.3508(4)	2.4170(5)
Ru-S2	2.2384(8)	2.2562(4)	2.3589(5)
Ru-S3	2.3940(8)	2.3532(4)	2.3068(5)
Ru-P1	2.3045(8)	2.3276(4)	
Ru-N1	2.188(3)	2.1746(13)	2.1351(16)
Ru-N2	2.190(3)	2.1852(14)	2.1639(16)
S2-O1	1.478(3)	1.4862(12)	
S2-O2	1.485(3)	1.4767(12)	
S3-O3		1.5197(14)	
P1-Ru-S2	90.24(3)	90.574(15)	
P1-Ru-S3	98.00(3)	91.759(15)	
S2-Ru-S3	93.94(3)	95.939(15)	85.762(17)

Experimental Methods

Materials and Methods

All chemicals were obtained from commercially available sources and were used as received unless otherwise noted. All glassware was flame-dried under vacuum three times prior to use and refilled with argon. All solvents were dried and freshly distilled using conventional techniques under a nitrogen atmosphere before degassing using freeze-pump-thaw techniques or purging with argon for 15 minutes. Standard Schlenk techniques under an argon atmosphere or an argon-filled glovebox were utilized for the manipulation of all compounds in this study. Oxygen (99.8%) was obtained from Welder's Supply, Louisville, KY. Oxygen addition was accomplished with the use of a 23 ml gas transfer bulb (fabricated by Wilmad-LabGlass) connected to a Schlenk line. 1-Thia-4,7-diazacyclononane (TASN) was prepared according to the method of Hoffmann et al (*39*). The ligand H_2(bmmp-TASN) was prepared as previously reported and stored as oil under argon in the glovebox (*28, 40*). $RuCl_2(PPh_3)_3$ was prepared according to literature methods (*41*). The metal complexes **1** (*16*), **2** (*17*), and **3** (*16*) were prepared by previously reported methods.

Phosphine Exchange Reactions

(bmmp-TASN)RuPPh₂CH₃ (5)

A solution of **1** (21.5 mg, 0.0314 mmol) was stirred in 22 mL of dry, degassed chlorobenzene in a 250 mL Schlenk flask. 29 μl (31.20 mg 0.16 mmol) of methyldiphenylphosphine was transferred to the solution in the glovebox using an adjustable volume pipette of 1000 μl. The flask containing **1** and methyldiphenylphosphine was heated at 100 °C for 36 hours using and oil bath. The solution was concentrated under air-free conditions at room temperature using the Schlenk line. The yellow solid was transferred to the glovebox inside the Schlenk flask. Yield 3.5 mg (0.0056 mol, 20%) ^{31}P NMR 34.64 ppm in CD$_3$OD referenced to PPh$_3$ -6.23 ppm at 25°C.

(bmmp-O₂-TASN)RuPPh₂CH₃ (6)

A solution of **2** (11 mg, 0.015 mmol) was stirred in 5 mL of dry, degassed methanol in a 50 mL Schlenk flask. 20.2 mg (0.10 mmol) of methyldiphenylphosphine was added to the solution inside the glovebox. The flask containing the solution was heated at 62 °C for 36 hours. The solvent was evacuated under air-free conditions at room temperature and the crude product was dissolved in degassed methanol under air-free conditions for crystallization using the solvent evaporation technique in a test tube inside the glovebox. After two days the solution was filtered. Crystals were collected from the walls of the tube after five days. Yield: 9.0 mg (0.013 mmol, 92%). (+)ESI-MS, m/z calcd for C$_{27}$H$_{41}$N$_2$O$_2$PRuS$_3$, [M+Na]$^+$ 676.85 found, 677.11. ^{31}P NMR 25.49 ppm in CD$_3$OD referenced to PPh$_3$ -6.23 ppm at 25°C. X-ray quality crystals of **6** were obtained by slow evaporation of a methanolic solution under air-free conditions.

(bmmp-O₃-TASN)RuPPh₂CH₃ (7)

A solution of **3** (12 mg, 0.016 mmol) was stirred in 5 mL of dry, degassed methanol in a 50 mL Schlenk flask. 15.0 mg (0.075 mmol) of methyldiphenylphosphine was added to the solution inside the glovebox. The flask containing the solution was heated at 62 °C for 36 hours. The solvent was evacuated under air-free conditions at room temperature and the crude product was dissolved in degassed methanol under air-free condition for crystallization by solvent evaporation. After two days the solution was filtered. Crystals were collected from the walls of the tube after five days. Yield: 10 mg (0.015 mmol, 94%). (+)ESI-MS, m/z calcd for C$_{27}$H$_{41}$N$_2$O$_3$PRuS$_3$, [M+Na]$^+$ 692.86 found, 692.86. ^{31}P NMR 22.54 ppm in CD$_3$OD referenced to PPh$_3$ -6.23 ppm at 25°C. X-ray quality crystals of **7** were obtained by slow evaporation of a methanolic solution under air-free conditions.

[(bmmp-TASN)Ru]₂ (8)

A suspension of 650 mg (2.51 mmol) of H_2(bmmp-TASN) in dry, degassed 20 mL of THF was cooled to -78 °C in an acetone bath and 1.20 mL (12.7 mmol) of *n*-butyl lithium (2M in hexane) was added using a glass syringe (Precaution; *n*-butyl lithium reacts violently with water). The solution was allowed to slowly warm to room temperature and then transferred via cannula to a Schlenk flask containing 1.50 g (5.72 mmol) of $RuCl_2(PPh_3)_3$. The resulting solution was stirred for 12 h and then refluxed for 15 minutes. An orange solid was observed on the walls of Schlenk flask within 30 minutes. The resulting orange solid was isolated by filtration due to its low solubility in THF and washed with diethyl ether under air-free conditions. The orange solid was then washed with 100 ml of H_2O and 3 portions of 20 ml of diethyl ether. The solid was dissolved in methanol for crystallization by ether vapor diffusion. After one month, the crystals of the bimetallic compound were recovered from the walls of the air-free reaction tube. Three attempts to reproduce the bimetallic product were unsuccessful. X-ray quality crystals of **8** were obtained by solvent diffusion of methanol-ether under air-free conditions.

Physical Methods

Elemental analyses were obtained from Midwest Microlab (Indianapolis, IN). Infrared spectra were recorded as KBr pellets using a Thermo Nicolet Avatar 360 spectrometer at 4 cm⁻¹ resolution. The infrared spectra were analyzed using OMNIC 6.0a Thermo Nicolet Corporation and the KnowItAll Informatic System 8.2. ³¹P NMR spectra were obtained on a Varian Inova 500 MHz spectrometer and the analysis of the data was performed using MestReNova Version 7.0.2-8636. NMR measurements were carried out under argon atmosphere in J-Young NMR tubes at 25°C, unless otherwise specified. Electrospray Ionization mass spectrometry (+ESI-MS) was performed by the Laboratory for Biological Mass Spectrometry at Texas A&M University.

Squarewave voltammetry was performed using a EG&G Princeton Applied Research potentiostat/galvanostat model 273 potentiostat with a three electrode cell (glassy carbon working electrode, platinum wire counter electrode, and Ag wire pseudo reference electrode) at room temperature in an argon-filled glovebox. All potentials were scaled to a ferrocene standard used as internal reference. The analysis and plot of electrochemistry data was performed in SigmaPlot for Windows Version 11.0. Electronic absorption spectra were recorded with an Agilent 8453 diode array spectrometer with an air-free 1 cm path length quartz cell or an air-free 0.5 cm path length quartz cell.

Crystallographic Data

Crystal data for **6**: triclinic, space group P-1; a = 9.7483(3) Å, b = 14.4088(5) Å, c = 24.8048(7) Å; α = 75.203(3)°, β = 80.610(3)°, γ = 81.507(3)°, V = 3302.85(19) Å³, ρ_{calcd} = 1.457 mg/m³, Z = 2. Data were collected on an Agilent

Technologies/Oxford Diffraction Gemini CCD diffractometer using MoKα radiation. For all 16098 unique reflections (R(int) = 0.0279), the final anisotropic full-matrix least-squares refinement on F^2 for 709 variables converged at R1 = 0.0532, wR2 = 0.1279 with a GOF of 1.085.

Crystal data for **7**: triclinic, space group P-1; a = 8.7767(3) Å, b = 15.8615(5) Å, c = 21.4533(7) Å; α = 89.823(3)°, β = 79.435(3)°, γ = 76.449(3)°, V = 2851.70(16) Å^3, ρ_{calcd} = 1.560 mg/m^3, Z = 4. Data were collected on an Agilent Technologies/Oxford Diffraction Gemini CCD diffractometer using MoKα radiation. For all 14094 unique reflections (R(int) = 0.0252), the final anisotropic full-matrix least-squares refinement on F^2 for 677 variables converged at R1 = 0.0289, wR2 = 0.0584 with a GOF of 1.051.

Crystal data for **8**: triclinic, space group P-1; a = 8.3441(10) Å, b = 11.7392(13) Å, c = 12.4768(14) Å; α = 111.689(2)°, β = 107.581(2)°, γ = 91.071(2)°, V = 1070.8 (2) Å^3, ρ_{calcd} = 1.977 mg/m^3, Z = 2. Data were collected on a Bruker SMART APEX CCD diffractometer using MoKα radiation. For all 4842 unique reflections (R(int) = 0.0136), the final anisotropic full-matrix least-squares refinement on F^2 for 223 variables converged at R1 = 0.0210, wR2 = 0.0490 with a GOF of 1.052.

CCDC# 940595 – 940597 contain full crystallographic data for **6 – 8**. Data can be obtained free of charge from The Cambridge Crystallographic Data Center via www.ccdc.cam.ac.uk/data_request.cif.

Acknowledgments

This research was supported in part by the National Science Foundation (CHE-0749965) and the University of Louisville Executive Vice President for Research (Competitive Enhancement Grant).

References

1. Mascharak, P. K. *Coord. Chem. Rev.* **2002**, *225*, 201–214.
2. Kovacs, J. A. *Chem. Rev.* **2004**, *104*, 825–848.
3. Yano, T.; Ozawa, T.; Masuda, H. *Chem. Lett.* **2008**, *37*, 672–677.
4. O'Toole, M., G.; Grapperhaus, C., A. In *Bioinorganic Chemistry*; American Chemical Society: 2009; Vol. 1012, pp 99-113.
5. Prasad, S.; Bhalla, T. C. *Biotechnol. Adv.* **2010**, *28*, 725–741.
6. Shigehiro, S.; Nakasako, M.; Dohmae, N.; Tsujimura, M.; Tokoi, K.; Odaka, M.; Yohda, M.; Kamiya, N.; Endo, I. *Nat. Struct. Biol.* **1998**, *5*, 347–351.
7. Miyanaga, A.; Fushinobu, S.; Ito, K.; Wakagi, T. *Biochem. Biophys. Res. Commun.* **2001**, *288*, 1169–1174.
8. Murakami, T.; Nojiri, M.; Nakayama, H.; Odaka, M.; Yohda, M.; Dohmae, N.; Takio, K.; Nagamune, T.; Endo, I. *Protein Sci.* **2000**, *9*, 1024–1030.
9. Tyler, L. A.; Noveron, J. C.; Olmstead, M. M.; Mascharak, P. K. *Inorg. Chem.* **2003**, *42*, 5751–5761.

10. Lee, C. M.; Chen, C. H.; Chen, H. W.; Hsu, J. L.; Lee, G. H.; Liaw, W. F. *Inorg. Chem.* **2005**, *44*, 6670–6679.

11. Rose, M. J.; Betterley, N. M.; Oliver, A. G.; Mascharak, P. K. *Inorg. Chem.* **2010**, *49*, 1854–1864.

12. Hashimoto, K.; Suzuki, H.; Taniguchi, K.; Noguchi, T.; Yohda, M.; Odaka, M. *J. Biol. Chem.* **2008**, *283*, 36617–36623.

13. Hopmann, K. H.; Guo, J. D.; Himo, F. *Inorg. Chem.* **2007**, *46*, 4850–4856.

14. Swartz, R. D.; Coggins, M. K.; Kaminsky, W.; Kovacs, J. A. *J. Am. Chem. Soc.* **2011**, *133*, 3954–3963.

15. Dey, A.; Jeffrey, S. P.; Darensbourg, M. Y.; Hodgson, K. O.; Hedman, B.; Solomon, E. I. *Inorg. Chem.* **2007**, *46*, 4989–4996.

16. Masitas, C. A.; Mashuta, M. S.; Grapperhaus, C. A. *Inorg. Chem.* **2010**, *49*, 5344–5346.

17. Masitas, C. A.; Kumar, M.; Mashuta, M. S.; Kozlowski, P. M.; Grapperhaus, C. A. *Inorg. Chem.* **2010**, *49*, 10875–10881.

18. Kumar, D.; Masitas, C. A.; Nguyen, T. N.; Grapperhaus, C. A. *Chem. Commun.* **2013**, *49*, 294–296.

19. Tsujimura, M.; Odaka, M.; Nakayama, H.; Dohmae, N.; Koshino, H.; Asami, T.; Hoshino, M.; Takio, K.; Yoshida, S.; Maeda, M.; Endo, I. *J. Am. Chem. Soc.* **2003**, *125*, 11532–11538.

20. Song, L.; Wang, M.; Yang, X.; Qian, S. *Biotechnol. J.* **2007**, *2*, 717–724.

21. Dilworth, J.; Zheng, Y.; Lu, S.; Wu, Q. *Transition Met. Chem.* **1992**, *17*, 364–368.

22. Buonomo, R. M.; Font, I.; Maguire, M. J.; Reibenspies, J. H.; Tuntulani, T.; Darensbourg, M. Y. *J. Am. Chem. Soc.* **1995**, *117*, 963–973.

23. Kung, I.; Schweitzer, D.; Shearer, J.; Taylor, W. D.; Jackson, H. L.; Lovell, S.; Kovacs, J. A. *J. Am. Chem. Soc.* **2000**, *122*, 8299–8300.

24. Aucott, S. M.; Milton, H. L.; Robertson, S. D.; Slawin, A. M. Z.; Walker, G. D.; Woollins, J. D. *Chem. Eur. J.* **2004**, *10*, 1666–1676.

25. Shearer, J.; Callan, P. E.; Amie, J. *Inorg. Chem.* **2010**, *49*, 9064–9077.

26. Dutta, A.; Flores, M.; Roy, S.; Schmitt, J. C.; Hamilton, G. A.; Hartnett, H. E.; Shearer, J. M.; Jones, A. K. *Inorg. Chem.* **2013**, *52*, 5236–5245.

27. Grapperhaus, C. A.; Patra, A. K.; Mashuta, M. S. *Inorg. Chem.* **2002**, *41*, 1039–1041.

28. Grapperhaus, C. A.; Li, M.; Patra, A. K.; Poturovic, S.; Kozlowski, P. M.; Zgierski, M. Z.; Mashuta, M. S. *Inorg. Chem.* **2003**, *42*, 4382–4388.

29. O'Toole, M. G.; Kreso, M.; Kozlowski, P. M.; Mashuta, M. S.; Grapperhaus, C. A. *J. Biol. Inorg. Chem.* **2008**, *13*, 1219–1230.

30. Grapperhaus, C. A.; O'Toole, M. G.; Mashuta, M. S. *Inorg. Chem. Commun.* **2006**, *9*, 1204–1206.

31. Shearer, J.; Callan, P. E.; Masitas, C. A.; Grapperhaus, C. A. *Inorg. Chem.* **2012**, *51*, 6032–6045.

32. Sellmann, D.; Hadawi, B.; Knoch, F.; Moll, M. *Inorg. Chem.* **1995**, *34*, 5963–5972.

33. Sellmann, D.; Engl, K.; Heinemann, F. W.; Sieler, J. *Eur. J. Inorg. Chem.* **2000**, 1079–1089.

34. Grapperhaus, C. A.; Poturovic, S.; Mashuta, M. S. *Inorg. Chem.* **2005**, *44*, 8185–8187.

35. Farmer, P. J.; Reibenspies, J. H.; Lindahl, P. A.; Darensbourg, M. Y. *J. Am. Chem. Soc.* **1993**, *115*, 4665–4674.

36. Farrugia, L. J. *J. Appl. Crystallogr.* **1997**, *30*, 565.

37. Sellmann, D.; Prakash, R.; Heinemann, F. W.; Moll, M.; Klimowicz, M. *Angew. Chem. Int. Ed.* **2004**, *43*, 1877–1880.

38. Sellmann, D.; Hein, K.; Heinemann, Frank W. *Eur. J. Inorg. Chem.* **2004**, *2004*, 3136–3146.

39. Hoffmann, P.; Steinhoff, A.; Mattes, R. *Z. Naturforsch., B: Chem. Sci.* **1987**, *42*, 867–873.

40. Grapperhaus, C. A.; Patra, A. K.; Mashuta, M. S. *Inorg. Chem.* **2002**, *41*, 1039–1041.

41. Parry, R. *Inorg. Synth.* **1970**, *12*, 237–240.

Chapter 5

Biochemistry of Nucleic Acids Functionalized with Sulfur, Selenium, and Tellurium: Roles of the Single-Atom Substitution

Manindar Kaur, Abdur Rob, Julianne Caton-Williams, and Zhen Huang[*]

Department of Chemistry, Georgia State University, Atlanta, Georgia 30303
[*]E-mail: Huang@gsu.edu.

Nucleic acids and their modifications are instrumental in discovering functional oligonucleotides as important biochemical and therapeutic agents. In the search of bioactive compounds, a variety of synthetic strategies have been developed to design novel analogs of the nucleosides, nucleotides, and nucleic acids. Since oxygen atoms are found in abundance in natural nucleic acids, the atom-specific replacement of these oxygen atoms with the heavier chalcogens sulfur, selenium, and tellurium facilitates generation of novel nucleic acids with distinctive properties, such as enhanced duplex stability, binding affinity, nuclease resistance, bioavailability, and base-pair fidelity. These structural alterations create a new archetype of nucleic acids with potential applications in therapeutics and drug development. Moreover, these engineered nucleic acids are useful biological tools for disease detection, molecular sensing, and fundamental understanding of nucleic acid structures and functions. This review outlines sulfur, selenium, and tellurium modifications to nucleic acids and their role in generating new classes of nucleic acids with tunable biochemical and physico-chemical properties. Such modifications offer structural, functional, and mechanistic probes for investigating the structures and biological functions of nucleic acids (DNA and RNA).

1. Introduction

Nucleic acids are fundamental biomacromolecules that function to store genetic information. Virtually omnipresent, nucleic acids function to replicate, transcribe and translate genetic information, andthe structure-function study of nucleic acids is essential to the knowledge of life. The significance of nucleosides and nucleotides in various biochemical processes has been amply demonstrated. Nucleosides and nucleotides are powerful building blocks for creating functional molecules with novel properties. They can be easily modeled and chemically functionalized at various positions with diverse groups. The modified analogues, being more stable, specific, and efficient, are potential candidates for novel drug therapies and effective biological tools for the fundamental understanding of diseases at the molecular and genetic level (*1–3*). Both chemical and enzymatic approaches have been utilized to synthesize nucleic acids with modified nucleobases, sugars or phosphodiester backbones. The modification of nucleic acids represents an enormous opportunity to tailor the properties and functions of these biomacromolecules, escalating their potential for biochemical and biomedical applications (*4–14*).

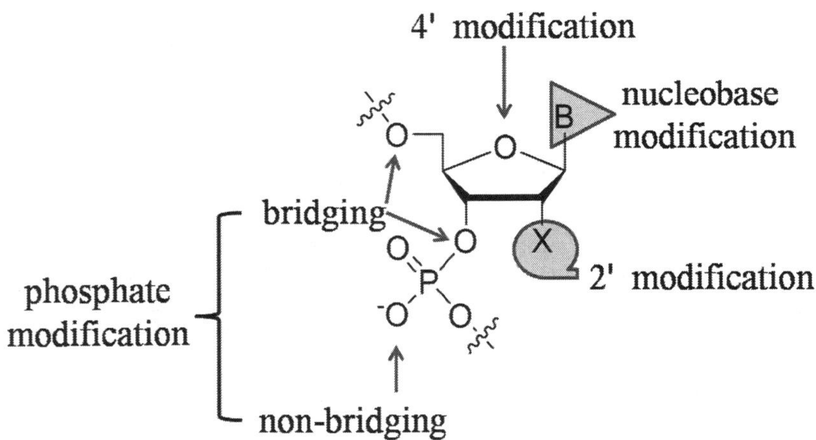

Figure 1. Probable sites for chemical modifications (with S, Se, and Te) in DNAs and RNAs. See Figure 2 for more details.

It is amazing how only five elements, namely oxygen, nitrogen, phosphorus, carbon, and hydrogen (as well as rarely sulfur and selenium), create nucleic acids. However, in comparison to proteins, which contain a combination of twenty-one amino acids with diverse side chains, the functions of nucleic acids have been limited by the combination of only five nucleotide building blocks. Therefore, chemical modifications of nucleosides and nucleotides have been developed to offer nucleic acids with tunable physico-chemical properties for applications in catalysis, molecular binding, and nucleic-acid labeling (*5,*

15–18). Oxygen atoms are abundant in nucleic acids. Therefore, atom-specific replacement of oxygen with sulfur, selenium, and tellurium, which belong to the oxygen family, is a useful and successful chemical strategy to improve and tune the biophysical and biochemical properties of nucleic acids (Figure 1). Compared to oxygen, these heavier chalcogens are polarizable and have larger atomic radii (O: 0.73 Å; S: 1.02 Å; Se: 1.16 Å; Te: 1.40 Å) and higher electron densities (*19*). These atomic features facilitate the tailor-made recognition and catalytic activities of the atom-specifically replaced nucleic acids, without significant structural perturbation. Moreover, an X-ray crystallographic study showed that thio- and seleno-nucleosides and -nucleotides are good structural mimics of the corresponding oxo-nucleosides and -nucleotides, respectively, yet they offer many distinctive and novel properties. This review briefly discusses structure-and-function studies and applications of S-, Se- and Te-modified nucleic acids. These modifications are presented in Figure 2 and discussed with an emphasis on their biochemical and therapeutic potentials.

2. Sulfur-Functionalized Nucleic Acids

Sulfur is an interesting element with divergent chemical properties. It is an essential element for all forms of life and is involved in various biochemical processes. The biological occurrence of sulfur was recognized about a century ago with the isolation of 5'-deoxy-5'-methylthioadenosine (MTA) from yeast (*20*). Since then, there have been many reports underscoring the physiological and pharmacological significance of thionucleosides. Incorporation of sulfur in nucleosides, nucleotides, and nucleic acids can afford exclusive biological properties such asincreased metabolic stability and bioavailability, nuclease resistance, high hybridization, and decreased toxicity. In search of bioactive compounds, a variety of modified oligonucleotides containing a sulfur atom on the sugar, the nucleobase, or the phosphoryl group have been synthesized. These sulfur-derivatized oligonucleotides are valuable tools in understanding protein-nucleic acid interactions, nucleic acid structure-function relationship, and oligonucleotide therapy.

2.1. Modifications

2.1.1. Sugar-Modified Analogues

Sulfur replacement of oxygen in nucleic acids has mostly focused on the 2', 3', 4', or 5' position of the sugar. The initial challenges in the synthesis of 2'-thionucleosides hampered the structure-function assessment of such a modification. The advancement in scientific techniques and tools led to the successful synthesis, characterization and structural evaluation of 2'-thionucleosides (*21*). The 2'-hydroxyl group plays a vital role in RNA hydrolysis. Therefore, the 2'-functionalization with groups such as -SH is helpful in elucidating the mechanism of RNA hydrolysis and ribozyme action and also bestows nucleic acids with tailored properties for use in therapeutics (*22, 23*).

The 3'-hydroxyl group is an important functionality dictating template-directed DNA polymerization. Thus, incorporation of a thiol group at the 3'-position generates analogs with attractive biological properties. For example, it has been demonstrated that the 2',3'-dideoxy-3'-thionucleosides (T, A, C, G) have potent antiviral activities and the corresponding 5'-triphosphates behave as effective DNA chain terminators (24). Sulfur derivatization of the furanose ring oxygen has facilitated the generation of novel nucleoside metabolites, commonly known as the 4'-thionucleosides. The 4'-thioribonucleosides are promising candidates for use as antiviral and antineoplastic agents, but synthetic challenges have limited their further development (25). Significant focus has also been given to the synthetic development of 2'-deoxy-4'-thionucleosides, further facilitating the biological evaluation of these metabolites (26, 27). Interestingly, incorporation of 4'-thio-2'-deoxyribonucleosides into DNA lowers the thermal stability of the DNA-RNA hybrid, whereas the incorporation of 4'-thiouridine into RNA enhances duplex stability (28). 4'-Thionucleosides exhibit lower host toxicity and enhanced stability towards viral and cellular enzymes. With the improvement in their synthetic strategies, the interest in biological activity of 4'-thionucleosides has rekindled (29).

2.1.2. Nucleobase-Modified Analogues

Sulfur is widely found in a broad array of natural biomolecules owing to its chemical and structural behavior. The presence of sulfur in certain amino acids, cofactors and in RNA thionucleosides has been well appreciated (30). The largest variety of naturally modified bases is found in tRNAs. Several of these modified nucleobases contain sulfur. These thionucleosides, isolated from the tRNA of *Escherichia coli, Bacillus subtilis*, yeast, and other higher organisms, improve translational efficacy and regulate cellular processes (31). The presence of 2-thiouridine (s^2U) derivatives at wobble position 34 in tRNA^{Glu}, tRNA^{Lys}, and tRNA^{Gln} is crucial for many biological functions of tRNA (32, 33). The s^2U modification is known to stabilize the 3'-endo sugar conformation and also participates in codon-anticodon interaction during translation. The s^2U undergoes further modification on C-5 to form 5-methylaminomethyl-2-thiouridine (mnm^5s^2U) in *Escherichia coli* and 5-methoxycarbonylmethyl-2-thiouridine (mcm^5s^2U) in humans. 4-Thiouridine (s^4U), with a thiol modification on C-4, is another type of uridine derivative found in tRNA. It has been proposed that s^4U is a major chromophore resulting in photo-induced growth delay (34). Biophysical and biochemical studies demonstrate that, in comparison to the non-modified RNA, s^2U-containing RNA is thermally stabilized, whereas s^4U-containing RNA is destabilized (35). The third thio-derivatized nucleobase found in nature is 2-thiocytidine (s^2C). The 2-thioribothymidine (s^2T), found at position 54 in the T-loop of tRNAs, has important contribution to the thermal stability of the tRNA's tertiary structure (36, 37). There is no report of the natural existence of 6-thioguanosine. To investigate the structure and diverse functions of DNA and RNA, many research groups have reported the synthesis of various thio-nucleobases and -nucleosides. 6-Thioguanine, an important anticancer and immunosuppressive agent, is an approved antitumor drug (*38*) by the US Food and Drug Administration (FDA). With the advancement in bioanalytical tools, the synthesis and structure-function analysis of various S-nucleobases have become so straightforward thatvarious thiolated phosphoramidites are commercially available.

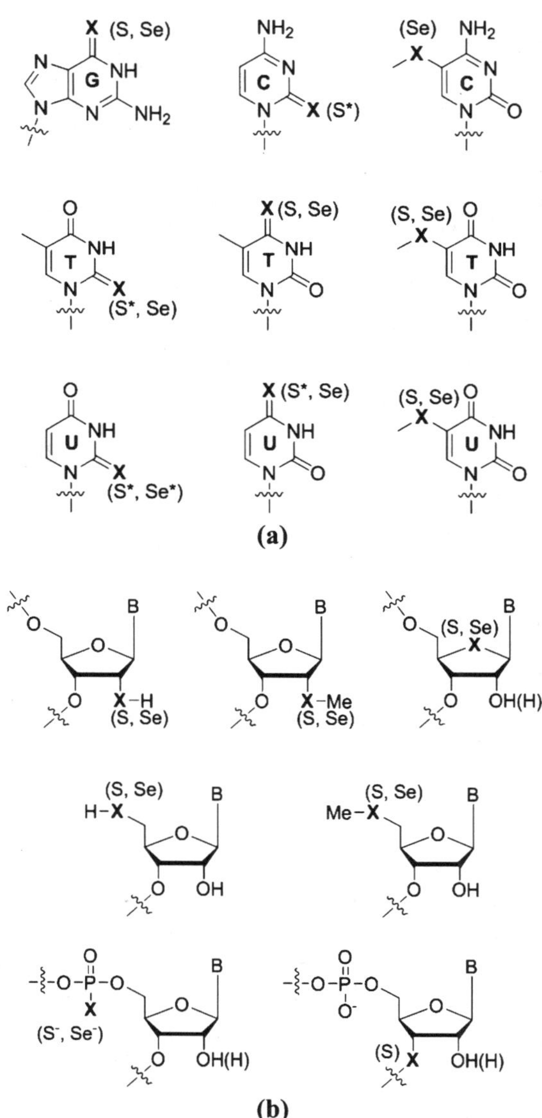

*Figure 2. Atom-specific replacement of oxygen with S and Se on (**a**) the nucleobase and (**b**) the sugar and phosphate backbone. The asterisk (*) indicates the naturally occurring modifications.*

The phosphodiester backbone ($-O-PO_2-O-$) connects two adjacent nucleotides. Several modifications have been incorporated to delineate their structural and functional impacts on the functionalized oligonucleotides (*9*). Nucleoside phosphorothioates are obtained by the substitution of an oxygen atom of the linking phosphate group with a sulfur atom. With minimal atomic replacement, phosphorothioates are an important class of biomolecules for structure-function studies. Phosphorothioates are good substrates for a variety of polymerases, kinases, ligases and restriction endonucleases. Phosphorothioate-containing nucleic acids, while maintaining the structural conformation of the natural nucleic acids, offer superior biochemical properties and applications, such as enhanced nuclease resistance, biochemical probes, and mutagenesis facilitation (*39–41*). Their stereochemical nature, due to the chirality of the thiophosphate linkage, has led to a number of applications such as antisense and small interfering RNAs (siRNAs) therapeutics (*42–50*), mechanistic probes for investigating enzyme catalyzed reactions (*39–41, 51*), DNA and RNA sequencing (*52, 53*), and site-directed mutagenesis facilitation (*54, 55*). Following Eckstein's pioneering work in synthesizing nucleoside 5′-phosphorothioates by condensation of unprotected 2′-deoxyribonucleosides or the protected ribonucleosides with triimidazolyl-1-phosphinsulfide in pyridine or dimethylformamide-tetrahydrofuran (*56, 57*), several other synthetic methodologies have been developed for creating phosphorothioates (*58*). These strategies embrace both enzymatic (*59–61*) and chemical solid-phase synthesis of phosphorothioates (*62, 63*). Strategic incorporation and hydrolysis of phosphorothioate linkages in nucleic acids have been successfully exploited for chemical sequencing of DNAs and RNAs (*53*). The incorporation of sulfur in place of one non-bridging oxygen provides a chiral center, giving rise to a mixture of diastereomeric phosphorothioates (denoted as R_P and S_P) (*63, 64*). This facilitates the stereochemical analyses of enzyme-catalyzed reactions (*65*). However, it is barely possible to control diastereomer formation via chemical synthesis. On the other hand, the DNA and RNA polymerases can recognize only one of the two diastereoisomers of deoxy- and ribo-nucleoside 5′-(α-P-thio)-triphosphates (α-S-dNTPs and α-S-NTPs), thus facilitating the stereoselective synthesis of diastereomerically pure PS-DNAs and PS-RNAs (*66–69*). These diastereomeric thio-triphosphates are useful in the preparation and analysis of oligonucleotides. They are important biochemical probes for elucidating the stereochemical course of the interactions of the triphosphates with cellular enzymes. Recently, Huang and co-workers have developed a novel strategy for convenient, one-pot synthesis of α-S-dNTPs and α-S-NTPs (*70*). Incorporation of sulfur at the 3′ and 5′ positions can also result in analogs with interesting biochemical properties (*71*). Oligonucleotides with the 3′-S-phosphorothiolate linkage can also act as mechanistic probes for investigating a variety of enzyme-catalyzed cleavage reactions (*72, 73*).

2.2. Applications

The rational design of drugs and therapeutic agents has become more plausible with our increasing understanding of biochemistry and molecular and cellular biology. Chemically modified nucleic acids, with their diverse functionalities, have drawn widespread attention from multidisciplinary research teams. Ample growth has occurred in developing optimized and automated methods for the preparation of modified nucleic acids, which serve as significant tools in fundamental research as structure-and-function probes, chemical probes, and mechanistic probes.

2.2.1. Mechanistic Probes and Biochemical Applications

Sulfur adds substantial functionality to nucleic acids, as a result of its structural and chemical versatility. For example, 4′-thio-RNAs, built by 4′-thioribonucleoside building blocks, exhibit improved thermostability and nuclease resistance (towards both endo- and exo-nucleases), high hybridization ability, and structural similarity to A-form RNA duplexes, affording a new generation of artificial RNA molecules for therapeutic applications (*74*). Alternatively, 4′-thionucleoside-5′-triphosphates (4′-thio-UTP and -CTP), fine substrates for the RNA polymerases, have been used effectively for isolating 4′-thio-RNA aptamers by SELEX (Systematic Evolution of Ligands by Exponential Enrichment) process and in developing THE siRNAs (*75, 76*). A detailed study has demostrated the preference of 4′-thio-DNAs to hybridize with RNA as well as their enhanced resistance towards nuclease digestion, especially to 3′-exonuclease hydrolysis. Interestingly, 4′-thio-DNAs structurally act as RNA-like molecules, adopting A-form helical structure (*77*).

2.2.2. Photolabels

Photo-crosslinking has emerged as a useful labeling technique for improved understanding and direct identification of RNA/DNA-protein interactions in complex nucleoprotein assemblies (*78*). Photo-crosslinking is an extremely powerful bioanalytical tool for probing the three-dimensional architecture of biomacromolecules in solution. In the last two decades, thionucleobases have drawn considerable interest as exceptional intrinsic photolabels for probing RNA folding and activity in solution, especially because their incorporation causes minimal structural perturbation. 4-Thiouracil, 4-thiothymine and 6-thioguanine have been investigated as potential photolabels. Thiolation of these nucleosides results in a red shift (λ_{max} = 330-340 nm) in the absorption spectra, well separated from the range of protein (λ_{max} = 280 nm) and native DNA/RNA (λ_{max} = 260 nm). This makes the selective photo-crosslinking of thionucleobases more viable and valuable. These photolabels have several notableattributes: they are (i) stable in dark, (ii) selectively photoactive at 330-370 nm, (iii) structurally similar to native nucleosides, and (iv) highly photoreactive towards nucleic acid and protein residues (*79*). 4-Thiouridine and 6-thioguanosine, when incorporated into RNA, are efficient structural probes for short range (approximately 3 Å) RNA-RNA cross-links (*80*). A systematic investigation of RNA-RNA crosslinks can be beneficial for three-dimensional conformational analysis of RNA. Strategic incorporation and placement of 2′-deoxy-4-thiouridine and 2′-deoxy-6-thioinosine in the ribozyme substrate have helped to decipher the conformational mobility of the hammerhead ribozyme domain (*81, 82*). A recent study suggests that DNA thiobases also act as UVA sensitizers, which warrants their importance in the clinical management of superficial tumors and non-malignant tumors such as psoriasis (*83*).

2.2.3. Therapeutic Development

Antisense technology embraces the class of nucleic acids with the exceptional ability of inhibiting gene-expression at mRNA level to knock-down the expression of disease-causing proteins. With the systematic development of modified nucleosides and nucleotides, antisense technology is experiencing a renaissance. Many modifications of the nucleobase, sugar and phosphate have been successfully introduced to convey resistance to nucleases and inhibit gene expression by escalating the efficiency of the hybridization with target mRNA. The rich and facile chemistry of phosphorothioates affords increased nuclease resistance, while maintaining RNase H activity. These typical characteristics make phosphorothioates prospective contenders for antisense therapeutics (*44, 84*). Vitravene (Fomivirsen), an S-oligonucleotide drug developed by ISIS pharmaceuticals, was approved by FDA in 1998 for the treatment of cytomegalovirus (CMV)-induced retinitis (*85*). It is an antisense 21-mer phosphorothioate oligonucleotide, meeting the safety and efficacy standards of an antiviral drug. It binds to the coding region of the CMV gene transcript and inhibits translation, thereby repressing viral replication and inhibiting viral adsorption to host cells (*86*). The discovery of antiviral Vitravene was a significant finding and paved the path to a number of other antisense drug candidates, most of which are still in different clinical trial phases (*87*).

A second-generation antisense inhibitor, Mipomersen, is an interesting phosphorothioate oligonucleotide with terminal 2′-*O*-methoxyethyl-riboses to resist cellular nucleases (*88*). Mipomersen is an apolipoprotein (apo) B synthesis inhibitor, and this drug was approved by FDA in 2013 for cholesterol reduction. During its initial phases of clinical study, it demonstrated significant reduction in apo B and low-density lipoprotein (LDL) in patients with heterozygous familial hypercholesterolemia, a genetic disorder arising due to defective LDL-receptor alleles, thus causing exceptionally high levels of LDL cholesterol (*89, 90*). Alicaforsen (ISIS-2302; Atlantic Healthcare), a 20-base-long antisense phosphorothioate oligonucleotide, suppresses the expression and message of ICAM-1 (*91*). Unlike Vitravene, it was unable to demonstrate efficacy in patients with Crohn's disease. Other phosphorothioate-based antisense drugs, currently in phase III clinical trials, are Trabedersen (Antisense Pharma), which targets transforming growth factor beta 2 (TGF-β2) and tumor necrosis factor ligand 13 and will be used in the treatment of glioblastoma and anaplastic astrocytoma; Genasense (Genta), which targets BCL-2 and will be used in the treatment of melanoma and various types of leukemia; and Aganirsen (Gene Signal), which targets insulin receptor substrate-1 and will be used to treat corneal graft rejection (*90*). However, due to the toxicological drawbacks associated with phosphorothioate-based therapeutics, focus has now shifted to the use of mixed-backbone analogues (*92*).

The siRNAs, recent additions to the wide repertoire of nucleic acid-based therapeutics, have broadened the horizons of antisense molecules. They are potent, sequence-specific, gene-silencing agents. siRNAs are designed to exclusively trigger the enzymatic destruction of specific transcripts via the RNA interference (RNAi) pathway. A typical siRNA construct consists of 19-30 nucleotide dsRNAwith 2 nucleotide overhangs at each terminus (*11*). In comparison with the antisense technology, S-containing siRNA therapeutics are clinically underutilized. The first fully tailored siRNA construct to silence a therapeutically relevant gene *in vivo* was reported in 2006 with a 2′-*O*-methyl sense strand and a phosphorothioate antisense strand (*93*).The intelligent combination of a 2′-*O*-methyl sense strand and the phosphorothioate modification provides this siRNA duplex with additional stability towards nucleases, along with extended serum retention, thereby augmenting its physicochemical properties. Phosphorothioate-modified siRNAs have been studied extensively for their usefulness as efficient and enduring pharmacological agents (*94*).

RNA triplex, a significant RNA tertiary structure motif, is formed by non-canonical base pairing between a duplex and a third strand. Recently, there has been tremendous focus on developing triplex-forming oligonucleotides (TFOs) as influential tools in gene expression/regulation and targeted mutagenesis and recombination (*95*). Various chemical modifications have facilitated the formation of stable triplexes (*96*). Oligonucleotides containing 2-thiouridine (s^{2}U), 2′-*O*-methyl-2-thiouridine (s^{2}Um) and 2-thioribothymidine (s^{2}T) derivatives have a rigid C*3′-endo* conformation, partially because of the steric repulsion between the 2-thiocarbonyl group and the 2′-hydroxyl group. The resulting strong stacking ability of the 2-thiocarbonyl group results in the stabilization of triplexes (*97*). Moreover, thermal analysis of TFOs has established that the incorporation of the 2-thiocarbonyl moiety does not reduce Hoogsteen base-pairing ability. 8-Thioxoadenine (s^{8}A) was also investigated as a potential modified nucleoside for improved hybridization of TFOs via enhanced stacking interactions. The thio-modification at C-8 of the adenine base favors the *syn* orientation around the glycosyl bond, making the formation of a triad entropically favorable (*98*, *99*). Due to these structural features, thionated TFOs are favorable molecules for antigene strategy. Further structural investigation, for better understanding of the parameters dictating triplex stability, would facilitate the discovery of novel diagnostic and therapeutic agents.

Aptamers, the nucleic acid analogues of antibodies, are typically 15-40 nucleotide DNA or RNA molecules that can adopt three-dimensional motifs facilitating specific binding to molecular targets. New aptamersare rapidly selected and generated by SELEX process (*100*). The design of the aptamer can be rationally engineered for high target specificity and affinity, biological efficacy, and exceptional pharmacokinetic properties (*101*). The design and development of aptamers for therapeutics rely comprehensively on nucleotide modifications. Using 4′-thio-UTP and 4′-thio-CTP, Kato et al. successfully isolated a high affinity, nuclease-resistant thio-RNA aptamer to human α-thrombin (*75*). Owing to their small size, ease of synthesis and modification, and low synthetic cost, aptamers offer versatile biological tools for intracellular and extracellular target validation (*102*).

3. Selenium-Functionalized Nucleic Acids

Selenium is an essential element and a vital micronutrient for humans with potential antioxidant properties. Although selenium was discovered in proteins in the mid-1950s (*103–106*), early investigations on selenium was limited by its noxious nature. It was not until 1973 that the roles played by Se in the body, involving in numerous physiological processes, were better understood. Today, selenium is known as an essential trace element that is present naturally in foods and soils, and it is available via dietary supplements as nutrients for the health of humans and other organisms (*107–109*). Selenium is biologically active *via* the functions of selenoproteins, which play important roles in thyroid hormone metabolism, redox signaling, antioxidant defense, immune responses, reproduction, DNA synthesis, and protection from oxidative damage and infection (*108*). Studies have shown that Se consumption may reduce the risk of certain types of cancer by reducing oxidative stress and DNA damage (*110*). It must be noted that both excess and deficiency of selenium from the diet leads to many fatal diseases. Among the many benefits of Se in selenoproteins, the role of Se in the prevention of cancer has drawn great attention (*111*). These positive influences of Se in the health of humans allow for the growth and development of interest in Se-related research (*111–117*). Despite its resemblance to its neighbors sulfur and tellurium, selenium displays exceptional chemical and biological activities owing to its unique electronegativity and atomic radius (*118, 119*). Selenium is the only trace element found in amino acids (selenocysteine and selenomethionine). Selenocysteine, the 21st amino acid (*120–122*), is the central component of selenoproteins and helps to regulate various enzymatic functions (*123, 124*). Analogous to sulfur, natural occurrence of selenium in tRNAs has been observed in certain bacteria and mammals. The selenium nucleoside, identified as 5-[(methylamino)methyl]-2-selenouridine (mnm^{5}se^{2}U), is often found at the first or wobble position of the anticodon loop of tRNA (*125, 126*). While the precise function of selenium at the C-2 position and the comprehensive utility of Se-derivatized tRNAs are not delineated, this modification is essential for catalysis and regulation and probably capable of augmenting translational efficiency and efficacy (*127*).

Selenium derivatization of nucleic acids generates the modified biopolymers with novel properties. Huang and co-workers have pioneered the synthesis and structure-function studies of selenium-derivatized (or -modified) nucleic acids (SeNA). Selenium functionalization at different positions in the nucleobase, sugar, and phosphate group has been achieved by Huang's research group and other groups (*128–148*). The presence of selenium atoms can facilitate crystallographic phasing as well as crystallization, and selenium derivatization of nucleic acids (*128, 129, 149*) has surpassed other known approaches (such as heavy-atom soaking, co-crystallization, and halogen derivatization) for nucleic acid X-ray crystallography. The nucleosides, nucleotides and nucleic acids obtained by atom-specific substitution of oxygen with selenium serve as useful biophysical, structural, and mechanistic probes. The Se-functionality, stability, and compatibility with chemical and enzymatic synthesis have made the introduction and incorporation of selenium at different positions a reliable and useful technique.

3.1. Selenium-Containing tRNAs and Selenoproteins

Transfer RNA (tRNA) is the link between messenger RNA (mRNA) and ribosomal RNA (rRNA) during protein synthesis. These RNAs are critical for life and each has its own unique function. tRNA has two important regions: a tri-nucleotide region containing the anticodon and an attachment site for a specific amino acid (*150–152*). Apart from oxygen and sulfur atoms being present in tRNAs, selenium in the form of selenocysteine (Sec) has been found in several charged tRNAs, such as tRNA(Sec). This Se-containing tRNA is the longest characterized mammalian tRNA (90 nt in length) and its anticodon is complementary to the stop codon UGA (*153, 154*). The secondary structures of tRNA(Sec) differ among the three domains of life: bacteria, eukarya and archaea (*155, 156*). Krol's group (*157*) has derived a three-dimensional structure model to demonstrate some unique aspects of secondary and tertiary structural features of eukaryotic tRNA(Sec). Furthermore, Yokoyama's group (*151, 152, 158*) has solved and elucidated the crystal structure of tRNA(Sec) in bacteria, eukarya, and archaea, and conducted an in-depth comparison on their secondary and tertiary structures to have a better understanding of the functions and properties of tRNA(Sec) in these three domains of life. Moreover, the presence of Se-modified nucleosides has been observed in certain tRNAs, such as seleno-tRNA in naturally-occurring 5-[(methylamino)methyl]-2-selenouridine (mnm^{5}Se^{2}U; Figure 3a) and 2-selenouridine (Figure 3b) derivatives (*116, 122, 125, 126, 159, 160*). Selenium in tRNAs and proteins provides additional advantages and functions, such as stability (*161*), catalytic efficiency (*162*), and balanced cellular environment (*107, 163*).

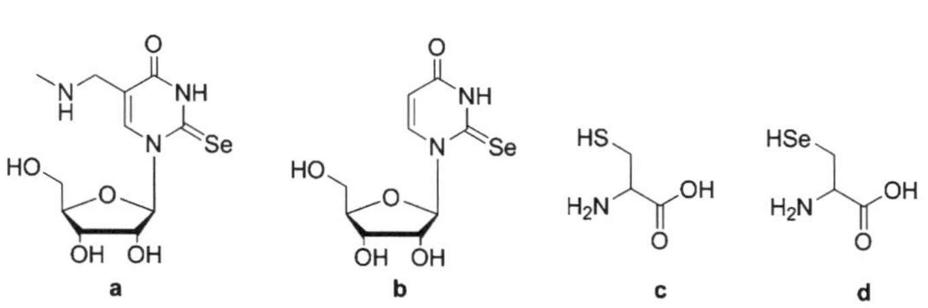

Figure 3. (*a*) *5-(methylamino)methyl-2-selenouridine,* (*b*) *2-Selenouridine,* (*c*) *cysteine, and* (*d*) *selenocysteine.*

Selenium also provides protein structures with many unique biochemical and biological functions and properties. Organic Se in the body is mainly found as the amino acids selenomethionine and selenocysteine (*154, 164*). Selenocysteine occurs as a highly specific and essential component of many selenium-dependent enzymes, including formate dehydrogenase, glycine reductase, hydrogenase of bacterial origin, and glutathione peroxidase (GSH-Px) from mammalian and avian sources (*165, 166*). GSH-Px enzymes, such as classical GPx1, gastrointestinal GPx2, plasma GPx3, phospholipid hydroperoxide GPx4, are a major class of functionally important selenoproteins and are the first to be characterized (*167*). GSH-Px is essential for protecting cellular membranes from being destroyed. The identification of Se in glutathione peroxidase has led to a clearer understanding of the importance of Se in nutrition and health (*168, 169*).Later, another class of Sec-containing enzymes, thioredoxin reductases (TR1 or Txnrd1 and TR3 or Txnrd2), which catalyze the NADPH-dependent reduction of thioredoxin, was identified (*170*). TR plays a regulatory role in its metabolic activity (*171*). TR1, GPx1, and GPx4 are three of the most studied selenoproteins (*172*). A study conducted with a mouse breast cancer cell line (EMT6) in the presence of these enzymes and popular antibiotics (Dox, Cp, and G418) demonstrated that antibiotic treatment can increase error of Sec insertion during selenoprotein synthesis, thereby affecting their functions (*117, 173*). On the other hand, selenomethionine is incorporated nonspecifically into proteins by bacteria, plants, and animals (*165*). Selenocysteine (Sec, or U; Figure 3d), an analog of cysteine (Cys; Figure 3c), has been co-translationally incorporated into proteins, and the UGA codon (typically a termination codon in protein synthesis) is used as a Sec codon by many organisms in the three domains of life: eubacteria, archaea, and eukaryotes (*121, 174, 175*). There are approximately 25 known Se-containing proteins in mammals (*121, 176*). However, they are not found in all organisms, such as fungi and higher plants (*140, 174, 177*). These organisms lack the Sec synthesis pathway, while lower organisms, such as green algae, kinetoplastida, and slime molds, have cysteine-containing homologs of selenoproteins (*178, 179*).

3.1.1. Different Properties of Selenium and Sulfur in Redoxcenters of Selenoproteins

In contrast to thiols, the aforementioned selenoenzymes (*165, 166*) contain selenol groups that function as a redox centers because of selenium's ionization potential at physiological pH. Although S and Se are in the same family of the periodic table, these redox-type enzymes (selenoproteins) take advantage of the special properties of selenium compared to sulfur. The S in Cys (Figure 3c) has remarkably different properties from the Se in Sec (Figure 3d) (*161, 162*). One of the most noticeable differences are their pK_a values. In Sec, the pK_a of Se-H group (5.3) is dramatically lower, compared to the pK_a of S-H (8.5) in Cys. Based on this factor, Sec possesses greater nucleophilicity and a more positive reduction potential than Cys at physiological pH (*163, 180, 181*). These differences are contributing factors for Sec participation in selenoprotein activity, where redox reaction rates can be enhanced up to 1000-fold (*182–184*) compared to sulfur-substituted analogs. Clearly, in many important redox enzymes, selenium plays an important role to catalytically maintain cellular redox balance in both eukaryotes and prokaryotes (*163, 172, 185*).Sulfur and selenium, therefore, contribute significantly to the proper functions of many natural tRNAs and proteins (*125, 186–189*).

3.1.2. 2-Selenouridine Derivatives

In addition to functioning in the active site of selenoproteins, selenium also plays critical role in some tRNAs of bacteria, archaea, eukaryotes, and plant species (*126, 159, 160, 164, 174, 190–193*). A tRNA containing a seleno-nucleobase was first discovered in *Escherichia coli* by Saelinger and his coworkers (*194*). These nucleoside modifications are found in both 5-[(methylamino)methyl]-2-selenouridine (mnm^{5}Se^{2}U) and 2-selenouridine derivatives (Figure 3a and 3b), of which mnm^{5}Se^{2}U occurs most frequently (60%). The remaining 40% are derivatives of 2-selenouridine (*191*). Mnm^{5}Se^{2}U has been observed mainly at the wobble position (the first position) of the anticodon in tRNA^{Glu}, tRNA^{Gln}, and tRNA^{Lys}, where tRNA^{Glu} accounts for 80% of the modification (*126, 128, 176, 193*). Because of the specific location of mnm^{5}Se^{2}U in tRNA, it has been theorized that the 2-seleno functionality may contribute to the enhancement of translation accuracy and efficiency in protein synthesis, similar to that proposed for the S counterpart (mnm^{5}S^{2}U) at the wobble position (*195, 196*).

3.1.3. Biosynthesis of tRNAs(Sec)

The amino acid Sec is biosynthesized on its tRNA, unlike any other known amino acid in eukaryotes (*170*). The biosynthesis of Sec was established in eubacteria in the early 1990s (*164*), and at a later time, the complete biosynthetic pathway of Sec in eukaryotes was established (*170*). Eubacteria, eukaryotes, and archaea all have similar pathways to biosynthesize Sec (*117*). However, while bacteria utilize Sec synthase (designated SelA) to synthesize Sec from Ser-tRNA(Sec), archaea and eukaryotes use SelS to synthesize Sec from phosphoserine (Sep) acylated to tRNA(Sec) (*151*). In bacteria, tRNA(Sec) is ligated first with serine by seryl-tRNA synthetase, followed by the conversion of Ser-to-Sec by Sec synthase (SelA). The incorporation of Sec into selenoproteins in eukaryotes was described comprehensively in a review by Allmang (*179*). Sec is synthesized on tRNA(Sec) in multiple steps, most often requiring four enzymes for its biosynthesis from serine. A comprehensive biosynthesis of tRNA(Sec) is described in the literature (*151, 152, 179, 197, 198*).

After the discovery of selenonucleoside in seleno-tRNA, the biosynthesis of the selenonucleoside (mnm^{5}Se^{2}U) in bacteria and archaea was partially characterized. Earlier experiments reported the presence of Se-modified nucleoside in tRNAs in mouse leukemia cells (*191*) and the presence of mnm^{5}Se^{2}U in bovine tRNA (*199*). In general, to introduce the Se atom at the modified uridines (Figure 3a and 3b) in the biosynthesis of seleno-tRNAs, the native uridine located at the wobble position 34 of the tRNA anticodon is first sulfurized by MnmA and IscS forming 2-thiouridine (*33, 200*). Subsequently, MnmE further modifies the 2-thiouridine to 5-methylaminomethyl-2-thiouridine (mnm^{5}S^{2}U) (*201*). The sulfur atom at C2in 2-thiouridine is then substituted by a Se atom using monoselenophosphate, the Se donor (*122, 125, 200, 202–204*). Monoselenophosphate is catalyzed by selenophosphatesynthetase (the *E. coli*selD gene product) from the selenide and ATP (*202, 203, 205*). Seleno-tRNAs are found in species such as *Escherichia coli, Clostridium sticklandii, Methanococcusvannielii*, and *Salmonella typhimurium* (*159, 160, 200, 202, 203, 206, 207*).

Study of the biosynthesis of Se-containing tRNAs in *Methanococcusvannielii* suggested that L-selenocysteine was also a Se donor. This study also revealed that the selenium incorporation into these nucleosides required ATP for the biosynthesis (*207*). Furthermore, Lacourciere and Stadtman (*205*) showed that the selenide derived from L-selenocysteine by *Escherichia coli* NIFS-like proteins can be used as the substrate for selenophosphate synthetase *in vitro*. Verse and coworkers discovered that a tRNA 2-selenouridine synthase utilized in the catalysis reaction was first found in *Salmonella typhimurium*cell extract (*202*). More recently, Wolfe and coworkers have also identified a tRNA 2-selenouridine synthase that was partially characterized as the product of the ybbB gene in *Escherichia coli* (*193, 206*). Despite this research, the mechanism of the Se modification in tRNAs is not yet well understood. Studying the presence and function of Se in proteins could provide much useful information for understanding Se-modifications and function in RNA. The presence of Se in natural RNA provides many useful and unique properties to RNAs to improve the function and survivability of organisms (*116, 125, 126, 159, 160*).

3.2. Selenium Modifications of Nucleic Acids

3.2.1. Sugar-Modified Analogues

The single-atom Se-functionalization of the 2′, 4′ and 5′ sugar and nucleobase positions has been accomplished by Huang's research group and other groups (*128, 129, 132–149, 208*). 16 years ago, Huang and co-workers pioneered and developed the first synthesis of the Se-labeled phosphoramidities (A, C, G, T and U) (*128*) for solid-phase nucleic acid synthesis. The 5′-position was first activated by good leaving groups (mesyl, Ms- and bromide, Br-), followed by nucleophilic substitution with sodium selenide. The 5′-seleno-nucleosides were found to be stable and compatible with solid-phase phosphoramidite chemistry, and were successfully used to synthesize oligonucleotides containing selenium at 5′-termini. A similar approach was used to introduce selenium at the 2′-α-position of the nucleoside (*129*). The 2′-methylseleno (Me-Se) nucleosides are stable and are used to synthesize the corresponding DNA and RNA analogues by standard solid-phase synthesis. The 2′-α-Me-Se-substituted furanoses retain the native C3′-*endo* conformation of A-form DNA and RNA molecules. The stability and compatibility of this modification with solid-phase synthesis facilitated large scale synthesis of Se-DNAs and Se-RNAs (*148, 209*). The successful synthesis of 2′-α-Me-Se-nucleosides (A, T, C, G and U) has allowed substantial flexibility in the modification within DNAs and RNAs, making 2′-Se-DNAs and 2′-Se-RNAs valuable derivatives for crystallographic phase determination (*131–137, 139, 142, 149, 210–212*). For example, the 2′-Se-RNA has been successfully used to determine the crystal structure of the Diels-Alder ribozyme (*146*). Atom-specific replacement of the 4′-oxygen in the ribose ring with selenium creates 4′-selenonucleosides. The synthesis and structural aspects of 4′-selenonucleosides were first reported by Matsuda's group (*213*) after they were successful in synthesizing 4′-selenocytidine and 4′-selenouridine via the stereoselective seleno-Pummer reaction. Later, there were reports from different groups using the modified procedures to synthesize and crystallize 4′-seleno purine and pyrimidine nucleosides (*145, 214*). With improved stability and binding affinity, the 4′-seleno oligonucleotides are promising candidates for crystallographic phasing and other biological applications.

3.2.2. Nucleobase-Modified Analogues

The substitution of oxygen with selenium in the nucleobases permits a search for novel aspects of nucleic acid base-pairing and stacking interactions at the atomic level (*131–136, 138, 140, 149, 208, 215, 216*). The first synthesis of the Se-modified nucleobase was recently reported by Huang and coworkers (*216*). A selenium functionality was introduced at the 4-position of thymidine (Se^{4}T). With the successful synthesis of Se^{4}T phosphoramidite, this modification was incorporated into oligonucleotides with high yield. Since then, Huang and coworkers have developed diverse building units for the incorporation of 2-Se-thymidine (Se^{2}T) (*140*), 6-Se-guanosine (Se^{6}G) (*132, 215*), and 2-Se-uridine (Se^{2}U) (*135*) into DNA and RNA by solid-phase synthesis. In the incorporation of the Se-functionality to the 2-position of thymidine, the key selenization reaction is accomplished by displacing the 2-methylsulfide activating group with sodium hydrogen selenide (*140*). This Se-incorporation reaction, which was finally developed after three years of hard work, is also called the Huang reaction. Crystallographic studies of the Se^{4}T-containing DNA (5′-G-dU$_{Se}$-G-^{Se}T-ACAC-3′) revealed thermally stable DNA with no significant structural perturbations. The Se^{6}G-containing DNA was co-crystallized as a ternary complex with RNA and RNase H. The structural analysis demonstrated its similarity to the native duplex, but with a slightly shifted base pair (approximately 0.3 Å) to accommodate the bulky selenium atom. The 2-Se-modification has a pronounced ability to discriminate wobble base-pairing. Structural and biophysical studies on Se^2T- and Se^2U-containing oligonucleotides highlight the increased fidelity of Se^2T/A and Se^2U/A base pairs, respectively (Figures 4 and 5). Interestingly, it was observed that atom-specific replacement of oxygen with selenium in the nucleobases imparts a bright yellow color to the otherwise colorless nucleoside (hence, the corresponding nucleotide and oligonucleotide). For instance, relative to the native dG (UV λ_{max}= 254 nm, ε = 1.22 x $10^4M^{-1}cm^{-1}$), the absorption maximum of Se^6dG (UV λ_{max}= 360 nm, ε = 2.3 x $10^4M^{-1}cm^{-1}$) is red-shifted by 100 nm, resulting in yellow DNA. This large shift in the absorption maximum is likely due to the greater delocalization of selenium electrons, thereby lowering the excitation energy of the nucleotide. Since Huang means yellow in Chinese, we refer these yellow nucleotides and nucleic acids as Huang reagents (or Huang nucleotides). In another interesting study, selenium was inserted at the 5-position of thymidine (5-Se-T) (*136, 177*) and cytosine (*217*). The structural and biophysical studies on 5-Se-T-and 5-Se-C-containing DNAs also revealed no significant structural perturbations.

3.2.3. Phosphate Backbone-Modified Analogs

Phosphoroselenoates (PSe) are oligonucleotides where one of the oxygen atoms of the linking phosphate is replaced with selenium. Initially, phosphoroselenoates were envisioned as model phosphorochalcogenoate derivatives with promising antisense and anti-HIV activity (*218*). But it was found that they are too unstable for *in vivo* applications. Egli and co-workers later demonstrated that on a crystallographic time scale, PSe-modified DNA was stable to oxidation and was a suitable candidate for structure determination by multi-wavelength anomalous dispersion (MAD) analysis (*141, 219*). The chemical preparation of phosphoroselenoates (PSe) oligonucleotides utilized potassium selenocyanate (KSeCN) as the selenizing reagent. The solid-phase synthetic approach results in the formation of P-diastereomers (R_P and S_P) (*141, 142*). Compared with the unmodified oligonucleotides and PS-oligonucleotides, the PSe-oligonucleotide diastereomeric mixture has reduced hybridization capabilities with complementary DNA and RNA (*218*). Recently, Stec and co-workers elaborated a novel method, based on the oxathiophospholane approach, for the stereo-controlled chemical synthesis of PSe-oligonucleotides (*220*). For short oligonucleotides containing a single phosphoroselenoate, diastereomerically-pure phosphoroselenoates can be obtained by chromatographic techniques (*141, 221*). However, for oligonucleotides containing more than one phosphoroselenoate group, separation of diastereomers is currently challenging and impractical.

The enzymatic synthesis of stereo-defined phosphoroselenoates was first developed by Huang's research group (*181, 222*). This enzymatic synthesis is based on the recognition of the S_P and/or R_P diastereomers of α-Se-dNTPs or α-Se-NTPs by the polymerases. For this study, diastereomeric monomers of α-Se-TTP were chemically synthesized and characterized. Unexpectedly, the DNA polymerase was able to recognize and incorporate both diastereomers of α-Se-TTP, thereby synthesizing diastereomerically-pure PSe-DNAs (*181*). On the other hand, only the S_P diastereoisomer of α-Se-ATP is efficiently recognized by T7 RNA polymerase, enabling synthesis of diastereomerically-pure PSe-RNAs (*222*). Later, the α-Se-NTPs (α-Se-ATP, α-Se-GTP, α-Se-CTP andα-Se-UTP) were used to transcribe hammerhead ribozyme to generate diastereomerically-pure PSe-RNAs for structural and mechanistic studies (*223*). Recently, Huang and coworkers have developed a novel, efficient, one-pot synthesis of α-Se-NTPs without protecting the nucleosides and used this method for the large-scale synthesis of α-Se-NTPs for PSe-RNA transcription (*130*).

3.3. Applications

Selenium is a trace essential element with valuable antioxidant, chemo-preventive, anti-inflammatory and antiviral properties (*224*). There is a rising interest in understanding the physiological functions of selenium for rational design of drugs and therapeutics. SeNAs could also serve as crucial tools in the structural, functional and mechanistic studies of nucleic acids, nucleic acid-protein interactions, DNA-drug complexes, DNAzymes and ribozymes, viral RNAs, regulation of transcription and translation, and DNA replication.

3.3.1. Crystallographic Studies

Selenium offers favorable scattering properties with a K absorption edge of 0.9795 Å, which is readily accessible with the synchrotron radiation (*225*). It is considered as the ideal anomalous scattering center for multiple wavelength anomalous dispersion (MAD) and single wavelength anomalous dispersion (SAD) analysis and has facilitated crystallographic analysis of bio-macromolecules. The use of selenium in protein structure determination was successfully demonstrated by incorporating selenomethionine (in place of methionine) and using the MAD technique (*225*). Sulfur replacement with selenium has been shown to cause minimal structural perturbation (*226*, *227*). Since that time, selenomethionine derivatization has become a standard strategy for protein crystal structure determination. Currently, selenomethionine derivatization and MAD phasing are now routine in protein crystallography and are used for more than two-thirds of new protein structures (*228*). Selenium has also been used to derivatize nucleic acids for crystal structure determination (*229*). In comparison with proteins, the structural database for nucleic acids is very limited. Since oxygen and selenium belong to the same family in the periodic table, in theory it is possible to replace any oxygen atom in a nucleotide with selenium. These Se-modifications are stable and do not cause significant structural and functional perturbations (*131, 135, 140, 149, 181, 215, 216*). Thus, the strategic and systematic replacement of specific oxygen atoms with selenium provides a versatile strategy for generating modified nucleic acids for X-ray crystallography (*131–138, 149, 230, 231*). SeNA, fine models for native DNA and RNA, could serve as a useful tool for visualizing the structural details (via crystallography) for comprehensive understanding of nucleic acids and protein-nucleic acid complexes (*143, 149, 232*).

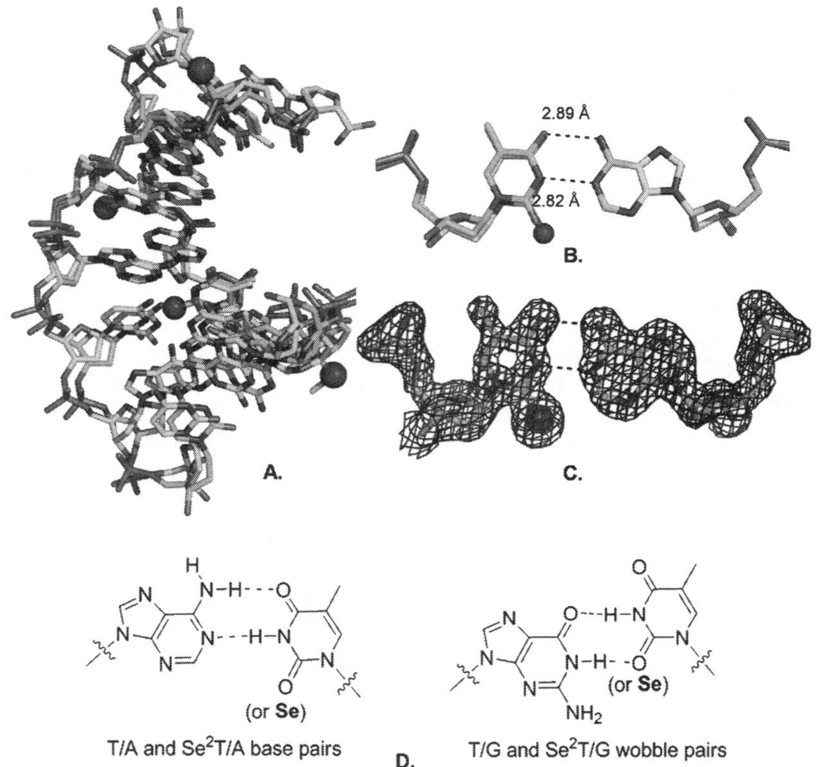

Figure 4. Global and local structures of the Se²T-DNA [(5'-GdU₂'-Se-G-SeT-ACAC-3')₂], with a resolution of 1.58 Å. (A) The superimposed comparison of the Se-DNA duplex (3HGD, in green) with its native counterpart (1D78, in cyan). The red balls represent the selenium atoms. (B) The superimposed comparison of the local SeT4/A5 (in green) and native T4/A5 (in cyan) base pairs. (C) The experimental electron density map of the SeT4/A5 base pair with σ = 1.0. (D) Native and Se-modified T/A base pair and T/G wobble pairs. (see color insert)

The classical DNA/RNA labeling approach for structural phasing of nucleic acids includes derivatization with a heavy atom or halogen (*233*). Due to their complex three-dimensional structures and charges, heavy atom derivatization of nucleic acids is difficult. Halogen (Br or I) derivatized nucleic acids are radiation (UV and X-ray) sensitive and limited to the incorporation of modifications at primarily 5-position of pyrimidines (Br⁵U, Br⁵C, and I⁵U), often causing structural perturbations and hampering crystallization (*146, 231*). In contrast, SeNAs are attractive alternatives with enhanced stability and functional diversity. Huang's research group has pioneered the chemical and enzymatic synthesis of Se-containing nucleotides (phosphoramidites and triphosphates) and nucleic acids for structure and function studies. The collaborative efforts of Huang's and Egli's research groups resulted in the first successful crystal structure determination of nucleic acids via selenium derivatization and MAD phasing (*129*). For the mentioned study, the 2′-seleno-uridine was synthesized and used to label oligonucleotides for phase and structure determination in X-ray crystallography. Over the past a few years, many crystal structures of DNA, RNA and their complexes with proteins and small molecules have been successfully determined by covalent selenium derivatization (*135, 136, 138, 149, 217*). The nucleic acid X-ray crystallography and function studies via direct selenium derivatization has been extensively reviewed and reported (*131, 133–138, 146, 149, 231, 232, 234*).

These nucleic acid crystal structures, obtained by the insertion of Se-nucleobase-derivatized nucleosides, provide valuable information on the structure and function of nucleic acids. Huang and co-workers have been successful in incorporating selenium at various positions in the nucleobases (Se^{4}T, Se^{2}T, Se^{4}U, Se^{2}U,Se^{6}dG and Se^{6}rG). The DNA/RNA crystals thus obtained have paved the way for further exploring base-pairing (dictated by the shape of the base and its hydrogen bonding capacity), duplex stability and flexibility, polymerase recognition, and replication and translation efficiency and efficacy. Structure and stability studies performed on 4-Se-T containing DNA reveal that this modification is thermally stable (*216*). The Se atom is well accommodated in the duplex structure *via* the formation of a novel hydrogen bond between Se and the adenosine 6-amino group (Se⋯HN), resulting in minimal structural perturbation. In another modification, selenium was introduced at the 6-position in guanosine (Se^{6}G) to explore the consequences of such a modification on duplex recognition and stability (*215*). It was observed that Se^{6}G base pairs with C via a Se mediated hydrogen bond. Although the Se^{6}G/C base pair is similar to the native G/C pair, the large size of Se causes the base pair to be shifted approximately 0.3 Å. The co-crystals of Se^{6}G-DNA with RNA and RNase H were the first protein-nucleic acid complex structure determined using MAD phasing of Se-derivatized nucleic acids. The 2-position in thymidine and uridine is decisive for the discrimination between Watson-Crick base pairs (T/A and U/A) and wobble base-pairs (T/G and U/G). The steric and electronic effects of this 2-exo position can be tailored by replacement of oxygen with selenium (Se^{2}T and Se^{2}U), offering a unique strategy for improving the base-pairing preference (Figures 4 and 5) (*135, 140*). Structural and biophysical studies of Se^{2}T-DNAs and Se^{2}U-RNAs indicate that they have very similar thermal stabilities compared to the native nucleic acids. With negligible structural impact, the 2-Se-modification is capable of strongly discriminating wobble basepairing, thereby increasing the fidelity of Se^{2}T/A and Se^{2}U/A base pairs. The insertion of selenium at the exo-5-position of thymidine and cytosine and incorporation of these seleno-bases into DNAs do not cause considerable structural perturbations (*177, 217*). Furthermore, the comparison of uridine with 5-O-Me, 5-S-Me, and 5-Se-Me substitutions indicates that only the 5-O-Me modification facilitates an interaction between the nucleobase and the 5′-phosphate group via a novel methyl/phosphate hydrogen bond (*136*). Specifically, this hydrogen bond is formed between the 5-methyl group and the 5′-phosphosphate (CH$_3$⋯O–PO$_3$), and this unique interaction might be further exploited to design inhibitors for backbone digestion. Although these modifications do not significantly alter the stability of the duplex, the atom-specific replacement of 5-O with S or Se abolishes the hydrogen bond between the CH$_3$ and 5′-phosphate R_P-oxygen.

These findings demonstrate the importance of such modifications in nucleic acids at the atomic level. Additional modifications on the nucleobases will result in the discovery of unique structural and functional features of the nucleic acids, especially non-coding RNA (ncRNA).

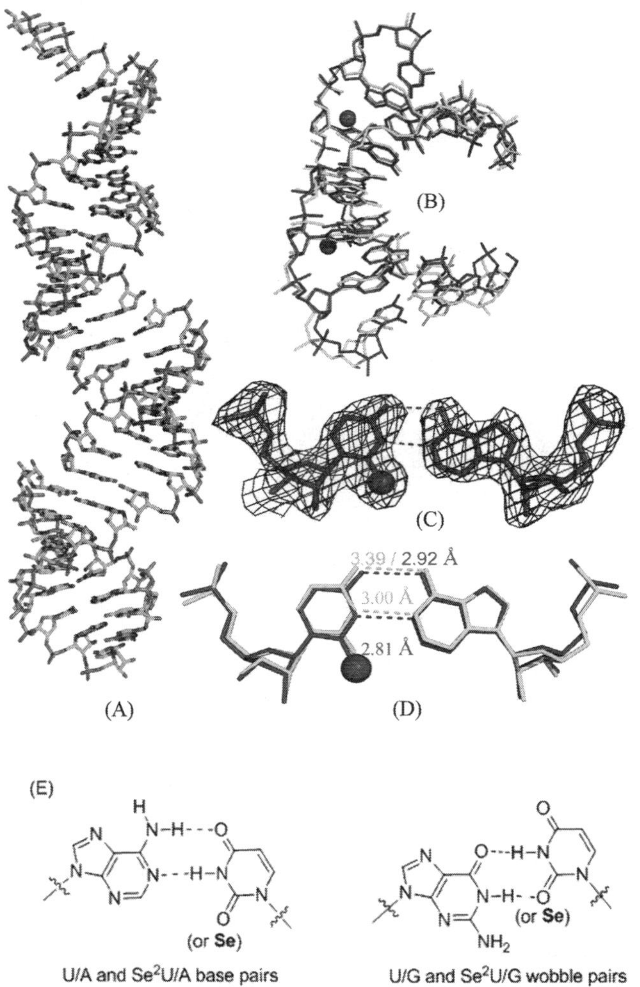

Figure 5. Global and local structures of the Se²U-containing RNA r[(5'-GUAUA(^{Se}U)AC-3']₂ with a resolution of 2.3 Å. (A) The overall structure of the duplex. (B)The superimposed comparison of one ^{Se}U-RNA duplex (red; PDB ID: 3S49) with its native counterpart [5'-r(GUAUAUA)-dC-3']₂ (cyan; PDB ID: 246D) with a RMSD value 0.55. The two red balls represent the selenium atoms. (C) The experimental electron density of ^{Se}U6/A11 basepair with σ=1.0. (D) The superimposed comparison of the local base pair ^{Se}U6/A11 (red) and the native U6/A11 (cyan). The numbers indicate distances between the corresponding atoms. (E) Native and Se-modified U/A base pair and U/G wobble pairs. (see color insert)

In a comparative study between Se (at 2′-position of the sugar) and Br (at the 5-position of the nucleobase dU) derivatization, it was observed that the introduction of selenium affects crystal packing, supporting high-quality crystal growth also broadening the buffer conditions for crystal formation (*210, 231*). The 2′-Se-DNAs crystallize much faster than the native and Br-derivatized DNAs. Since the synthesis and purification of SeNAs are much easier than those of proteins, successful Se-derivatization of nucleic acids and structure analysis by MAD and SAD phasing will promote structure-function studies of nucleic acids and their complexes with proteins and pave the way for additional development and applications of SeNA. Thus, it is meaningful to further explore the possible benefits of selenium as a facilitator of crystallization.

3.3.2. Biochemical Applications

Selenium-derivatized nucleic acids entail unique structural and functional features that make them strong contenders for biochemical and biophysical research. Owing to the chirality of the seleno-phosphate, synthesis of nucleoside triphosphate analogues with selenium at the non-bridging α-position of phosphate results in the formation of R_P and S_P diastereomers. As a typical example, thymidine 5′-(α-P-seleno)triphosphate (α-Se-TTP) was synthesized and both the diastereomers formed were subjected to enzymatic synthesis of PSe-DNAs (*181*). It was observed that the DNA polymerase is capable of recognizing both the diastereomers to the same capacity as the natural TTP. Similarly, PSe-RNAs were enzymatically synthesized by the *in vitro* transcription of adenosine 5′-(α-P-seleno)triphosphate (α-Se-ATP) with T7 RNA polymerase (*222*). However, only the S_P diastereomer is efficiently recognized by the RNA polymerase; the R_P diastereomer is neither a substrate nor an inhibitor of the RNA polymerase. This disparity in recognition of the α-Se-NTPs and α-Se-dNTPs can be utilized to study the catalysis and substrate specificity of the DNA and RNA polymerases. The S_P and R_P diastereomers of the Se-derivatized nucleotides are instrumental for the fundamental understanding of the structure and biophysical chemistry of the polymerases, nucleases, kinases and phosphatases. The site-specifically Se-derivatized hammerhead ribozymes are useful models for investigating ribozyme structure, activity and catalytic mechanisms and identifying highly conserved sequences in functional RNAs. These Se-hammerhead ribozymes, transcribed using all the α-Se-NTPs (A, C, G and U), are important biochemical tools for probing the structure and catalytic behavior of ribozymes (*223*). They demonstrate the catalytic activity at different levels: hammerhead ribozymes transcribed with α-Se-UTP and α-Se-CTP have the same level of activity as the native ribozyme, while hammerhead ribozymes transcribed with α-Se-ATP and α-Se-GTP have 1000-fold and 3-fold activity reduction, respectively. These observations highlight the significance of Se-modification and its participation in ribozyme-catalysis and intramolecular interactions. This Se-NTP tool can be used to facilitate the rapid screening of ribozymes and other non-coding RNAs. Moreover, Se-modification renders the

PSe-DNAs and PSe-RNAs resistant to nucleases, thereby making them potential therapeutic agents. Furthermore, we have demonstrated that the synthesized [Se]UTP is stable and recognizable by T7 RNA polymerase (39). Under the optimized conditions, the transcription yield of [Se]U-RNA can reach up to 85% of the corresponding native RNA. Furthermore, the transcribed [Se]U-hammerhead ribozyme has the similar activity as the corresponding native (Figure 6), which suggests usefulness of [Se]U-RNAs in function and structure studies of non-coding RNAs.

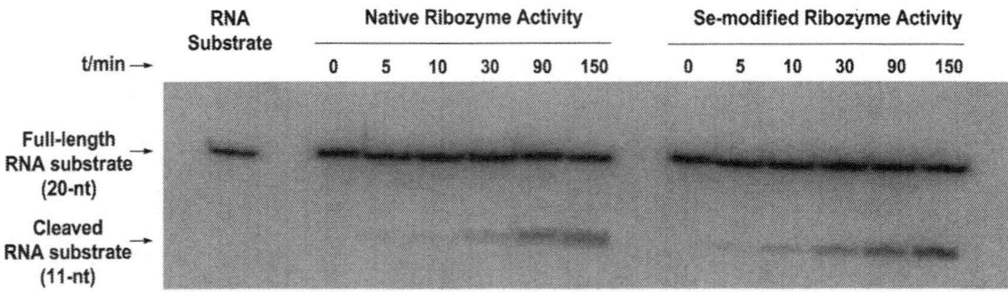

Figure 6. The catalytic activity of the Se-modified ribozyme. The time-course experiment of the 5'-^{32}P-RNA substrate digested with the non-self-cleaving native and Se-modified hammerhead ribozymes under the same conditions.

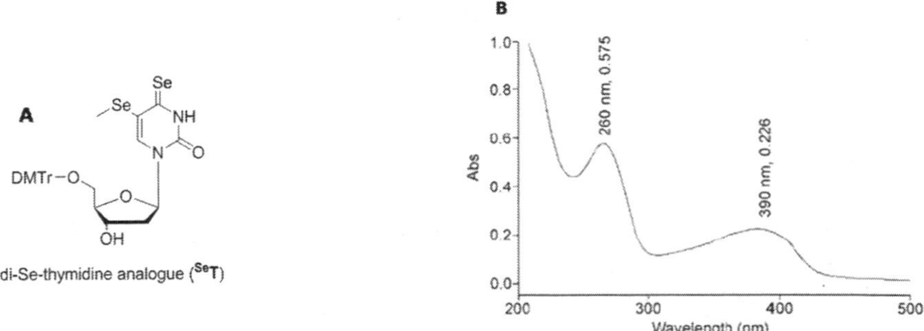

Figure 7. (A) Chemical structure of the di-Se-thymidine analogue 5-methylseleno-4-(2-cyanoethyl)seleno-5'-O-(4,4-dimethoxytrityl)-thymidine (^{Se}T). (B) UV spectrum of the T^{Se}TT trimer.

The 2'-methylseleno-modification is amongst the most widely explored selenium modifications of RNAs and successful incorporation of this modification led to structure determination of several RNAs (210, 235). To further utilize the potential of 2'-Se-RNAs, 2'-Se-NTPs were synthesized to accomplish the enzymatic synthesis of longer modified RNAs (*236, 237*).   Interestingly, 2′-MeSe-NTPs are good substrates for T7 RNA polymerase and are efficiently incorporated into RNA. These results could be valuable in the design of engineered RNA polymerases. Moreover, the Se functionality can be inserted into 5-methylcytosine (m5C) without significant perturbation (*217*).   The structural and biophysical studies on SeC and SeC-DNAs indicate that this modification does not affect DNA duplex structure and stability.   Such a modification encourages the further exploration of the epigenetic mechanism of m5C.

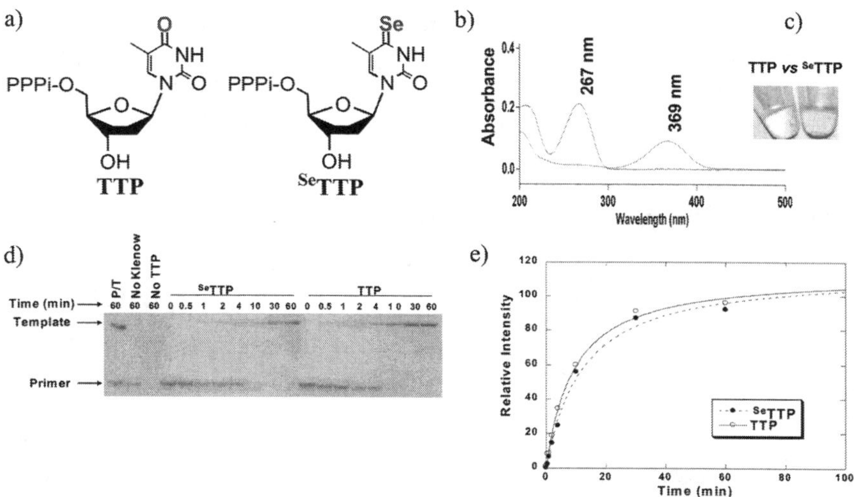

*Figure 8.   The color of SeTTP and its Se-DNA. **a)** TTP and SeTTP chemical structures; **b)** UV profiles of TTP and SeTTP; **c)** colorless TTP and yellow SeTTP; **d)** Gel analysis of DNA polymerization with TTP and SeTTP; **e)** Plots of the incorporation of TTP and SeTTP into DNA. (see color insert)*

3.3.3. Labeling and Detection

Nucleic acid-based detection and visualization are essential for many biochemical, biophysical and biomedical applications (*238–243*).   The oxygen atoms in nucleic acids are present in diverse chemical environments. Therefore, strategic substitution of these oxygen atoms with selenium atoms provides novel opportunities to tailor the properties of nucleic acids. According to our previous work, selenium derivatization of the nucleobase imparts a strong yellow color to the otherwise colorless nucleotides and nucleic acids, resulting in strongly colored DNAs and RNAs. Incorporation of a single Se atom in the nucleobases can cause over a 100 nm red-shift in the absorption maximum of the nucleobases (*215, 216, 244*). This distinctive property is very useful in nucleic acid-based detection and visualization. The spectroscopic and functional repertoire of the Se-derivatized nucleobases was further improved by the synthesis of a thymidine analog containing two selenium atoms (Figure 7) (*134*). Due to the electron-donating effect of two selenium atoms on the same nucleobase, the absorption maximum of the di-Se-thymidine is further red-shifted to 390 nm with $\varepsilon = 6.2 \times 10^3$ M^{-1}cm^{-1}. This double-selenium modification greatly enables phasing to investigate the structural dynamics of nucleic acids and protein-nucleic acid complexes. These di-Se-T-DNAs are stable and structurally similar to the native DNAs. With their unique spectroscopic properties, the di-Se-T-DNAs have great potential in detection, visualization, and profiling of nucleic acids and their complexes. Synthesis of a yellow 4-Se-thymidine-5′-triphosphate ([Se]TTP) from the 4-Se-thymidine nucleoside was also successful (*244*). This modified [Se]TTP acts as a good substrate for DNA polymerase (Figure 8) and is recognized with virtually the same efficiency as the native TTP. The enzymatic incorporation of [Se]TTP also results in yellow DNA. These colored Se-DNAs could serve as unique visual probes for enzymatic assays.

4. Tellurium-Functionalized Nucleic Acids

Tellurium is a non-essential trace element that is toxic to humans and does not naturally occur in any biomolecule. Tellurium has a lower electronegativity (Te: 2.0; Se: 2.55; S: 2.58) and greater atomic radius (Te: 1.37 Å; Se: 1.17 Å;S: 1.04 Å) than its fellow chalcogens, which gives it distinctive chemical properties in comparison with sulfur and selenium (*245*) as well as some metallic character. Tellurium was discovered in 1782 by a Transylvanian chemist, Franz-Joseph Mueller von Reichenstein (1742–1825), who was studying gold-containing ores. A typical human body (a 70-kg person) contains approximately 0.7 mg of Te. The reported level of Te is 5 ng/mL in blood and 50 ng/mL in urine (*246*). Tellurium exists in several oxidation states, including +4 and +6, which have comparable stability and tellurium compounds have a wide range of biological functions (*247*), including exceptional antioxidant properties. Since it is not naturally occurring, the biological functions of tellurium have not been as extensively explored as selenium. The toxic nature of tellurium, which was first reported nearly a century ago and was later studied in a few toxicological research projects (*248–251*), perhaps prevented tellurium from being closely examined. However recently, due to its unique physical, chemical, and spectroscopic properties, tellurium and its compounds have attracted appreciable interest from researchers in drug development and diagnostics, and in protein and nucleic acid structure determination.

4.1. Tellurium in Proteins

Since the chemical properties of tellurium are similar to that of sulfur, it is not surprising that tellurium could be incorporated into amino acids, such as cysteine and methionine, yielding tellurocysteine and telluromethionine (TelMet), respectively. After successful isosteric replacement of naturally-occurring Met by SeMet and its effective usefulness in X-ray structure analysis, several attempts have been initiated to express TelMet-protein in bacterial, yeast, and mammalian systems (*245, 252, 253*). TelMet is chemically very reactive, which makes its synthesis very challenging. This higher reactivity makes it toxic to the expression systems and leads to misfolding of the expressed TelMet-proteins. By optimizing the expression systems and selecting methionine-auxotrophic *Escherichia coli* strains, incorporation of telluromethionine with higher rates and yields has been reported for a series of proteins, such as dihydrofolate reductase (*254*), human recombinant annexin V, human mitochondrial transamidase, arabidopsis glutathione-S-transferase, and the N-terminal domain of *Salmonella* tailspike adhesion protein (*255, 256*). During TelMet-protein expression, the TelMet analog is activated by the Met-tRNA synthatase and charged properly onto t-RNA[Met], which is also accepted by all other host translational machinery. For many proteins, including the recombinant proteins successively used for X-ray diffraction data collection and analysis, TelMet can be incorporated into the polypeptide chain at every position of Met, including N-terminal positions.

Tellurium offers similar advantages to selenium in crystallographic structure determination by MAD phasing, such as a convenient labeling procedure, high isomorphism with the parent molecule, stable amino acid analogues, strong phasing power, and relative abundance and mobility of target sites (*257*). In addition, as a heavy-atom, tellurium provides clear signals in both isomorphous and anomalous difference Patterson maps in the crystallographic analysis of the proteins at commonly used CuKα wavelengths. However, tellurium is not as suitable for MAD experiments as selenium due its larger absorption edge, and as a phasing strategy, Te performs better as a heavy atom for isomorphous replacement. Although the C-Te single bond is 0.20 and 0.33 Å longer than C-Se and C-S single bonds, respectively, the three-dimensional structure of Te-proteins were found to be similar to the native forms, probably because molecular plasticity (the ability of a protein to retain its native form) tolerates these atomic modifications. A thorough analysis of the X-ray crystallography data for recombinant proteins reveals that incorporated TelMet may have a significant effect on the structure of TelMet-protein (Figure 9). Due to the higher reactivity of tellurium atom, the side chains of TelMet residues on the surface or near surface may be exposed and reactive, causing misfolding, exclusion body formation, rapid chemical modification, oxidation, and degradation (*254, 256*). During diffraction data analysis, it is very hard to assign electron density peaks for the modified TelMet residues. Fortunately, TelMet residues located in the hydrophobic core of protein often do not have these problems.

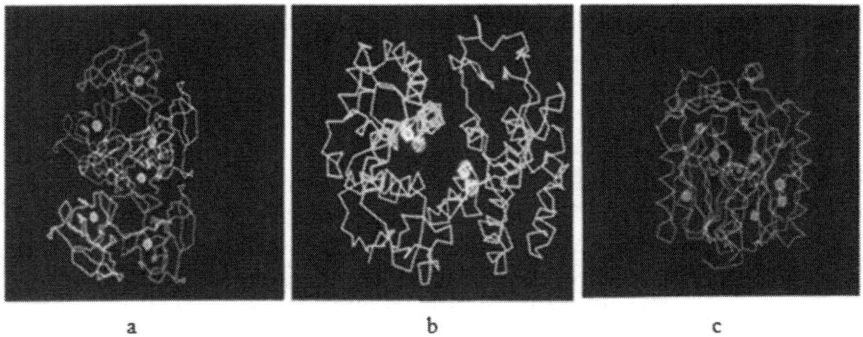

Figure 9. C^{α} -representation of TelMet proteins. Electron difference maps between sulphur and tellurium at the side-chain sites of replacement (δ) shown by difference Fourier maps (F_{TeMet} - F_{Nati}) contoured at 6σ. (**a**) N-terminal domain of tailspike protein; (**b**) dimeric gluthatione-S-transferase from Arabidopsis thaliana; (**c**) human recombinant transamidase (at this contouring level it is only possible to find 9 of 11 TelMet proteins, because the remaining two are solvent-exposed) (255).

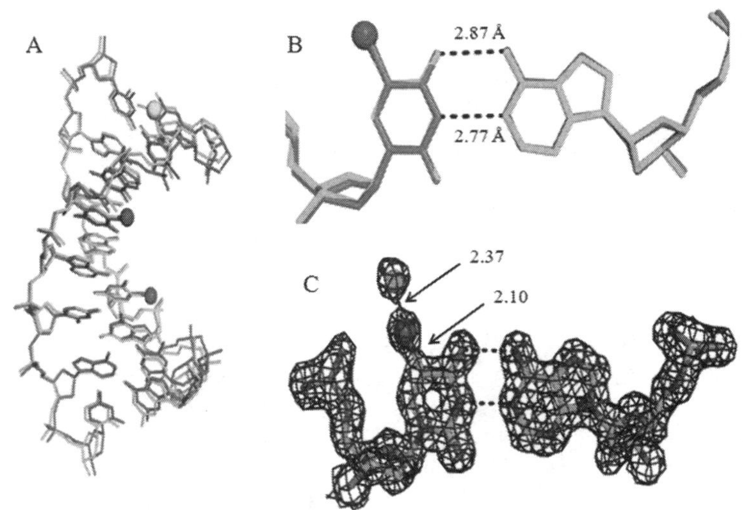

Figure 10. The global and local structures of the Te-DNA duplex [5'-G(2'-SeMe-dU)G($^{5\text{-}Ph\text{-}Te}$T)ACAC-3']₂. The red and yellow balls represent Te and Se atoms, respectively. (**A**) Te-dsDNA structure (in cyan; 1.50 Å resolution; PDB ID: 3FA1) superimposed over the native structure (in magenta; PDB ID: 1DNS). (**B**) Local structure of the $^{5\text{-}Ph\text{-}Te}$T /A base pair (in cyan) superimposed over the native T/A base pair (in magenta). (**C**) The experimental electron density $(2|F_o|\_|F_c|$ map) of the $^{5\text{-}Ph\text{-}Te}$T /A base pair (σ =1.0). The green ball represents approximately one-sixth of a phenyl group (one carbon). (see color insert)

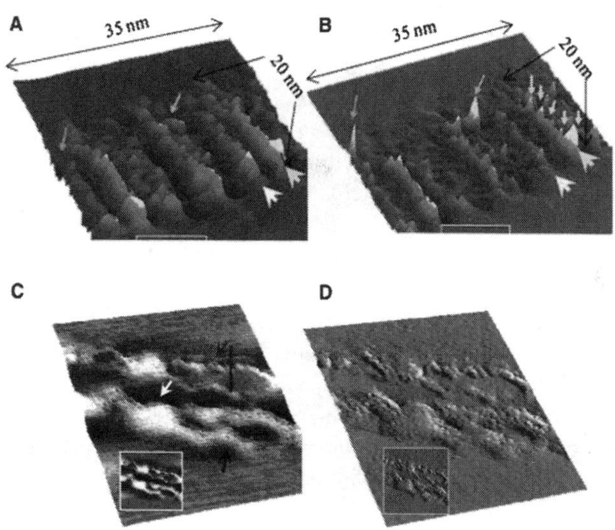

*Figure 11. STM images of the Te-modified DNA duplex [5'-ATGG(TeT)-GCTC-3'
and 5'-(GAGCACCAT)$_6$-3'] and its counterpart native duplex on HOPG. The
arrows indicate the edges or current peaks of the measured molecules. (**A**)
Topographic image of the Te-duplex; (**B**) Current image of Te-duplex; (**C**)
Topographic image of native duplex; (**D**) Current image of native duplex. In
comparison to the Te-modified DNA duplex, the native sample does not show
obvious conductivity. The insets are 2D images corresponding to the 3D images.
For these experiments, the sample bias was 0.50 V and the current set point
was 100 pA.*

4.2. Tellurium in Nucleic Acids

Successful atom-specific introduction of selenium into various positions
of the sugar, nucleobases, and backbones of DNA and RNA has encouraged
the atom-specific incorporation of tellurium into nucleic acids. This research
area was pioneered by Huang and co-workers (*258*). Tellurium functionality
has been introduced into different positions of the nucleobases and sugar
moieties, such as the 5-position of thymidine (5-phenyl-tellurium thymidine,
[5-Ph-Te]T), the 2'-position of thymidine (2'-[Te]T), and the 2'-position of deoxyuridine
(2'-d[Te]U) (*149*). These modified nucleosides were converted to the corresponding
Te-phosphoramides first, followed by synthesizing Te-nucleic acids via
solid-phase methods. Thermal stability and crystal structure studies reveal
that DNA duplexes containing 5-phenyl-tellurium ([5-Ph-Te]T-DNA) have similar
stabilities compared to the native duplexes (*130, 215*). The superimposed X-ray
crystal structures indicate that the [5-Ph-Te]T-DNA and the native DNA structures
are virtually identical, and that the [5-Ph-Te]T/A base pair is formed similarly to

the native T/A pair (Figure10A-C) (*149*). Furthermore, tellurium possesses unique optical and electrical properties thatare being explored by quantum dot nanotechnology for various biological applications (*259*). DNA duplexes are relatively electron-deficient, therefore, the presence of an electron-rich tellurium atom in a DNA base is likely to donate electrons and facilitate electron delocalization. This phenomenon helps to direct DNA imaging without structural perturbation. A STM imaging study with Te-DNA suggests that the Te-DNA duplexes have higher visibility and conductivity under STM, and exhibit stronger topographic and current peaks than the corresponding native duplexes (Figure 11). The topographic and current images of the native and Te-DNA duplexes under STM also reveal that the native and Te-modified duplexes assemble with different orientations on the graphite surface (*149*). Thus, by the virtue of their unique opto-electronic properties, Te-modified nucleic acids will open novel research avenues for nucleic acids, including mechanistic and functional studies, STM imaging and nano-electronic materials.

5. Perspective

The strategic atom-specific substitution of oxygen with heavier chalcogens (S, Se, and Te) has greatly enhanced the structural and functional repertoire of nucleic acids and allowed the emergence of novel nucleic acid paradigms. Chemical modification of nucleic acids is of paramount importance in nucleic acid research. These structural and functional variations have caused a renaissance of nucleic acid chemistry and biology, and they has paved the way for fresh avenues in nucleic acid research, including therapeutic expansion, nucleic acid biology, structure-function studies, catalytic and mechanistic analysis, and material science and nanotechnology. Among all the chalcogen derivatives, the thiol-modified nucleic acids have been largely found in nature and have demonstrated great potential for nucleic acid therapeutics. Owing to their nuclease stability, bioavailability, efficient delivery, low toxicity, high hybridization ability, and high specificity, S-modified oligonucleotides have been widely utilized in drug design and therapeutic development. Currently, many S-modified therapeutic oligonucleotides are in clinical trials, and some are already available as drugs.

Selenium modifications of nucleic acids offer additional valuable advantages by facilitating crystallization and phasing in X-ray crystallography for 3D structural determination of nucleic acids and their complexes with proteins, small molecules, and/or metal ions. Similar to the thiol-derivatives, these Se-modified nucleic acids exhibit increased resistance to nucleases with minimal structural perturbation, supporting a better understanding and probing of the highly diversified secondary and tertiary structures of nucleic acids, such as bulges, junctions, triplexes, i-motifs and quadruplexes. As discussed in this chapter, by facilitating structure determination, selenium-derivatized nucleic acids will create paradigm shifts in the functional studies, structure-based drug design and discovery, and development of novel materials with tailored properties. For instance, strong Se-Au interactions may be useful in gold-nanoparticle stabilization and property tailoring, thus opening new avenues for SeNA in material science and nanotechnology. These findings will further guide the design of novel therapeutics with high specificity, selectivity, potency and low toxicity.

The incorporation of tellurium into nucleic acids offers similar advantages as sulfur and selenium, such as convenient labeling and enhanced duplex thermal stability. In a symbiotic relationship, the electron-deficient nucleobases stabilize the Te-functionality and consequently, the electron-rich tellurium increases electron delocalization in the nucleobases, thereby improving nucleic acid conductivity. This renders exceptional optical and electronic properties to Te-modified nucleic acids. These unique properties make Te-modifications extremely beneficial to nucleic acid-based structure-function studies, molecular imaging, and nanotechnology. The single-atom substitution is undeniably a convenient and viable strategy for tailoring the functionality of nucleic acids, and is advantageous in designing novel nucleic acid derivatives with customized properties, such as nuclease resistance, base-pairing specificity, drug bioavailabilty, and increased duplex thermostability, while virtually preserving structural integrity. In the near future, this atom-specific modification strategy will generate valuable novel materials for both fundamental research and the therapeutic development of nucleic acids.

Acknowledgments

This work was financially supported by NSF (MCB-0824837) and the Georgia Cancer Coalition (GCC) Distinguished Cancer Clinicians and Scientists.

References

1. DiPaolo, J. A.; Alvarez-Salas, L. M. *Expert Opin. Biol. Ther.* **2004**, *4*, 1251–1264.
2. Gitlin, L.; Andino, R. *J. Virol.* **2003**, *77*, 7159–7165.
3. Opalinska, J. B.; Gewirtz, A. M. *Nat. Rev. Drug Discovery* **2002**, *1*, 503–514.
4. Kuwahara, M.; Sugimoto, N. *Molecules* **2010**, *15*, 5423–5444.
5. Verma, S.; Jäger, S.; Thum, O.; Famulok, M. *Chem. Rec.* **2003**, *3*, 51–60.
6. Hollenstein, M. *Molecules* **2012**, *17*, 13569–13591.
7. Bittker, J. A.; Phillips, K. J.; Liu, D. R. *Curr. Opin. Chem. Biol* **2002**, *6*, 367–374.
8. Breaker, R. R. *Nature* **2004**, *432*, 838–845.
9. Freier, S. M.; Altmann, K.-H. *Nucleic Acids Res.* **1997**, *25*, 4429–4443.
10. Venkatesan, N.; Seo, Y. J.; Bang, E. K.; Park, S. M.; Lee, Y. S.; Kim, B. H. *Bull. Korean Chem. Soc.* **2006**, *27*, 613–630.
11. Dorsett, Y.; Tuschl, T. *Nat. Rev. Drug Discovery* **2004**, *3*, 318–329.
12. Proske, D.; Blank, M.; Buhmann, R.; Resch, A. *Appl. Microbiol. Biotechnol.* **2005**, *69*, 367–374.
13. Jepsen, J. S.; Sørensen, M. D.; Wengel, J. *Oligonucleotides* **2004**, *14*, 130–146.
14. Potaman, V. N. *Expert Rev. Mol. Diagn.* **2003**, *3*, 481–496.

15. Polsky, R.; Gill, R.; Kaganovsky, L.; Willner, I. *Anal. Chem.* **2006**, *78*, 2268–2271.

16. Lee, S. E.; Sidorov, A.; Gourlain, T.; Mignet, N.; Thorpe, S. J.; Brazier, J. A.; Dickman, M. J.; Hornby, D. P.; Grasby, J. A.; Williams, D. M. *Nucleic Acids Res.* **2001**, *29*, 1565–1573.

17. Micklefield, J. *Curr. Med. Chem.* **2001**, *8*, 1157–1179.

18. Jäschke, A. *Curr. Opin. Struct. Biol.* **2001**, *11*, 321–326.

19. Hecht, S. M.; Zhang, L.-H.; Xi, Z.; Chattopadhyaya, J. *Medicinal Chemistry of Nucleic Acids*; Wiley: New York, 2011; Vol. 16.

20. Williams-Ashman, H. G.; Seidenfeld, J.; Galletti, P. *Biochem. Pharmacol.* **1982**, *31*, 277–288.

21. Patel, A. D.; Schrier, W. H.; Nagyvary, J. *J. Org. Chem.* **1980**, *45*, 4830–4834.

22. Dantzman, C. L.; Kiessling, L. L. *J. Am. Chem. Soc.* **1996**, *118*, 11715–11719.

23. Johnson, R.; Reese, C. B.; Zhang, P.-Z. *Tetrahedron* **1995**, *51*, 5093–5098.

24. Chambert, S.; Gautier-Luneau, I.; Fontecave, M.; Decout, J.-L. *J. Org. Chem.* **2000**, *65*, 249–253.

25. Yoshimura, Y.; Kuze, T.; Ueno, M.; Komiya, F.; Haraguchi, K.; Tanaka, H.; Kano, F.; Yamada, K.; Asami, K.; Kaneko, N. *Tetrahedron Lett.* **2006**, *47*, 591–594.

26. Secrist, J. A., III; Tiwari, K. N.; Riordan, J. M.; Montgomery, J. A. *J. Med. Chem.* **1991**, *34*, 2361–2366.

27. Van Draanen, N. A.; Freeman, G. A.; Short, S. A.; Harvey, R.; Jansen, R.; Szczech, G.; Koszalka, G. W. *J. Med. Chem.* **1996**, *39*, 538–542.

28. Haeberli, P.; Berger, I.; Pallan, P. S.; Egli, M. *Nucleic Acids Res.* **2005**, *33*, 3965–3975.

29. Gunaga, P.; Moon, H. R.; Choi, W. J.; Shin, D. H.; Park, J. G.; Jeong, L. S. *Curr. Med. Chem.* **2004**, *11*, 2585–2637.

30. Mueller, E. G. *Nat. Chem. Biol.* **2006**, *2*, 185–194.

31. Ajitkumar, P.; Cherayil, J. D. *Microbiol. Rev.* **1988**, *52*, 103–113.

32. Agris, P. F.; Soell, D.; Seno, T. *Biochemistry* **1973**, *12*, 4331–4337.

33. Kambampati, R.; Lauhon, C. T. *Biochemistry* **2003**, *42*, 1109–1117.

34. Thomas, G.; Favre, A. *Eur. J. Biochem.* **1980**, *113*, 67–74.

35. Kumar, R. K.; Davis, D. R. *Nucleic Acids Res.* **1997**, *25*, 1272–1280.

36. Davanloo, P.; Sprinzl, M.; Watanabe, K.; Albani, M.; Kersten, H. *Nucleic Acids Res.* **1979**, *6*, 1571–1581.

37. Watanabe, K. *Bull. Chem. Soc. Jpn.* **2007**, *80*, 1253–1267.

38. Wang, H.; Wang, Y. *Biochemistry* **2009**, *48*, 2290–2299.

39. Yoshida, A.; Sun, S.; Piccirilli, J. A. *Nat. Struct. Biol.* **1999**, *6*, 318–321.

40. Piccirilli, J. A.; Vyle, J. S.; Caruthers, M. H.; Cech, T. R. *Nature* **1993**, *361*, 85–88.

41. Eckstein, F.; Gish, G. *Trends Biochem. Sci.* **1989**, *14*, 97–100.

42. Padmapriya, A.; TANG, J.; AGRAWAL, S. *Antisense Res. Dev.* **1994**, *4*, 185–199.

43. Agrawal, S.; Temsamani, J.; Tang, J. Y. *Proc. Natl. Acad. Sci. U.S.A.* **1991**, *88*, 7595–7599.

44. Metelev, V.; Lisziewicz, J.; Agrawal, S. *Bioorg. Med. Chem. Lett.* **1994**, *4*, 2929–2934.

45. Stein, C. A.; Goel, S. *Clin. Cancer Res.* **2011**, *17*, 6369–6372.

46. Marshall, W.; Caruthers, M. *Science* **1993**, *259*, 1564–1570.

47. Tillman, L. G.; Geary, R. S.; Hardee, G. E. *J. Pharm. Sci.* **2008**, *97*, 225–236.

48. Seth, P. P.; Jazayeri, A.; Yu, J.; Allerson, C. R.; Bhat, B.; Swayze, E. E. *Mol. Ther. Nucleic Acids* **2012**, *1*, e47.

49. Detzer, A.; Sczakiel, G. *Curr. Top. Med. Chem.* **2009**, *9*,, 1109–1116.

50. Koller, E.; Vincent, T. M.; Chappell, A.; De, S.; Manoharan, M.; Bennett, C. F. *Nucleic Acids Res.* **2011**, *39*, 4795–4807.

51. Eckstein, F. *Acc. Chem. Res.* **1979**, *12*, 204–210.

52. Nakamaye, K. L.; Gish, G.; Eckstein, F.; Vosberg, H.-P. *Nucleic Acids Res.* **1988**, *16*, 9947–9959.

53. Gish, G.; Eckstein, F. *Science* **1988**, *240*, 1520–1522.

54. Carter, P. *Biochem. J.* **1986**, *237*, 1–7.

55. Potter, B.; Eckstein, F. *J. Biol. Chem.* **1984**, *259*, 14243–14248.

56. Eckstein, F. *J. Am. Chem. Soc.* **1966**, *88*, 4292–4294.

57. Eckstein, F. *J. Am. Chem. Soc.* **1970**, *92*, 4718–4723.

58. Cook, A. F. *J. Am. Chem. Soc.* **1970**, *92*, 190–195.

59. Eckstein, F. *Annu. Rev. Biochem.* **1985**, *54*, 367–402.

60. Tang, J.; Roskey, A.; Li, Y.; Agrawal, S. *Nucleosides, Nucleotides Nucleic Acids* **1995**, *14*, 985–990.

61. Łazewska, D.; Guranowski, A. *Nucleic Acids Res.* **1990**, *18*, 6083–6088.

62. Iyer, R. P.; Guo, M.-J.; Yu, D.; Agrawal, S. *Tetrahedron Lett.* **1998**, *39*, 2491–2494.

63. Almer, H.; Stawinski, J.; Strömberg, R. *Nucleic Acids Res.* **1996**, *24*, 3811–3820.

64. Knowles, J. R. *Annu. Rev. Biochem.* **1980**, *49*, 877–919.

65. Brautigam, C. A.; Steitz, T. A. *J. Mol. Biol.* **1998**, *277*, 363.

66. Romaniuk, P. J.; Eckstein, F. *J. Biol. Chem.* **1982**, *257*, 7684–7688.

67. Eckstein, F.; Jovin, T. M. *Biochemistry* **1983**, *22*, 4546–4550.

68. Bartlett, P.; Eckstein, F. *J. Biol. Chem.* **1982**, *257*, 8879–8884.

69. Yang, Z.; Sismour, A. M.; Benner, S. A. *Nucleic Acids Res.* **2007**, *35*, 3118–3127.

70. Caton-Williams, J.; Fiaz, B.; Hoxhaj, R.; Smith, M.; Huang, Z. *Sci. China: Chem.* **2012**, *55*, 80–89.

71. Alefelder, S.; Patel, B. K.; Eckstein, F. *Nucleic Acids Res.* **1998**, *26*, 4983–4988.

72. Gaynor, J. W.; Piperakis, M. M.; Fisher, J.; Cosstick, R. *Org. Biomol. Chem.* **2010**, *8*, 1463–1470.

73. Meena; Sam, M.; Pierce, K.; Szostak, J. W.; McLaughlin, L. W. *Org. Lett.* **2007**, *9*, 1161–1163.

74. Hoshika, S.; Minakawa, N.; Matsuda, A. *Nucleic Acids Res.* **2004**, *32*, 3815–3825.

75. Kato, Y.; Minakawa, N.; Komatsu, Y.; Kamiya, H.; Ogawa, N.; Harashima, H.; Matsuda, A. *Nucleic Acids Res.* **2005**, *33*, 2942–2951.

76. Keefe, A. D.; Cload, S. T. *Curr. Opin. Chem. Biol* **2008**, *12*, 448–456.

77. Inoue, N.; Minakawa, N.; Matsuda, A. *Nucleic Acids Res.* **2006**, *34*, 3476–3483.

78. Favre, A.; Fourrey, J.-L. *Acc. Chem. Res.* **1995**, *28*, 375–382.

79. Favre, A.; Saintomé, C.; Fourrey, J.-L.; Clivio, P.; Laugâa, P. *J. Photochem. Photobiol., B* **1998**, *42*, 109–124.

80. Harris, M. E.; Christian, E. L. *Methods Enzymol.* **2009**, *468*, 127–146.

81. Saintomé, C.; Clivio, P.; Fourrey, J.-L.; Woisard, A.; Laugâa, P.; Favre, A. *Tetrahedron* **2000**, *56*, 1197–1206.

82. Woisard, A.; Favre, A.; Clivio, P.; Fourrey, J. L. *J. Am. Chem. Soc.* **1992**, *114*, 10072–10074.

83. Massey, A.; Xu, Y.-Z.; Karran, P. *Curr. Biol.* **2001**, *11*, 1142–1146.

84. De Mesmaeker, A.; Haener, R.; Martin, P.; Moser, H. E. *Acc. Chem. Res.* **1995**, *28*, 366–374.

85. Crooke, S. T. *Antisense Nucleic Acid Drug Dev.* **1998**, *8*, vii–viii.

86. Tafech, A.; Bassett, T.; Sparanese, D.; Lee, C. H. *Curr. Med. Chem.* **2006**, *13*, 863–881.

87. Persidis, A. *Nat. Biotechnol.* **1999**, *17*, 403–404.

88. Bell, D. A.; Hooper, A. J.; Burnett, J. R. *Expert Opin. Invest. Drugs* **2011**, *20*, 265–272.

89. Akdim, F.; Stroes, E. S.; Sijbrands, E. J.; Tribble, D. L.; Trip, M. D.; Jukema, J. W.; Flaim, J. D.; Su, J.; Yu, R.; Baker, B. F. *J. Am. Coll. Cardiol.* **2010**, *55*, 1611–1618.

90. Kling, J. *Nat. Biotechnol.* **2010**, *28*, 295–297.

91. Yamamoto-Furusho, J. K. *World J. Gastroenterol.* **2007**, *13*, 1893–1896.

92. Akhtar, S.; Hughes, M. D.; Khan, A.; Bibby, M.; Hussain, M.; Nawaz, Q.; Double, J.; Sayyed, P. *Adv. Drug Delivery Rev.* **2000**, *44*, 3–21.

93. Kraynack, B. A.; Baker, B. F. *RNA* **2006**, *12*, 163–176.

94. Wilson, C.; Keefe, A. D. *Curr. Opin. Chem. Biol* **2006**, *10*, 607–614.

95. Cobb, A. J. *Org. Biomol. Chem.* **2007**, *5*, 3260–3275.

96. Zhou, Y.; Kierzek, E.; Loo, Z. P.; Antonio, M.; Yau, Y. H.; Chuah, Y. W.; Geifman-Shochat, S.; Kierzek, R.; Chen, G. *Nucleic Acids Res.* **2013**, *41*, 6664–6673.

97. Okamoto, I.; Seio, K.; Sekine, M. *Bioorg. Med. Chem. Lett.* **2006**, *16*, 3334–3336.

98. Miyata, K.-i.; Tamamushi, R.; Tsunoda, H.; Ohkubo, A.; Seio, K.; Sekine, M. *Org. Lett.* **2009**, *11*, 605–608.

99. Ohkubo, A.; Nishino, Y.; Yokouchi, A.; Ito, Y.; Noma, Y.; Kakishima, Y.; Masaki, Y.; Tsunoda, H.; Seio, K.; Sekine, M. *Chem. Commun.* **2011**, *47*, 12556–12558.

100. Dausse, E.; Da Rocha Gomes, S.; Toulmé, J.-J. *Curr. Opin. Pharmacol.* **2009**, *9*, 602–607.

101. Pendergrast, P. S.; Marsh, H. N.; Grate, D.; Healy, J. M.; Stanton, M. *J. Biomol. Tech.* **2005**, *16*, 224–234.

102. Ni, X.; Castanares, M.; Mukherjee, A.; Lupold, S. E. *Curr. Med. Chem.* **2011**, *18*, 4206–4214.

103. Flohe, L.; Gunzler, W. A.; Schock, H. H. *FEBS Lett.* **1973**, *32*, 132–134.

104. Rotruck, J. T.; Pope, A. L.; Ganther, H. E.; Swanson, A. B.; Hafeman, D. G.; Hoekstra, W. G. *Science* **1973**, *179*, 588–590.

105. Gladyshev, V. N.; Boyington, J. C.; Khangulov, S. V.; Grahame, D. A.; Stadtman, T. C.; Sun, P. D. *J. Biol. Chem.* **1996**, *271*, 8095–8100.

106. Lacourciere, G. M.; Mihara, H.; Kurihara, T.; Esaki, N.; Stadtman, T. C. *J. Biol. Chem.* **2000**, *275*, 23769–23773.

107. Stadtman, T. C. *Science* **1974**, *183*, 915–922.

108. Sunde, R. A. Selenium. In *Modern Nutrition in Health and Disease*; Ross A. C., Caballero, B., Cousins, R. J., Tucker, K. L., Ziegler, T. R, Eds.; Lippincott Williams & Wilkins: Baltimore, MD, 2012; pp 225–237.

109. Tsuji, P. A., Davis, C. D., Milner, J. A. Berry, M., Gladyshev, V., Hatfield, D. In *Selenium: Its molecular biology and role in human health*, 3rd ed.; Hatfield, D. L., Berry, M. J., Gladyshev, V. N., Eds.; Springer: New York, 2012

110. Almondes, K. G.; Leal, G. V; Cozzolino, S. M; Philippi, S. T; Rondó, P. H. *Rev. Assoc. Med. Bras.* **2010**, *6*, 484–488.

111. Hudson, T. S.; Carlson, B. A.; Hoeneroff, M. J.; Young, H. A.; Sordillo, L.; Muller, W. J.; Hatfield, D. L.; Green, J. E. *Carcinogenesis* **2012**, *33*, 1225–1230.

112. Schwarz, E. *Dtsch. Gesundheitswes.* **1957**, *12*, 97–104.

113. Shamberger, R. J.; Frost, D. V. *Can. Med. Assoc. J.* **1969**, *100*, 682.

114. Schrauzer, G. N.; White, D. A.; Schneider, C. J. *Bioinorg. Chem.* **1977**, *7*, 35–56.

115. Foster, L. H.; Sumar, S. *Crit. Rev. Food. Sci. Nutr.* **1997**, *37*, 211–228.

116. Hoffman, J. L.; McConnell, K. P. *Biochim. Biophys. Acta* **1974**, *366*, 109–113.

117. Rees, K.; Hartley, L.; Day, C.; Flowers, N.; Clarke, A.; Stranges, S. *Cochrane Database Syst. Rev.* **2013**, *1*, CD009671.

118. Boyd, R. *Nat. Chem.* **2011**, *3*, 570–570.

119. Xu, H.; Cao, W.; Zhang, X. *Acc. Chem. Res.* **2013**, *46*, 1647–1658.

120. Birringer, M.; Pilawa, S.; Flohe, L. *Nat. Prod. Rep.* **2002**, *19*, 693–718.

121. Hatfield, D. L.; Gladyshev, V. N. *Mol. Cell. Biol.* **2002**, *22*, 3565–3576.

122. Lacourciere, G. M.; Levine, R. L.; Stadtman, T. C. *Proc. Natl. Acad. Sci. U.S.A.* **2002**, *99*, 9150–9153.

123. Rayman, M. P. In *Proceedings of the Nutrition Society of London*; Cambridge University Press: New York, 2005; Vol. 64, p 527.

124. Rayman, M. P. *Lancet* **2000**, *356*, 233–241.

125. Ching, W. M.; Alzner-DeWeerd, B.; Stadtman, T. C. *Proc. Natl. Acad. Sci. U.S.A.* **1985**, *82*, 347–350.

126. Ching, W.-M. *Arch. Biochem. Biophys.* **1986**, *244*, 137–146.

127. Stadtman, T. C. *FASEB J.* **1987**, *1*, 375–379.

128. Carrasco, N.; Ginsburg, D.; Du, Q.; Huang, Z. *Nucleosides Nucleotides Nucleic Acids* **2001**, *20*, 1723–1734.

129. Du, Q.; Carrasco, N.; Teplova, M.; Wilds, C. J.; Egli, M.; Huang, Z. *J. Am. Chem. Soc.* **2002**, *124*, 24–25.

130. Lin, L.; Caton-Williams, J.; Kaur, M.; Patino, A. M.; Sheng, J.; Punetha, J.; Huang, Z. *RNA* **2011**, *17*, 1932–1938.

131. Sheng, J.; Gan, J.; Huang, Z. *Med. Res. Rev.* **2013**.

132. Salon, J.; Gan, J.; Abdur, R.; Liu, H.; Huang, Z. *Org. Lett.* **2013**, *15*, 3934–3937.

133. Sun, H.; Jiang, S.; Caton-Williams, J.; Liu, H.; Huang, Z. *RNA* **2013**, *19*, 1309–1314.

134. Zhang, W.; Hassan, E. A.; Huang, Z. *Sci. China: Chem.* **2013**, *56*, 273–278.

135. Sun, H.; Sheng, J.; Hassan, A. E.; Jiang, S.; Gan, J.; Huang, Z. *Nucleic Acids Res.* **2012**, *40*, 5171–5179.

136. Sheng, J.; Zhang, W.; Hassan, A. E.; Gan, J.; Soares, A. S.; Geng, S.; Ren, Y.; Huang, Z. *Nucleic Acids Res.* **2012**, *40*, 8111–8118.

137. Lin, L.; Huang, Z. In *Recombinant and In Vitro RNA Synthesis*; Springer: New York, 2012; p 213–225.

138. Sheng, J.; Hassan, A. E.; Zhang, W.; Zhou, J.; Xu, B.; Soares, A. S.; Huang, Z. *Nucleic Acids Res.* **2011**, *39*, 3962–3971.

139. Salon, J.; Sheng, J.; Gan, J.; Huang, Z. *J. Org. Chem.* **2010**, *75*, 637–641.

140. Hassan, A. E.; Sheng, J.; Zhang, W.; Huang, Z. *J. Am. Chem. Soc.* **2010**, *132*, 2120–2121.

141. Wilds, C. J.; Pattanayek, R.; Pan, C.; Wawrzak, Z.; Egli, M. *J. Am. Chem. Soc.* **2002**, *124*, 14910–14916.

142. Pallan, P. S.; Egli, M. *Nat. Protoc.* **2007**, *2*, 640–646.

143. Olieric, V.; Rieder, U.; Lang, K.; Serganov, A.; Schulze-Briese, C.; Micura, R.; Dumas, P.; Ennifar, E. *RNA* **2009**, *15*, 707–715.

144. Hobartner, C.; Micura, R. *J. Am. Chem. Soc.* **2004**, *126*, 1141–1149.

145. Watts, J. K.; Johnston, B. D.; Jayakanthan, K.; Wahba, A. S.; Pinto, B. M.; Damha, M. J. *J. Am. Chem. Soc.* **2008**, *130*, 8578–8579.

146. Wachowius, F.; Höbartner, C. *ChemBioChem* **2010**, *11*, 469–480.

147. Tram, K.; Wang, X.; Yan, H. *Org. Lett.* **2007**, *9*, 5103–5106.

148. Hobartner, C.; Rieder, R.; Kreutz, C.; Puffer, B.; Lang, K.; Polonskaia, A.; Serganov, A.; Micura, R. *J. Am. Chem. Soc.* **2005**, *127*, 12035–12045.

149. Lin, L.; Sheng, J.; Huang, Z. *Chem. Soc. Rev.* **2011**, *40*, 4591–4602.

150. Quigley, G. J.; Rich, A. *Science* **1976**, *194*, 796–806.

151. Itoh, Y.; Brocker, M. J.; Sekine, S.; Hammond, G.; Suetsugu, S.; Soll, D.; Yokoyama, S. *Science* **2013**, *340*, 75–78.

152. Itoh, Y.; Sekine, S. I.; Suetsugu, S.; Yokoyama, S. *Nucleic Acids Res.* **2013**, *41*, 6729–6738.

153. Leinfelder, W.; Zehelein, E.; Mandrand-Berthelot, M. A.; Bo¨ck, A. *Nature* **1988**, *331*, 723–725.

154. Leinfelder, W.; Stadtman, T. C.; Bock, A. *J. Biol. Chem.* **1989**, *264*, 9720–9723.

155. Baron, C.; Sturchler, C.; Wu, X. Q.; Gross, H. J.; Krol, A.; Bock, A. *Nucleic Acids Res.* **1994**, *22*, 2228–2233.

156. Abe, T.; Ikemura, T.; Sugahara, J.; Kanai, A.; Ohara, Y.; Uehara, H.; Kinouchi, M.; Kanaya, S.; Yamada, Y.; Muto, A.; Inokuchi, H. *Nucleic Acids Res.* **2011**, *39*, D210–213.

157. Sturchler, C.; Westhof, E.; Carbon, P.; Krol, A. *Nucleic Acids Res.* **1993**, *21*, 1073–1079.

158. Itoh, Y.; Sekine, S. I.; Yokoyama, S. *Acta Crystallogr., Sect F* **2012**, *68*, 678–682.

159. Chen, C. S.; Stadtman, T. C. *Proc. Natl. Acad. Sci. U.S.A.* **1980**, *77*, 1403–1407.

160. Chen, C. S.; Wen, T. N.; Tuan, H. M. *Biochim. Biophys. Acta* **1982**, *699*, 92–97.

161. Starks, C. M.; Francois, J. A.; MacArthur, K. M.; Heard, B. Z.; Kappock, T. J. *Protein Sci.* **2007**, *16*, 92–98.

162. Metanis, N.; Keinan, E.; Dawson, P. E. *J. Am. Chem. Soc.* **2006**, *128*, 16684–16691.

163. Allmang, C.; Krol, A. *Biochimie* **2006**, *88*, 1561–1571.

164. Bock, A.; Forchhammer, K.; Heider, J.; Leinfelder, W.; Sawers, G.; Veprek, B.; Zinoni, F. *Mol. Microbiol.* **1991**, *5*, 515–520.

165. Sliwkowski, M. X.; Stadtman, T. C. *J. Biol. Chem.* **1987**, *262*, 4899–4904.

166. Bock, A.; Stadtman, T. C. *Biofactors* **1988**, *1*, 245–250.

167. Forstrom, J. W.; Zakowski, J. J.; Tappel, A. L. *Biochemistry* **1978**, *17*, 2639–2644.

168. Barbezat, G. O.; Casey, C. E.; Reasbeck, P. G.; Robinson, M. F.; Thomson, C. D. *Current Topics in Nutrition*; Alan, R. Liss, Inc.: New York, 1984; Vol. 12.

169. Combs, G. F., Combs, S. B. In *The Role of Selenium in Nutrition*; Academic Press: New York, 1986.

170. Ganichkin, O. M.; Xu, X. M.; Carlson, B. A.; Mix, H.; Hatfield, D. L.; Gladyshev, V. N.; Wahl, M. C. *J. Biol. Chem.* **2008**, *283*, 5849–5865.

171. Brown, K. M.; Arthur, J. R. *Public Health Nutr.* **2001**, 593–599.

172. Hatfield, D. L., Berry, M. J., and Gladyshev, V. N. *Selenium: Its Molecular Biology and Role in Human Health*, 3rd ed.; Springer: New York, 2011.

173. Baudin-Baillieu, A.; Fabret, C.; Liang, X. H.; Piekna-Przybylska, D.; Fournier, M. J.; Rousset, J. P. *Nucleic Acids Res.* **2009**, *37*, 7665–7677.

174. Axley, M. J.; Bock, A.; Stadtman, T. C. *Proc. Natl. Acad. Sci. U.S.A.* **1991**, *88*, 8450–8454.

175. Kryukov, G. V.; Gladyshev, V. N. *Genes Cells* **2000**, *5*, 1049–1060.

176. Jameson, R. R.; Diamond, A. M. *RNA* **2004**, *10*, 1142–1152.

177. Hassan, A. E.; Sheng, J.; Jiang, J.; Zhang, W.; Huang, Z. *Org. Lett.* **2009**, *11*, 2503–2506.

178. Gladyshev, V. N.; Kryukov, G. V. *Biofactors* **2001**, *14*, 87–92.

179. Allmang, C.; Wurth, L.; Krol, A. *Biochim. Biophys. Acta* **2009**, *1790*, 1415–1423.

180. Krief, A., Hevesi, L. J. *Organoselenium chemistry I: functional group transformations*; Springer-Verlag: Berlin, 1988; pp 1–11.

181. Carrasco, N.; Huang, Z. *J. Am. Chem. Soc.* **2004**, *126*, 448–449.

182. Zhong, L.; Arner, E. S.; Holmgren, A. *Proc. Natl. Acad. Sci. U.S.A.* **2000**, *97*, 5854–5859.

183. Lee, S. R.; Bar-Noy, S.; Kwon, J.; Levine, R. L.; Stadtman, T. C.; Rhee, S. G. *Proc. Natl. Acad. Sci. U.S.A.* **2000**, *97*, 2521–2526.

184. Castellano, S.; Andres, A. M.; Bosch, E.; Bayes, M.; Guigo, R.; Clark, A. G. *Mol. Biol. Evol.* **2009**, *26*, 2031–2040.

185. Aboul-Fadl, T. *Curr. Med. Chem. Anticancer Agents* **2005**, *5*, 637–652.

186. Stadtman, T. C. *Fundam. Appl. Toxicol.* **1983**, *3*, 420–423.

187. Becker, H. F.; Motorin, Y.; Florentz, C.; Giege, R.; Grosjean, H. *Nucleic Acids Res.* **1998**, *26*, 3991–3997.

188. Sprinzl, M.; Vassilenko, K. S. *Nucleic Acids Res.* **2005**, *33*, D139–140.

189. McCloskey, J. A.; Rozenski, J. *Nucleic Acids Res.* **2005**, *33*, D135–138.

190. Flohe, L. *Biochim. Biophys. Acta* **2009**, *1790*, 1389–1403.

191. Ching, W. M. *Proc. Natl. Acad. Sci. U.S.A.* **1984**, *81*, 3010–3013.

192. Wittwer, A. J. *J. Biol. Chem.* **1983**, *258*, 8637–8641.

193. Anantharaman, V.; Iyer, L. M.; Aravind, L. *Mol. Biosyst.* **2012**, *8*, 3142–3165.

194. Saelinger, D. A.; Hoffman, J. L.; McConnell, K. P. *J. Mol. Biol.* **1972**, *69*, 9–17.

195. Takai, K.; Yokoyama, S. *Nucleic Acids Res.* **2003**, *31*, 6383–6391.

196. Sundaram, M.; Crain, P. F.; Davis, D. R. *J. Org. Chem.* **2000**, *65*, 5609–5614.

197. Turanov, A. A.; Xu, X. M.; Carlson, B. A.; Yoo, M. H.; Gladyshev, V. N.; Hatfield, D. L. *Adv. Nutr.* **2011**, *2*, 122–128.

198. Hatfield, D. L.; Carlson, B. A.; Xu, X. M.; Mix, H.; Gladyshev, V. N. *Prog. Nucleic Acid Res. Mol. Biol.* **2006**, *81*, 97–142.

199. Mizutani, T.; Watanabe, T.; Kanaya, K.; Nakagawa, Y.; Fujiwara, T. *Mol. Biol. Rep.* **1999**, *26*, 167–172.

200. Mihara, H.; Kato, S.; Lacourciere, G. M.; Stadtman, T. C.; Kennedy, R. A.; Kurihara, T.; Tokumoto, U.; Takahashi, Y.; Esaki, N. *Proc. Natl. Acad. Sci. U.S.A.* **2002**, *99*, 6679–6683.

201. Elseviers, D.; Petrullo, L. A.; Gallagher, P. J. *Nucleic Acids Res.* **1984**, *12*, 3521–3534.

202. Kim, I. Y.; Stadtman, T. C. *Proc. Natl. Acad. Sci. U.S.A.* **1994**, *91*, 7326–7329.

203. Glass, R. S.; Singh, W. P.; Jung, W.; Veres, Z.; Scholz, T. D.; Stadtman, T. C. *Biochemistry* **1993**, *32*, 12555–12559.

204. Gladyshev, V. N.; Khangulov, S. V.; Axley, M. J.; Stadtman, T. C. *Proc. Natl. Acad. Sci. U.S.A.* **1994**, *91*, 7708–7711.

205. Lacourciere, G. M.; Stadtman, T. C. *J. Biol. Chem.* **1998**, *273*, 30921–30926.

206. Wolfe, M. D.; Ahmed, F.; Lacourciere, G. M.; Lauhon, C. T.; Stadtman, T. C.; Larson, T. J. *J. Biol. Chem.* **2004**, *279*, 1801–1809.

207. Politino, M.; Tsai, L.; Veres, Z.; Stadtman, T. C. *Proc. Natl. Acad. Sci. U.S.A.* **1990**, *87*, 6345–6348.

208. Sheng, J.; Gan, J.; Soars, A. S.; Salon, J.; Huang, Z. *Nucleic Acids Res.* **2013**, *41*, 1723–1734.

209. Sheng, J.; Salon, J.; Gan, J.; Huang, Z. *Sci. China: Chem.* **2010**, *53*, 78–85.

210. Moroder, H.; Kreutz, C.; Lang, K.; Serganov, A.; Micura, R. *J. Am. Chem. Soc.* **2006**, *128*, 9909–9918.

211. Puffer, B.; Moroder, H.; Aigner, M.; Micura, R. *Nucleic Acids Res.* **2008**, *36*, 970–983.

212. Sheng, J.; Jiang, J.; Salon, J.; Huang, Z. *Org. Lett.* **2007**, *9*, 749–752.

213. Inagaki, Y.; Minakawa, N.; Matsuda, A. In *Nucleic Acids Symposium Series*; Oxford University Press: New York, 2007; Vol. 51, pp 139–140.

214. Jeong, L. S.; Tosh, D. K.; Kim, H. O.; Wang, T.; Hou, X.; Yun, H. S.; Kwon, Y.; Lee, S. K.; Choi, J.; Zhao, L. X. *Org. Lett.* **2008**, *10*, 209–212.

215. Salon, J.; Jiang, J.; Sheng, J.; Gerlits, O. O.; Huang, Z. *Nucleic Acids Res.* **2008**, *36*, 7009–7018.

216. Salon, J.; Sheng, J.; Jiang, J.; Chen, G.; Caton-Williams, J.; Huang, Z. *J. Am. Chem. Soc.* **2007**, *129*, 4862–4863.

217. Zhang, W.; Sheng, J.; Hassan, A. E.; Huang, Z. *Chem. Asian J.* **2012**, *7*, 476–479.

218. Mori, K.; Boiziau, C.; Cazenave, C.; Matsukura, M.; Subasinghe, C.; Cohen, J.; Broder, S.; Toulme, J.; Stein, C. *Nucleic Acids Res.* **1989**, *17*, 8207–8219.

219. Egli, M.; Pallan, P. S.; Pattanayek, R.; Wilds, C. J.; Lubini, P.; Minasov, G.; Dobler, M.; Leumann, C. J.; Eschenmoser, A. *J. Am. Chem. Soc.* **2006**, *128*, 10847–10856.

220. Guga, P.; Maciaszek, A.; Stec, W. J. *Org. Lett.* **2005**, *7*, 3901–3904.

221. Guga, P. *Curr. Top. Med. Chem.* **2007**, *7*, 695–713.

222. Carrasco, N.; Caton-Williams, J.; Brandt, G.; Wang, S.; Huang, Z. *Angew. Chem., Int. Ed.* **2006**, *45*, 94–97.

223. Brandt, G.; Carrasco, N.; Huang, Z. *Biochemistry* **2006**, *45*, 8972–8977.

224. Papp, L. V.; Lu, J.; Holmgren, A.; Khanna, K. K. *Antioxid. Redox Signaling* **2007**, *9*, 775–806.

225. Hendrickson, W. A.; Horton, J. R.; LeMaster, D. M. *EMBO J.* **1990**, *9*, 1665–1672.

226. Deacon, A. M.; Ealick, S. E. *Structure* **1999**, *7*, R161–166.

227. Doublié, S. *Methods Enzymol.* **1997**, *276*, 523–530.

228. Hendrickson, W. A.; Pähler, A.; Smith, J. L.; Satow, Y.; Merritt, E. A.; Phizackerley, R. P. *Proc. Natl. Acad. Sci. U.S.A.* **1989**, *86*, 2190–2194.

229. Ferre-D'Amare, A. R.; Zhou, K.; Doudna, J. A. *Nature* **1998**, *395*, 567–574.

230. Sheng, J.; Huang, Z. *Int. J. Mol. Sci.* **2008**, *9*, 258–271.

231. Jiang, J.; Sheng, J.; Carrasco, N.; Huang, Z. *Nucleic Acids Res.* **2007**, *35*, 477–485.

232. Sheng, J.; Huang, Z. *Chem. Biodiversity* **2010**, *7*, 753–785.

233. Holbrook, S. R.; Kim, S. H. *Biopolymers* **1997**, *44*, 3–21.

234. Zhang, W.; Sheng, J.; Huang, Z. *Med. Chem. Nucleic Acids* **2011**, *16*, 101.

235. Serganov, A.; Keiper, S.; Malinina, L.; Tereshko, V.; Skripkin, E.; Hobartner, C.; Polonskaia, A.; Phan, A. T.; Wombacher, R.; Micura, R.; Dauter, Z.; Jaschke, A.; Patel, D. J. *Nat. Struct. Mol. Biol.* **2005**, *12*, 218–224.

236. Siegmund, V.; Santner, T.; Micura, R.; Marx, A. *Chem. Sci.* **2011**, *2*, 2224–2231.

237. Santner, T.; Siegmund, V.; Marx, A.; Micura, R. *Bioorg. Med. Chem.* **2012**, *20*, 2416–2418.

238. Dueymes, C.; Décout, J. L.; Peltié, P.; Fontecave, M. *Angew. Chem., Int. Ed.* **2002**, *41*, 486–489.

239. Simon, P.; Dueymes, C.; Fontecave, M.; Décout, J. L. *Angew. Chem., Int. Ed.* **2005**, *44*, 2764–2767.

240. Wilson, J. N.; Kool, E. T. *Org. Biomol. Chem.* **2006**, *4*, 4265–4274.

241. Silverman, A. P.; Kool, E. T. *Trends Biotechnol.* **2005**, *23*, 225–230.

242. Kim, S. J.; Kool, E. T. *J. Am. Chem. Soc.* **2006**, *128*, 6164–6171.

243. Kim, H.-K.; Liu, J.; Li, J.; Nagraj, N.; Li, M.; Pavot, C. M.-B.; Lu, Y. *J. Am. Chem. Soc.* **2007**, *129*, 6896–6902.

244. Caton-Williams, J.; Huang, Z. *Angew. Chem.* **2008**, *120*, 1747–1749.

245. Moroder, L. *J. Pept. Sci.* **2005**, *11*, 187–214.

246. Schroeder, H. A.; Buckman, J.; Balassa, J. J. *J. Chronic Dis.* **1967**, *20*, 147–161.

247. You, Y.; Ahsan, K.; Detty, M. R. *J. Am. Chem. Soc.* **2003**, *125*, 4918–4927.

248. Franke, K. W.; Moxon, A. L. *J. Pharmacol. Exp. Ther.* **1937**, *61*, 89–102.

249. De Meio, R. *J. Ind. Hyg. Toxicol.* **1946**, *28*, 229–232.

250. Amdur, M. *AMA Arch. Ind. Health* **1958**, *17*, 665–667.

251. Sailer, B. L.; Liles, N.; Dickerson, S.; Sumners, S.; Chasteen, T. G. *Toxicol. In Vitro* **2004**, *18*, 475–482.

252. Knapp, F. F., Jr.; Ambrose, K. R.; Callahan, A. P. *J. Med. Chem.* **1981**, *24*, 794–797.

253. Kolar, Z. *Int. J. Appl. Radiat. Isot.* **1974**, *25*, 330–331.

254. Boles, J. O.; Lewinski, K.; Kunkle, M.; Odom, J. D.; Dunlap, B.; Lebioda, L.; Hatada, M. *Nat. Struct. Biol.* **1994**, *1*, 283–284.

255. Budisa, N.; Karnbrock, W.; Steinbacher, S.; Humm, A.; Prade, L.; Neuefeind, T.; Moroder, L.; Huber, R. *J. Mol. Biol.* **1997**, *270*, 616–623.

256. Budisa, N.; Steipe, B.; Demange, P.; Eckerskorn, C.; Kellermann, J.; Huber, R. *Eur. J. Biochem.* **1995**, *230*, 788–796.

257. Fanchon, E.; Hendrickson, W. A. *Acta Crystallogr., Sect. A* **1990**, *46*, 809–820.

258. Sheng, J.; Hassan, A. E.; Huang, Z. *Sci. China: Chem.* **2009**, *15*, 10210–10216.

259. Kang, Q.; Yang, L.; Chen, Y.; Luo, S.; Wen, L.; Cai, Q.; Yao, S. *Anal. Chem.* **2010**, *82*, 9749–9754.

Chapter 6

⁷⁷Se NMR Spectroscopy of Selenoproteins

Sharon Rozovsky*

Department of Chemistry and Biochemistry, University of Delaware, Newark, Delaware 19716, United States
***E-mail: rozovsky@udel.edu. Phone: 302-831-7028. Fax: 302-831-6335.**

Selenium is a cardinal contributor to cellular antioxidative defense by way of selenoproteins, a family of enzymes that contains the reactive amino acid selenocysteine. This unique group of proteins can be studied by ⁷⁷Se NMR spectroscopy, a versatile spectroscopic method capable of recording the electronic structure of reactive enzymatic centers. ⁷⁷Se is a spin 1/2 nucleus with a pronounced chemical shielding response that renders it highly sensitive to its local environment. This review discusses the application of ⁷⁷Se NMR spectroscopy to studies of native and nonnative selenoproteins, sample preparation, and data interpretation. We also discuss perspectives for the routine use of ⁷⁷Se in investigations of biological systems and the prospects of using selenium as a spectroscopic surrogate for sulfur, an abundant biological nuclei that is not easily detected by NMR spectroscopy.

Introduction

In biological systems, selenium's strong nucleophilicity, redox activity, metal binding capabilities, and low pKa are utilized to catalyze chemical reactions (*1*). Most health benefits of selenium are mediated by selenoproteins, a specialized group of enzymes that contains the rare amino acid selenocysteine (Sec, U). This family of proteins participates in cellular antioxidative functions and regulation of redox pathways. As a result, they play a cardinal role in chemopreventive, anti-inflammatory, and antiviral defense, as well as in the aging process. Several excellent reviews discuss the biological roles of selenoproteins (*2, 3*), the genetic coding and Sec incorporation into proteins *in vivo* (*4–6*), as well as the unique features that arise from the utilization of Sec in macromolecules (*1, 7*). This review focuses on studies of the selenoprotein family through the use of [77]Se nuclear magnetic resonance (NMR) spectroscopy.

NMR is an essential biophysical technique, due to its unsurpassed sensitivity to molecules' electronic and chemical structure. However, while NMR detection of the biological nuclei proton, carbon, nitrogen, and phosphorus is routine, the study of selenium in biological systems is uncommon and has only recently reemerged as potentially beneficial for these systems. This review discusses the underlying reasons for the sparse utilization of biological selenium NMR, recent advances, and its future in studies of proteins containing Sec.

Selenium's NMR Properties

The nuclear spin number of the only NMR active isotope of selenium, [77]Se, is I=1/2 (*8*), and its resonance frequency, at an external field of 2.3 T where ^{1}H is 100 MHz, is 19.07 MHz. In comparison, at the same field strength, carbon is 25.14 MHz, and nitrogen is 10.13 MHz. It is easily detected with conventional hardware and experiments (*9–11*). As illustrated in Figure 1, it has a wide chemical shift range, spanning over 5800 ppm, an order of magnitude larger than that exhibited by carbon (approximately 300 ppm). [77]Se is, therefore, highly sensitive to changes in the electronic environment, a characteristic that renders it an excellent spectroscopic reporter on bonding, geometry, and electronic structure. [77]Se is routinely employed in studies of materials and of organic and inorganic compounds, but it is not regularly used in studies of biological systems because of several limitations. First, selenium resonances are broad, since as the number of electrons increase, so does the chemical shielding anisotropy and potential relaxation pathways. The short T_2 relaxation time is reflected in solution-state NMR selenoproteins' resonances' linewidths, which range from tens of Hertz to several hundreds of Hertz (see 'Investigation of [[77]Se]-Sec in biological systems'). The short T_2 relaxation time reduces detection sensitivity and limits the use of multidimensional NMR experiments. Second, since [77]Se's natural abundance is 7.63%, samples need to be isotopically enriched. The preparation of selenoproteins at high yield and enriched in [77]Se remains an active field of development (see '[77]Se enrichment of biological samples'). Third, data interpretation is hindered by the limited information about the NMR properties of selenium in biological systems due to the paucity of relevant studies. Certainly, our perception of how the protein environment influences selenium's NMR parameters is just emerging (see 'Interpretation of ^{77}Se NMR data'). In the following sections, we review the progress made in addressing these challenges and in establishing biological selenium NMR as a powerful technique to study native and nonnative selenoproteins.

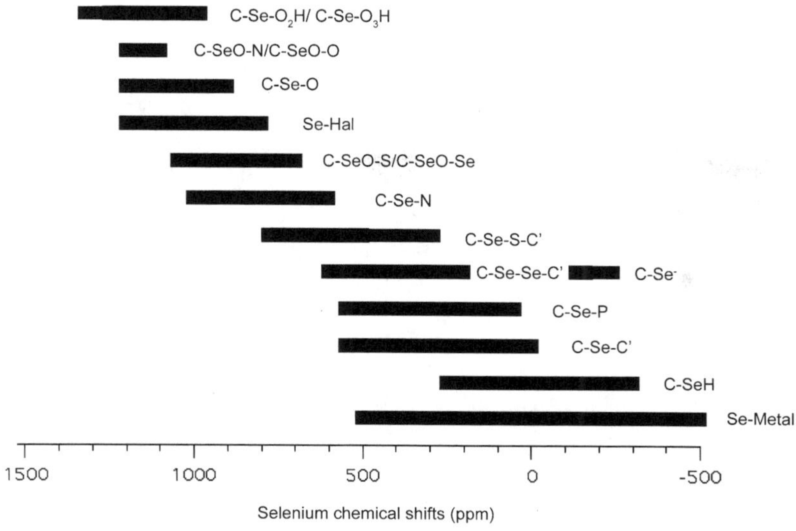

Figure 1. Chemical shifts of biologically relevant selenium chemical groups. The compilation is based on references (9–11).

^{77}Se Enrichment of Biological Samples

NMR experiments require a large amount of protein (between 0.1 to 2 μmoles of isotopically enriched protein). Isotopic enrichment with ^{77}Se is required because its natural abundance is only 7.63% and its broad line widths reduce detection sensitivity. In recent years, several routes for isotopic enrichment of selenoproteins were developed, thereby allowing the production of ^{77}Se-labeled proteins at high yield and low cost. There are two general isotopic enrichment schemes: insertion of a unique Sec in a specific location and insertion of Sec in more than one position, often without specificity (*12*). In some cases, it is desirable to label a specific Sec residue in the protein to either simplify the assignment or to preserve function for detailed mechanistic examination. In other cases, replacement and labeling of all sulfur-containing residues with selenium is beneficial for uses such as recording connectivity of diselenide bonds or simply charting the behavior of several sites of interest simultaneously.

In the first case involving applications that label a specific Sec residue, genetic incorporation of Sec using *E. coli*'s native selenium incorporation machinery is commonly employed. This approch relies on Sec biosynthesis *in vivo* on its own tRNA and the use of an mRNA loop, which designates the UGA codon as a Sec insertion codon (see below). *E. coli*'s selenium incorporation machinery is well understood and is most commonly used for the genetic incorporation of Sec in recombinant proteins (*4, 13*). The key for the successful expression of selenoproteins in *E. coli* is coexpressing the desired recombinant protein, along with the selenium production proteins (*14, 15*). Even then, the typical incorporation ratio of Sec is below 10%, with the majority of the protein being the truncated form, in which the UGA was misrecognized as a termination signal. It is possible to further separate the full-length and truncated forms of the protein if the presence of truncated protein interferes with the experiments. Isotopic enrichment with [77]Se is possible by providing [[77]Se]-sodium selenite in the growth media (*16*). Recently, we have developed a protocol based on defined growth media that has high yield and minimal background (*17*).

Because genetic incorporation allows for the labeling of a unique Sec residue at a specified position, it minimizes structural perturbation to the protein and simplifies interpretation of the NMR spectra. A significant disadvantage of genetic incorporation in bacteria, such as *E. coli*, is that unless the Sec is located at the C-terminus, the requirements for Sec coding would introduce mutations into the native sequence. This is because the UGA codon is immediately followed by a highly conserved mRNA stem-loop structure that mediates the recognition of the selenium incorporation proteins (*13*). When this Sec insertion sequence (SECIS) element is present in the coding segment, the protein of interest must include the amino acids it encodes. Recently, the Soll group overcame this limitation by engineering an artificial genetic incorporation machinery that bypasses the requirement for a coding segment in the mRNA (*18*). An alternative approach is the use of purified Sec insertion proteins in an *in vitro* cell-free wheat germ system (*19*). Since the eukaryotic system is employed, the Sec-encoding mRNA segment can be placed outside of the coding region. These techniques have great promise for future characterization of selenoproteins, because they allow for the biochemical synthesis of any native and nonnative selenoprotein, regardless of the position of the Sec.

A different path for incorporation of Sec(s) in a unique position is chemical synthesis including solid state peptide synthesis (*20*), the use of intein reactions (*21*), and chemical ligation (*22*). All three approaches rely on solid state synthesis of Sec-containing-peptides using protected [77]Se-labeled amino acids. These methods enjoy high selectivity in the insertion site but require expertise and familiarity with both synthesis and reactivity of Sec-containing peptides. Nevertheless, while these methods are not as commonly utilized as the previous approaches, they do offer great flexibility in the insertion site.

When specific incorporation of Sec is not necessary, there are several other methods to enrich the protein with [77]Se. Typically, instead of genetically encoding Sec with a UGA codon, a Cys (TGT/TGC) is employed. Subsequently, all Cys are converted to Sec *in vivo* by saturating the cellular sulfur pathways with selenium. In essence, random incorporation of [[77]Se] into proteins is carried out by first depleting all sulfur sources in *E. coli* cells' growth media and then, once protein expression is induced, supplementing the media with [[77]Se]-selenite (*23*). The ratio of selenium-to-sulfur incorporation into the target protein can be governed by changing the molar ratio of selenite to sulfate in the growth media. To eliminate nonspecific incorporation of selenomethionine, the methionine auxotroph strain B834(DE3) can be used with similar yield as that of proteins produced in conventional *E. coli* strains such as BL21(DE3). By definition, this method cannot be used to label specific positions in the protein; however, it is both simple and cost effective. It is compatible with protocols for additional partial and uniform isotopic enrichment of recombinant selenoproteins with ^{13}C and [15]N.

Similar to this method, it is also possible to uniformly substitute all Cys to Sec in *E. coli* by supplementing the growth media of a cysteine auxotroph strain (*24*) with selenocystine (*25, 26*). The overwhelming abundance of Sec results in misloading cysteyl-tRNA with Sec and its subsequent nonspecific incorporation, instead of Cys. While this method is nontoxic and has been employed for the purpose of crystallographic phasing with negligible yield loss, it does require [[77]Se]-selenocystine. Currently, this compound is not commercially available and must be prepared by organic or enzymatic synthesis (*27*). Also, for the method to become cost effective, the optimal Cys/Sec concentrations in the media require further investigation, due to the cost of isotopes.

Overall, the past few years have presented several new paths, not only for cheap and simple enrichment of samples with [77]Se, but also for improved chemical biology procedures to specifically incorporate Sec in proteins. Hence, previous limitations on sample preparation and enrichment are gradually being resolved. In addition, several exciting new applications promise to advance the ease and flexibility of specific incorporation of Sec, without the need for expert knowledge. Among these developments is a newly engineered *E. coli* cell, where release factor 1, which catalyzes translational termination at UAA and UAG codons, was deleted (*28*). Another frontier is the continuing improvement of the modified Sec tRNA engineered by the Soll group to optimize genetic incorporation in *E. coli* (*18*). A recent report described higher efficiency of Sec incorporation using engineered rRNA (*29*). Collectively, these advances show promise in ameliorating the limitations on the preparation of native and non-native selenoproteins in the near future. With the limitation on sample preparation and isotopic labeling well on the way to resolution, we turn to the description of the development in ^{77}Se NMR studies.

Investigation of [[77]Se]-Sec in Biological Systems

In this review, our focus is studies of the 21st amino acid Sec in proteins. The NMR properties of the amino acid in isolation, (e.g. the isotropic chemical shifts, coupling constants, and relaxation times) have been recorded by several groups (*30–32*). Due to solubility, the dimeric form, selenocystine, is better studied. The [1]H-[77]Se J couplings reported for selenocystine range from 23.6 to -0.97 Hz (*32*). The [77]Se chemical shift tensor of Sec was not reported, but measurements from our group show that L-selenocystine spans Ω=601.5 ppm (unpublished results). The chemical shift tensor of Sec can easily span 1000 ppm (*33*), which will be manifested in broad resonance lines and lower sensitivity.

The first report of ^{77}Se NMR spectroscopy conducted on a biological system, as opposed to model compounds, was published by Odom and Dunlap's research groups (*34*). In this study, the cysteines of two abundant proteins - ribonuclease-A (RNase A) and lysozyme - were reacted with [77]Se-labeled selenium compounds and the modified proteins were then examined. The authors strove to demonstrate how sensitive ^{77}Se NMR spectroscopy would be to the local protein environment. Unfortunately, due to low sensitivity, the proteins were studied under denaturing conditions. Even though the proteins were unfolded, the selenium chemical shifts still spanned a range of 15 ppm, which indeed demonstrates selenium's sensitivity to the local environment. Subsequently, Gettins and coworkers reported the spectra of a non-denatured protein labeled again with selenium compounds (*33*). They recorded the ^{77}Se NMR chemical shifts for folded and partially unfolded hemoglobin, as well as denatured RNase A. The relaxation properties of the denatured proteins were used to estimate that the chemical shift anisotropy is close to 890 ppm. This landmark paper served to establish the linewidths of resonances in folded and unfolded proteins, the ratio of T_1 and T_2 relaxation times, and the leading role of the chemical shift anisotropy in the relaxation mechanism.

These initial studies naturally led to investigations of selenoproteins, which represented a shift from studies demonstrating feasibility to those focusing on the special chemistry catalyzed by Sec. Gettins attempted to record the NMR properties in studies of the reaction mechanism of the highly efficient selenoenzyme glutathione peroxidase 1 (GPx1), which is responsible for scavenging free oxygen radicals *in vivo* (*35*). This unique endeavor involved feeding lambs for five months with a [[77]Se]-sodium selenite enriched diet and further reacting the purified protein with [77]Se-labeled selenocompounds to increase the likelihood of NMR detection. However, since resonances in the folded proteins were not detected, the protein had to be denatured. A similar ambition motivated Odom, Dunlap and Hilvert's report on different chemical forms of Sec in the serine protease subtilisin (*36*). Subtilisin was converted into a selenosubtilisin, gaining a peroxidase function by reacting the active site serine with [[77]Se]-hydrogen selenide. This work reported the chemical shifts of selenolate (E-Se⁻), seleninic acid (E-SeO$_2$H), and even of a selenylsulfide (E-Se-S-R) bond in the context of a protein (Table 1).

Dramatic improvements in the last decade in hardware and in the strengths of magnetic fields resulted in increased sensitivity that allowed further growth in the scope of the investigations. Recently, the assignments of diselenide bonds in a 37-residue peptide containing four diselenide bonds were described (*20*). The study combined 2D [1]H–[77]Se Heteronuclear Multiple Bond Correlation (HMBC), [1]H–[77]Se Heteronuclear Multiple Quantum Correlation (HMQC), and COrrelation SpectroscopY (COSY) to preform the assignments (*37*). The use of multidimensional NMR was unprecedented, but as will be detailed in the section about multidimensional NMR experiments, is not always applicable to larger selenoproteins. In this case, the peptide's small molecular weight and the relative rigidity of diselenide bonds led to narrow resonance widths that allowed for ms-long mixing times. A follow-up study measured the pKa of two Sec residues in a nine-residue hormone peptide (*38*). The pKas of the two residues were 3.3 and 4.3, respectively. The peptides for both those studies were prepared using solid-state peptide synthesis.

While this groundbreaking body of work established not only the feasibility of biological selenium NMR, but also the wide span of chemical shifts and how they can be used to study protein function, it left several questions unanswered: What is the feasibility of detecting Sec in a mid-size protein (>30 KDa)? Can multiple sites of Sec be resolved in a given protein? Can Sec be detected in mobile sites, such as is often the case for selenoproteins? Can multidimensional NMR experiments be used to study such sites? What are the advantages and disadvantages of high magnetic fields? Recently, our group set to address those open questions by studying a variety of mid-sized native and non-native, folded and active selenoproteins.

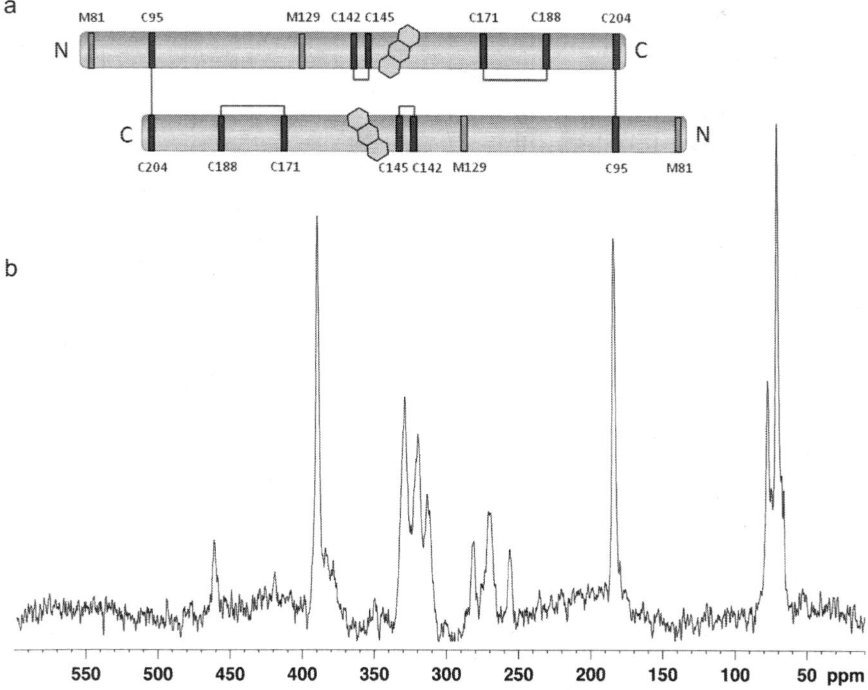

Figure 2. a) Depiction of the form of human ALR used in these studies. Each sub-unit of the covalent homodimer has 2 Met and 6 Cys (green and blue bars, respectively). The catalytic cysteines C142-C145 interact with the flavin cofactor, shown in yellow. b) Proton-decoupled ^{77}Se spectra of ALR with 80% selenium enrichment acquired at a high magnetic field of 19.97 T (^{1}H frequency of 850 MHz). The region from 150 to 500 ppm contains the Sec resonances. At least ten individual peaks are observed, corresponding to both diselenide and selenylsulfide bonds (unpublished data).

Our first focus was a Cys-rich sulfhydryl oxidase that employs a CXXC motif, in conjunction with a flavin cofactor, to catalyze the formation of disulfide bonds (*39*). Augmenter of liver regeneration (ALR) is a 32 kDa covalent homodimer that contains 6 Cys (and 2 Met) per subunit. All the sulfur sites in Cys and Met were enriched with selenium, using a defined media with a prefixed ratio of Se/S (*23*). The structure, stability, and activity of the seleno-ALR were extensively characterized to determine the ramifications of the substitution of sulfur for selenium. The seleno-ALR was found to be nearly identical structurally to wild type ALR by high resolution X-ray crystallography. Figure 2 shows the NMR spectrum at a high magnetic field of 19.97 T (^{1}H frequency of 850 MHz). Ten individual resonances attributed to Sec were resolved, corresponding to both diselenide (Se-Se) and selenylsulfide (Se-S) bonds. Since the protein was enriched with ^{77}Se, an overnight acquisition was sufficient to record the spectrum presented in Figure 2.

These encouraging results led our group to test whether ^{77}Se NMR would be a sensitive reporter of the oxidation state of the active site's flavin cofactor FAD. The FAD is reduced with sodium dithionite, a reagent that is capable of directly reducing the flavin but does not interact with disulfide, selenylsulfide or diselenide groups (*40*). Figure 3 displays the spectra of ALR before, during, and following reduction of the FAD. Several resonances change position and intensity and two are altogether absent upon reduction of FAD. A new resonance appears downfield close to 650 ppm. The changes in resonance positions are in large part reversed once the reducing agent is removed. Hence, the Sec is acting as a sensitive probe of the reaction progress.

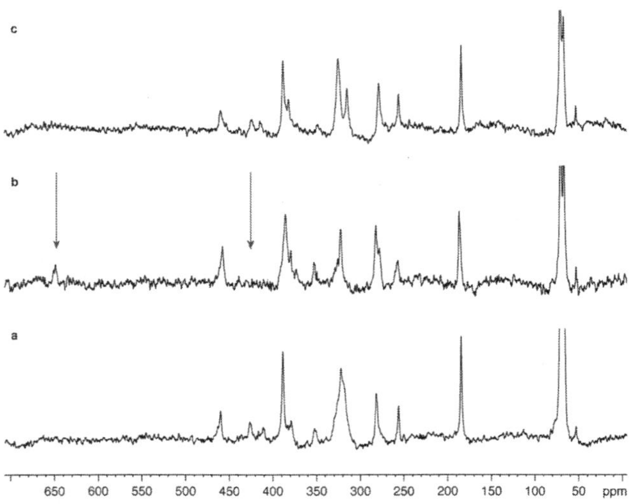

Figure 3. ^{77}Se NMR spectroscopic characterization of flavin reduction in ^{77}Se-labeled ALR. Proton-decoupled ^{77}Se spectra of ALR with 50% selenium enrichment acquired at 14.1 T (^{1}H frequency of 600 MHz). (a) Spectrum of ^{77}Se-labeled ALR with oxidized FAD (selenomethionine resonances in the 0-150 ppm region are truncated for clarity). (b) Spectrum of ^{77}Se-labeled ALR with reduced FAD under anaerobic conditions. Upon addition of a reducing agent, the resonance peaks at 412 and 426 disappear, while a new resonance peak at 651 ppm appears (highlighted with an arrow). (c) Same sample shown in (b), following exposure to air and removal of the reducing agent. Reproduced with permission from reference (23). Copyright 2013 Elsevier.

Table 1. Selenium-containing proteins that have been studied by ^{77}Se NMR

Protein	Labeling method	Chemical shifts (ppm)	Reference
Ribonuclease-A (dnatured)	Chemical labeling	S-Se: 580.2, 578.3, 573.9, 570.4, 566.2	(33, 34)
Lysozyme (denatured)	Chemical labeling	S-Se: 580.2, 578.3, 573.9, 570.4, 566.2	(34)
Hemgloblin	Chemical labeling	S-Se: 574	(33)
Glutathione peroxidase (denatured)	Chemical labeling	S-Se: 377, 576, 586 Se-: -212 Se-CH$_2$CONH$_2$: 195	(35)
Subtilisin	Semisynthesis	S-Se: 389 Se-: -215 SeO$_2$H: 1188, 1190	(36)
Spider toxin (k-ACTX-Hv1c)	Peptide synthesis	Se-Se: ~ 320, 334	(20)
Argenine vasopressin	Peptide synthesis	Se-Se: 250, 359 Se-: -220 SeH: ~ -60	(38)
Augmenter of liver regeneration	Random incorporation of Sec by heterologous expression	Se-Se and S-Se: 184.4, 256.0, 281.7, 322.0, 363.0, 380.0, 388.0, 411.2, 426.7, 460.1	(23)
Thioredoxin with various Sec-containing redox motifs	Genetic incorporation of Sec by heterologous expression	S-Se: 279.4, 290.0, 432.4, 433.7	(17)

While this study provides evidence that multiple selenium sites can be recorded and resolved in mid-sized proteins, even at high field strengths where T_2 relaxation may become pronounced, the assignment of individual peaks is not possible unless site-specific mutants are prepared and examined. Ultimately, this potential barrier to information about specific sites could impede the use of ^{77}Se NMR spectroscopy. The NMR data needs to become more readily interpretable. In order to address this challenge, a subsequent study in our group examined a library of selenoproteins in which the local environment of a single Sec residue was systematically altered (17). This new study focuses on Sec-containing redox motifs, in which a nearby Cys forms a selenenylsulfide bond with the Sec. The selenenylsulfide bond generates a ring, whose steric strain is determined by the number and identity of spacing residues and, to a lesser extent, flanking residues. The steric strain in the oxidized ring is directly related to the tendency of the ring to be reduced. The redox motifs SCUS, GCUG, GCAUG, and GCGAUG were positioned at the C-terminus of a Cys-less thioredoxin used as a scaffold protein. This position is used by a variety of selenoproteins to readily expose the active site's Sec to approaching substrates (*41*). In our study, [^{77}Se]-Sec was inserted using the *E. coli* genetically encoded selenium insertion machinery, in order to specifically incorporate the Sec at only one location. Preliminary results suggest that the resonances fall in the 280-460 ppm range, larger than that previously reported for the selenenylsulfide bond. The resonances' line widths span 100-500 Hz at a field strength of 14.1 T. Another notable trend is the shift of the Sec resonance downfield, in conjunction with the increase in ring strain. It remains to be seen whether the trend can be correlated to the motif's redox potential. Finally, the spectra of the reduced proteins show no obvious relation to the immediate amino acid sequence. They may reflect the stabilizing effects of nearby hydrogen bonds' donors, such as the Cys, charged residues, or weak interactions with the peptide backbone.

Quantum Calculations in Aiding Interpretation of ^{77}Se NMR Data

Despite the renaissance in biological selenium NMR, the number of reported spectra and identified resonances is still too sparse to form a comprehensive view of how the protein environment governs the NMR parameters. This is in striking contrast to the case of the commonly employed nuclei carbon and nitrogen, with their large data depository (http://www.bmrb.wisc.edu). For those nuclei, ab initio calculations are frequently used to predict their local environments. To gain better insight as to how the selenium NMR parameters relate to the protein environment, it is essential to expand the available chemical shift data, in combination with ab initio calculations of the magnetic shielding tensor in selenoproteins.

Prediction of a protein's spectroscopic properties is rapidly advancing, in direct proportion to the availability of increased computational power (*42*). It is now practical to employ ab initio phasing algorithms in solving protein structures by X-ray crystallography and in calculating chemical shift tensors in structural determination (*43, 44*). Calculations of the magnetic shielding tensors are increasingly improving in accuracy (*45*), and the magnetic shielding tensors of relatively heavy atoms, such as the transition metal ^{57}Fe, were calculated for proteins (*46, 47*). Also, both the quadrupolar coupling tensors and magnetic shielding tensors of ^{51}V, ^{59}Co, ^{65}Cu, and ^{67}Zn in a number of proteins have been determined (*48–51*). Regarding selenium, the magnetic shielding tensors were calculated for small organoselenium compounds using density functional theory, yielding close agreement (5-15%) with experiments (*52–54*). Calculations of isotropic chemical shifts for selenoproteins and selenol amino acids, as well as magnetic shielding tensors of bioactive selenium compounds (*55*), were also found to be in relatively good agreement with experimental data (*56*), although presently there are not many examples of terminal anionic selenium compounds (*57*). In summary, there is no conceptual or hardware barrier to extending studies of small compounds to those of selenoproteins.

Prospects for Utilizing Multidimensional NMR for Investigations of Selenoproteins

Another extension of ^{77}Se NMR is the use of multidimensional experiments, which offer several distinct advantages. In addition to theoretically increased sensitivity and resolution, they may also be used to probe specific sites in large macromolecules and protein complexes that are not easily amenable for structural determination. Dipole-dipole interactions between selenium and its neighboring atoms (such as carbon, nitrogen, etc.) can be used to unmask the surrounding local environment (such as enzymes' active sites or other regions of interest in macromolecules with convoluted spectrum) or measure site-specific dynamics. The magnitude of selenium dipole-dipole interactions is about 8.5% lower than that of carbon. It allows, assuming a weak dipolar coupling limit of 30 Hz, measurements of internuclear distances of up to 6.8 Å for selenium-phosphorus, 5.8 Å for selenium-carbon, 5.3 Å for selenium-selenium, and 4.3 Å for selenium-nitrogen. Furthermore, ^{77}Se introduces a large chemical shift in directly bonded atoms, allowing their immediate identification. For example, in selenomethionine-enriched calmodulin, the span of ^{77}Se chemical shifts is approximately 70 ppm, and ^{13}C chemical shifts of the selenomethionine's methyl are upfield shifted by as much as 11 ppm compared to those of methionine (*58*). The large ^{77}Se-^{77}Se scalar coupling can be exploited, possibly even in larger proteins, to study connectivities of diselenide bonds (*37*).

Currently, there are no reports of multidimensional experiments carried out for proteins above 6 kDa. What are the prospects and possibilities for employing multidimensional experiments for selenoproteins? It appears that the option to conduct multidimensional selenium NMR of biological systems depends on the specifics of the system under investigation. Typically, the fast T_2 relaxation rate is a considerable impediment for executing multi-dimensional experiments. We were only able to record ^{1}H-^{77}Se-Heteronuclear Multiple Bond Coherence (HMBC) for the relatively rigid eight-membered rings SCUS and GCUG (Figure 4) but not of the more flexible eleven and fourteen-membered rings GCAUG and GCGAUG. However, these redox motifs are located at the flexible C-terminus of thioredoxin. It is likely that the NMR properties of Sec positioned in rigid sites could resemble those of the eight-membered rings. Hence, it remains to be seen whether particular selenium sites in proteins are more amenable for multidimensional experiments.

Another consideration for ^{77}Se NMR experiments is the optimal magnetic field strength one should use. In general, T_2 relaxation rates scale with the square of the magnetic field strength as shown by Gettins for two proteins (*33*). For the subset of ^{77}Se containing proteins studied in our lab, the line widths range from 50 Hz to as much as 600 Hz at 14.1 T (the later reflecting protein dynamics, unpublished data). However, there has been no systematic study of the relation between the field strength, the relaxation times and the sensitivity. Clearly, this aspect of routine biological selenium NMR requires additional characterization and study.

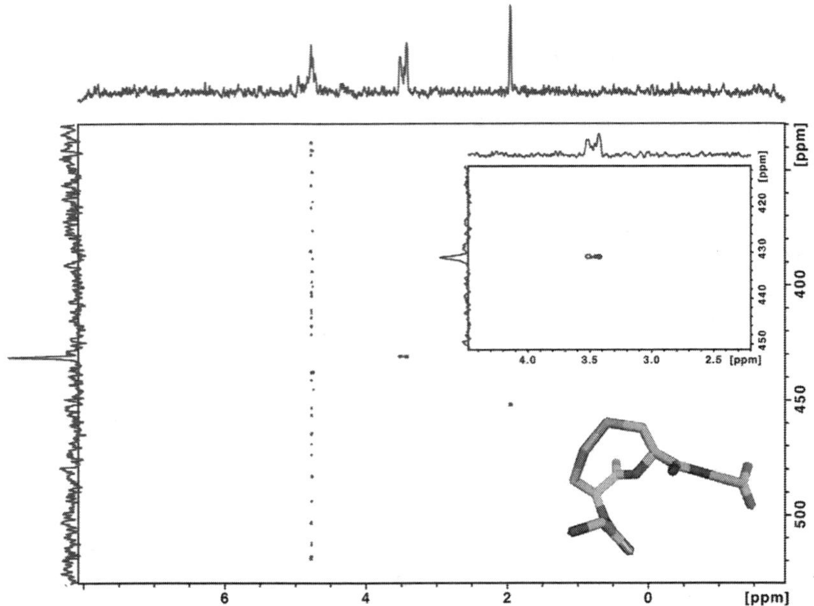

Figure 4. (^{1}H-^{77}Se)-Heteronuclear Multiple Bond Coherence (HMBC) spectra of ^{77}Se enriched Thioredoxin-Gly-Cys-Sec-Gly (12 kDa). The spectrum was acquired at 14.1 Tesla using 0.4 mM ^{77}Se-Trx-Gly-Cys-Sec-Gly in 20 mM phosphate buffer, pH 7.5; the acquisition time was one day. The evolution period was optimized for a 25 Hz long-range scalar coupling. Cross peaks are observed only for the 432 ppm conformation (the peak at 450 ppm does not originate from Sec). The insets contain an expansion of the 432 ppm region and an illustration of the selenenylsulfide bond formed between vicinal Cys and Sec (unpublished data).

Prospects for Utilizing Solid-State NMR for Investigations of Selenoproteins

Solid-state NMR spectroscopy is a powerful technique for studies of large proteins, protein dynamics, and temperature-sensitive samples such as enzyme-ligand and enzyme intermediates (*59, 60*). It is an important experimental technique that can probe conformational dynamics on all time scales relevant for enzymatic function (10^{-9} to 10^2 seconds) (*61*). Theoretically, selenium should be an excellent probe for molecular motion, due to its high sensitivity to the magnetic field and local environment. Its large chemical shift tensor can be used to detect motions that occur on the microseconds time scale. Dynamical information from such a local reporter present in enzyme active sites will enhance our understanding of conformational dynamics in promoting chemical reactivity. Our group has been able to record the solid-state spectra of proteins containing selenomethionine and is now attempting to record that of proteins containing Sec. The success of these studies will depend on T_2 values of specific Sec residues, as mentioned above in the section on multidimensional NMR.

138

Future Perspectives - Selenium as a Surrogate for Sulfur

Studies of selenium in biological systems are not limited to macromolecules in which selenium naturally occurs. Selenium may well be the closest spectroscopic surrogate to sulfur. While selenium is a trace element in humans, sulfur is the eighth most abundant element in the human body and the fifth most abundant in proteins. Sulfur-containing amino acids have distinct cellular specialization: Cys is among the most reactive amino acids and hence a regular participant in enzymes' active sites. Cys also governs a protein's three-dimensional architecture by forming disulfide bonds in selected proteins. Cys and Met are the second and third most frequent coordinators of Cu, Fe, and Ni (Cys is an excellent metal ligand in general). Functional analysis of proteins' functions will therefore greatly benefit from a direct method to measure sulfur's properties, such as bonding, pKa, and electronic structure. It is not surprising that the low sensitivity of the sulfur's only NMR active nuclei (^{33}S, spin I=3/2, 0.76% natural abundance) has long been perceived as a major limitation (*62*). Because ^{33}S, is a quadrupolar nucleus with a low frequency, it suffers from low sensitivity and broad lines. Despite technical advances such as high magnetic field strengths, its detection in biological systems remains a distant goal (*63*). Since sulfur cannot be used to study macromolecules, selenium could act as a spectroscopic surrogate for sulfur. Selenium is located directly below sulfur in the periodic table and shares many physicochemical properties with it (*64*). The substitution of sulfur by selenium typically does not interfere with protein function. Such substitutions can for example be exploited to investigate enzyme active sites, conformational dynamics and in favorable cases, the nature of the reaction intermediate.

Like every technique, ^{77}Se NMR is particularly well suited to answer certain questions and maybe less so for others. Specifically, it could be very useful to study site-specific dynamics because ^{77}Se NMR is sensitive to molecular motions at the low millisecond to high microseconds time scales. Furthermore, it is well suited to monitor oxidative damage and post-translational modifications. On the other hand, as discussed earlier, not every protein may be suitable for ^{77}Se NMR studies. Hence, ^{77}Se NMR is not expected to replace well-established characterization techniques, but rather to complement them. However, considering the eminent role of sulfur in cellular functions, this is a particularly exciting avenue with a potentially wide impact on biological NMR spectroscopy.

Summary

The considerable sensitivity of ^{77}Se NMR to its chemical environment holds great promise to make it one of the most powerful biophysical methods to characterize native and non-native selenoproteins. Recent advances in the preparation and isotopic enrichment of selenoproteins have significantly removed restrictions on and bottlenecks in sample preparation. Data interpretation is rapidly maturing, via the recent surge of publications on selenopeptides and selenoproteins and by developments in DFT calculations of selenium's magnetic shielding tensor. Applications of selenium NMR spectroscopy to biological systems are showing great promise to expand the NMR toolbox to include this diverse nucleus as a probe of both sulfur and selenium sites.

Acknowledgments

The author thanks Fei Li for assistance with figure preparation and acknowledges support from the National Institute of General Medical Sciences Grant P30 GM103519, and from a National Science Foundation MRI-R2 Grant 0959496. This material is based upon work supported by the National Science Foundation under Grant No. MCB-1054447 "CAREER: Reactivity of Selenoproteins".

References

1. Arner, E. S. J. *Exp. Cell Res.* **2010**, *316*, 1296–1303.
2. Lu, J.; Holmgren, A. *J. Biol. Chem.* **2009**, *284*, 723–727.
3. Kasaikina, M. V.; Hatfield, D. L.; Gladyshev, V. N. *Biochem. Biophys. Acta* **2012**, *1823*, 1633–1642.
4. Yoshizawa, S.; Bock, A. *Biochim. Biophys. Acta* **2009**, *1790*, 1404–1414.
5. Donovan, J.; Copeland, P. R. *Antioxid. Redox Signaling* **2010**, *12*, 881–892.
6. Allmang, C.; Wurth, L.; Krol, A. *Biochim. Biophys. Acta* **2009**, *1790*, 1415–1423.
7. Hondal, R. J.; Marino, S. M.; Gladyshev, V. N. *Antioxid. Redox Signaling* **2013**, *18*, 1675–1689.
8. Davis, S. P.; Jenkins, F. A. *Phys. Rev.* **1951**, *83*, 1269–1269.
9. Duddeck, H. *Prog. Nucl. Magn. Reson. Spectrosc.* **1995**, *27*, 1–323.
10. Duddeck, H. *Annu. Rev. NMR Spectrosc.* **2004**, *52*, 105–166.
11. Demko, B. A.; Wasylishen, R. E. *Prog. Nucl. Magn. Reson. Spectrosc.* **2009**, *54*, 208–238.
12. Johansson, L.; Gafvelin, G.; Arner, E. S. J. *Biochim. Biophys. Acta* **2005**, *1726*, 1–13.
13. Böck, A.; Rother, M.; Leibundgut, M.; Ban, N. In *Selenium: Its molecular biology and role in human health*; Hatfield, D. L., Berry, M. J., Gladyshev, V. N., Eds.; Springer: Boston, 2006; p 9–28.
14. Arner, E. S. J.; Sarioglu, H.; Lottspeich, F.; Holmgren, A.; Bock, A. *J. Mol. Biol.* **1999**, *292*, 1003–1016.
15. Rengby, O.; Arner, E. S. J. *Appl. Environ. Microbiol.* **2007**, *73*, 432–441.
16. Gladyshev, V. N.; Khangulov, S. V.; Stadtman, T. C. *Proc. Natl. Acad. Sci. U.S.A.* **1994**, *91*, 232–236.
17. Li, F.; Lutz, P. B.; Pepelyayeva, Y.; Arnér, E. S. J.; Bayse, C. A.; Rozovsky, S. unpublished work.
18. Aldag, C.; Brocker, M. J.; Hohn, M. J.; Prat, L.; Hammond, G.; Plummer, A.; Soll, D. *Angew. Chem.* **2013**, *52*, 1441–1445.
19. Gupta, N.; Demong, L. W.; Banda, S.; Copeland, P. R. *J. Mol. Biol.* **2013**, *425*, 2415–2422.

20. Mobli, M.; de Araujo, A. D.; Lambert, L. K.; Pierens, G. K.; Windley, M. J.; Nicholson, G. M.; Alewood, P. F.; King, G. E. *Angew. Chem.* **2009**, *48*, 9312–9314.

21. Eckenroth, B.; Harris, K.; Turanov, A. A.; Gladyshev, V. N.; Raines, R. T.; Hondal, R. J. *Biochemistry* **2006**, *45*, 5158–5170.

22. Metanis, N.; Keinan, E.; Dawson, P. E. *J. Am. Chem. Soc.* **2006**, *128*, 16684–16691.

23. Schaefer, S. A.; Dong, M.; Rubenstein, R. P.; Wilkie, W. A.; Bahnson, B. J.; Thorpe, C.; Rozovsky, S. *J. Mol. Biol.* **2013**, *425*, 222–231.

24. Muller, S.; Senn, H.; Gsell, B.; Vetter, W.; Baron, C.; Bock, A. *Biochemistry* **1994**, *33*, 3404–3412.

25. Strub, M. P.; Hoh, F.; Sanchez, J. F.; Strub, J. M.; Bock, A.; Aumelas, A.; Dumas, C. *Structure* **2003**, *11*, 1359–1367.

26. Salgado, P. S.; Taylor, J. D.; Cota, E.; Matthews, S. J. *Acta Crystallogr., Sect. D* **2011**, *67*, 8–13.

27. Wessjohann, L. A.; Schneider, A. *Chem. Biodiversity* **2008**, *5*, 375–388.

28. Heinemann, I. U.; Rovner, A. J.; Aerni, H. R.; Rogulina, S.; Cheng, L.; Olds, W.; Fischer, J. T.; Soll, D.; Isaacs, F. J.; Rinehart, J. *FEBS Lett.* **2012**, *586*, 3716–3722.

29. Thyer, R.; Filipovska, A.; Rackham, O. *J. Am. Chem. Soc.* **2013**, *135*, 2–5.

30. Pan, W. H.; Fackler, J. P. *J. Am. Chem. Soc.* **1978**, *100*, 5783–5789.

31. Tan, K. S.; Arnold, A. P.; Rabenstein, D. L. *Can. J. Chem.* **1988**, *66*, 54–60.

32. Salzmann, M.; Stocking, E. M.; Silks, L. A.; Senn, H. *Magn. Reson. Chem.* **1999**, *37*, 672–675.

33. Gettins, P.; Wardlaw, S. A. *J. Biol. Chem.* **1991**, *266*, 3422–3426.

34. Luthra, N. P.; Costello, R. C.; Odom, J. D.; Dunlap, R. B. *J. Biol. Chem.* **1982**, *257*, 1142–1144.

35. Gettins, P.; Crews, B. C. *J. Biol. Chem.* **1991**, *266*, 4804–4809.

36. House, K. L.; Dunlap, R. B.; Odom, J. D.; Wu, Z. P.; Hilvert, D. *J. Am. Chem. Soc.* **1992**, *114*, 8573–8579.

37. Mobli, M.; King, G. F. *Toxicon* **2010**, *56*, 849–854.

38. Mobli, M.; Morgenstern, D.; King, G. F.; Alewood, P. F.; Muttenthaler, M. *Angew. Chem.* **2011**, *50*, 11952–11955.

39. Farrell, S. R.; Thorpe, C. *Biochemistry* **2005**, *44*, 1532–1541.

40. Fox, J. L. *FEBS Lett.* **1974**, *39*, 53–55.

41. Kryukov, G. V.; Castellano, S.; Novoselov, S. V.; Lobanov, A. V.; Zehtab, O.; Guigo, R.; Gladyshev, V. N. *Science* **2003**, *300*, 1439–1443.

42. Oldfield, E. *Philos. Trans. R. Soc. London, Ser. B* **2005**, *360*, 1347–1361.

43. Mulder, F. A. A.; Filatov, M. *Chem. Soc. Rev.* **2010**, *39*, 578–590.

44. Wishart, D. S. *Prog. Nucl. Magn. Reson. Spectrosc.* **2011**, *58*, 62–87.

45. Casabianca, L. B.; De Dios, A. C. *J. Chem. Phys.* **2008**, *128*, 052201–052211.

46. Godbout, N.; Havlin, R.; Salzmann, R.; Debrunner, P. G.; Oldfield, E. *J. Phys. Chem. A* **1998**, *102*, 2342–2350.

47. Godbout, N.; Sanders, L. K.; Salzmann, R.; Havlin, R. H.; Wojdelski, M.; Oldfield, E. *J. Am. Chem. Soc.* **1999**, *121*, 3829–3844.

48. McDermott, A.; Polenova, T. *Curr. Opin. Struct. Biol.* **2007**, *17*, 617–622.

49. Lipton, A. S.; Heck, R. W.; de Jong, W. A.; Gao, A. R.; Wu, X. J.; Roehrich, A.; Harbison, G. S.; Ellis, P. D. *J. Am. Chem. Soc.* **2009**, *131*, 13992–13999.

50. Lipton, A. S.; Heck, R. W.; Staeheli, G. R.; Valiev, M.; De Jong, W. A.; Ellis, P. D. *J. Am. Chem. Soc.* **2008**, *130*, 6224–6230.

51. Pooransingh-Margolis, N.; Renirie, R.; Hasan, Z.; Wever, R.; Vega, A. J.; Polenova, T. *J. Am. Chem. Soc.* **2006**, *128*, 5190–5208.

52. Demko, B. A.; Eichele, K.; Wasylishen, R. E. *J. Phys. Chem. A* **2006**, *110*, 13537–13550.

53. Sutrisno, A.; Lo, A. Y. H.; Tang, J. A.; Dutton, J. L.; Farrar, G. J.; Ragogna, P. J.; Zheng, S. H.; Autschbach, J.; Schurko, R. W. *Can. J. Chem.* **2009**, *87*, 1546–1564.

54. Griffin, J. M.; Knight, F. R.; Hua, G. X.; Ferrara, J. S.; Hogan, S. W. L.; Woollins, J. D.; Ashbrook, S. E. *J. Phys. Chem. C* **2011**, *115*, 10859–10872.

55. Bayse, C. A.; Antony, S. *Main Group Chem.* **2007**, *6*, 185–200.

56. Bayse, C. A. *Inorg. Chem.* **2004**, *43*, 1208–1210.

57. Bayse, C. A. *J. Chem. Theory Comput.* **2005**, *1*, 1119–1127.

58. Zhang, M. J.; Vogel, H. J. *J. Mol. Biol.* **1994**, *239*, 545–554.

59. Schmidt-Rohr, K.; Spiess, H. W. *Multidimensional solid-state NMR and polymers*; Academic Press: San Diego, 1994.

60. Duer, M. J. *Introduction to solid-state NMR spectroscopy*; Wiley-Blackwell: Oxford, 2005.

61. Palmer, A. G.; Williams, J.; McDermott, A. *J. Phys. Chem.* **1996**, *100*, 13293–13310.

62. Musio, R. *Annu. Rev. NMR Spectrosc.* **2009**, *68*, 1–88.

63. O'Dell, L. A.; Moudrakovski, I. L. *J. Magn. Reson.* **2010**, *207*, 345–347.

64. Wessjohann, L. A.; Schneider, A.; Abbas, M.; Brandt, W. *Biol. Chem.* **2007**, *388*, 997–1006.

Chapter 7

Investigations of New Types of Glutathione Peroxidase Mimetics

Thomas G. Back*

**Department of Chemistry, University of Calgary,
Calgary, Alberta, Canada, T2N 1N4
*E-mail: tgback@ucalgary.ca.**

Cyclic seleninate esters, spirodioxyselenuranes and conformationally constrained diselenides are three classes of small-molecule organoselenium compounds that function as effective mimetics of the selenoenzyme glutathione peroxide. Their catalytic activity in the reduction of peroxides with thiols is described, along with a comparison of their associated redox mechanisms and the role of substituent and other structural effects upon their antioxidant properties.

Introduction

Respiration and aerobic metabolism are accompanied by the production of reactive oxygen species (ROS) such as hydrogen peroxide, lipid hydroperoxides, superoxide radical anion and hydroxyl radical. These byproducts contribute to oxidative stress, which is in turn implicated in inflammation, mutagenesis, cardiovascular and neurological damage and possibly the aging process. Higher organisms are protected from these deleterious effects by both exogenous antioxidants present in various dietary sources and by endogenous ones that catalytically destroy ROS to minimize their harmful effects. Endogenous antioxidants include the glutathione peroxidase (GPx) family (*1–10*) of selenoenzymes, which catalyze the reduction of hydrogen peroxide and lipid hydroperoxides to water and alcohols with the tripeptide thiol glutathione. This process proceeds via the catalytic cycle shown in Scheme 1 (*5*), where the selenol moiety of a selenocysteine residue (EnzSeH) serves to reduce a molecule of the peroxide substrate while generating the corresponding selenenic acid (EnzSeOH). The latter is in turn converted to the corresponding selenenyl sulfide (EnzSeSG) with glutathione (GSH), followed by reaction with a second equivalent of the thiol, thus affording the corresponding disulfide (GSSG) and restoring the original selenol. It is therefore the redox properties of the selenium atom that bestow this protective function upon GPx and enable it to detoxify harmful peroxides that would otherwise accumulate. However, selenols and their conjugate bases (selenolates) can also assume the opposite function by reducing dioxygen to ROS, which under certain circumstances can play a beneficial role in human physiology by destroying pathogens and cancer cells (*11–13*). Thus, while GPx mimetics serve as antioxidants with respect to their peroxide-destroying ability, they are better classified as redox modulators when they also contribute to the generation of ROS. Clearly, optimum health requires the maintenance of a delicate redox balance in which selenium plays a critical role. In some situations, exceptionally high levels of oxidative stress overwhelm the protective effects of GPx, resulting in extensive damage to surrounding cells and tissues. A particularly noteworthy example is the cardiovascular and neurological damage that accompanies the ischemic reperfusion of heart attack and stroke victims, respectively (*14–19*). In such cases, the removal of an arterial obstruction with clot-busting drugs or by surgical intervention causes the release of substantial amounts of ROS from neutrophils that are part of the immune system as they invade the previously hypoxic tissue downstream from the site of the original obstruction (*20*). In such cases, it might be beneficial to augment temporarily the antioxidant activity of GPx by the administration of small-molecule selenium compounds designed to mimic the catalytic properties of the selenoenzyme prior to ischemic reperfusion.

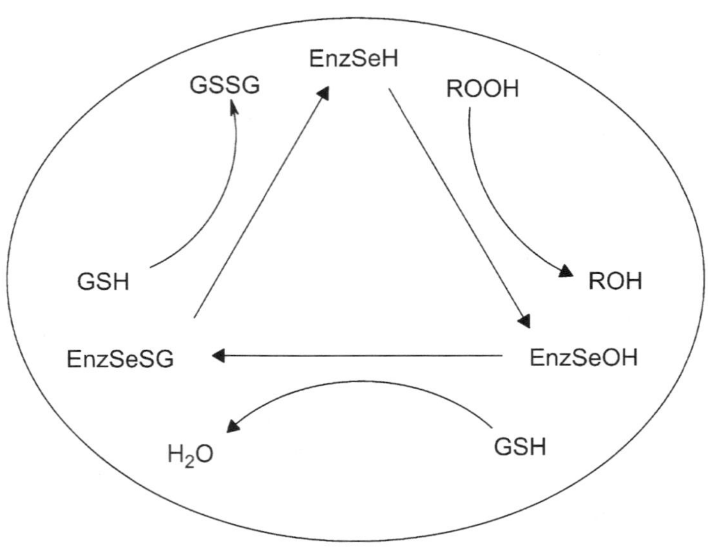

Scheme 1. Catalytic cycle of glutathione peroxidase

Numerous attempts to design such GPx mimetics have been reported and several reviews of this subject have appeared (*21–25*). Figure 1 shows a few representative examples of compounds that have been investigated in this context (*26–35*). Many of them contain a nitrogen atom either covalently bonded to selenium or in a pendant amino group capable of coordinating with selenium at various stages of the catalytic cycle, which is often beneficial to catalytic activity. Several tellurium compounds have also been investigated in this context. Two selenium compounds, ALT 2074 (formerly known as BXT 51072; Synvista Inc.) (*20, 36, 37*) and ebselen (*38–49*) (Daiichi-Sankyo Inc.), have undergone clinical trials for various medicinal applications of their antioxidant properties.

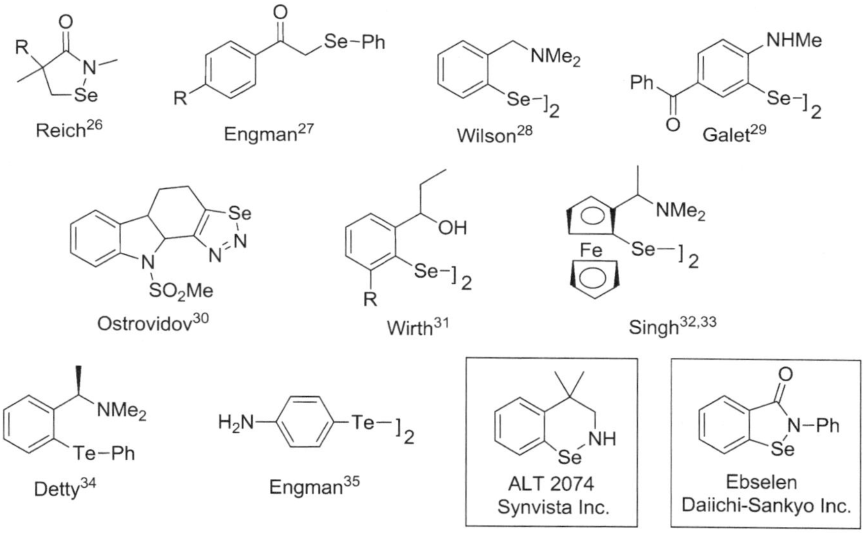

Figure 1. Representative examples of earlier types of GPx mimetics.

Our entry into this field was somewhat serendipitous. In the 1990s we were engaged in the development of camphorseleno derivatives for use as chiral auxiliaries in various types of electrophilic selenium reactions. One compound that was prepared during the course of this work was the cyclic selenenamide **1**, which appeared to have superficial similarities to ebselen and other Se-N compounds that were popular targets at the time. We therefore decided to test it for GPx-like activity (*50*). Toward this effect, we developed a simple assay (Scheme 2) in which the oxidation of benzyl thiol to its corresponding disulfide with excess *t*-butyl hydroperoxide or hydrogen peroxide was monitored in the presence of a catalytic amount of selenenamide **1** in dichloromethane-methanol (95:5). We chose benzyl thiol instead of the more customary benzenethiol for several reasons. Benzyl thiol contains both a chromophore and a characteristic methylene signal to facilitate analysis by both HPLC and NMR techniques, respectively. It is less acidic than aryl thiols and so more closely resembles glutathione and other alkyl thiols in its chemical properties. While aqueous conditions and the use of glutathione itself would have provided more realistic conditions, the catalyst **1** and many that we subsequently developed were poorly water-soluble and necessitated an organic solvent to maintain homogeneity. Since we later observed that the rate of disulfide formation in the presence of various other catalysts was often nonlinear, we typically compared their activity by the time required for 50% completion of the thiol to disulfide conversion ($t_{1/2}$).

$$2\ \text{BnSH} + t\text{-BuOOH or } H_2O_2 \xrightarrow[\substack{CH_2Cl_2\text{-MeOH} \\ (95:5) \\ 18\ ^\circ C}]{\substack{\text{Catalyst} \\ 10\ \text{mol \%}}} \text{BnSSBn} + t\text{-BuOH} + H_2O$$

(excess)

$t_{1/2}$ = time required for conversion of 50% of BnSH to BnSSBn

Scheme 2. In vitro assay conditions for evaluating GPx mimetics

These initial studies (*50*) revealed that selenenamide **1** was not the true catalyst in this process. Instead, it was irreversibly consumed by the thiol in the first step to produce the selenenyl sulfide **2**. From there it followed a cycle via selenol **3** and selenenic acid **4**, closely resembling that of GPx itself (Scheme 3). This prompted further efforts and, since then, our group has studied three different classes of GPx mimetics: cyclic seleninate esters, spirodioxyselenuranes and conformationally restricted diselenides. These are described in subsequent sections and are exemplified by the simple prototypes **5-7**, respectively, as shown in Figure 2.

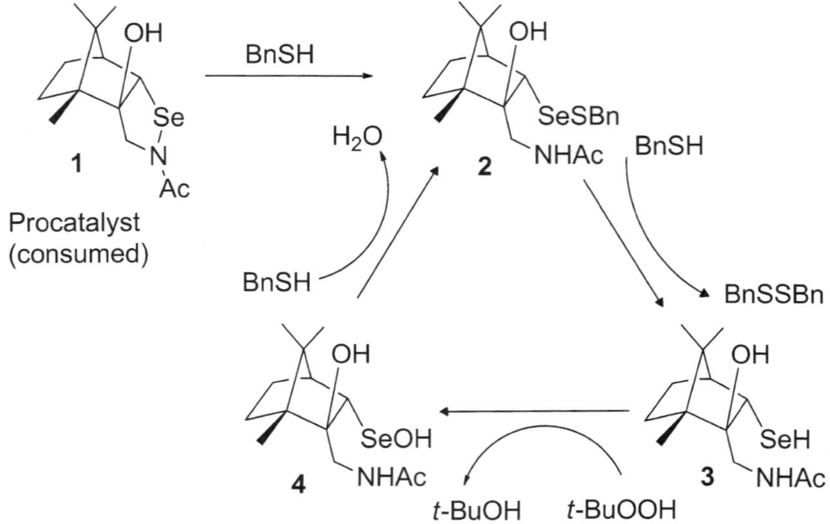

Scheme 3. Catalytic cycle initiated by selenenamide 1

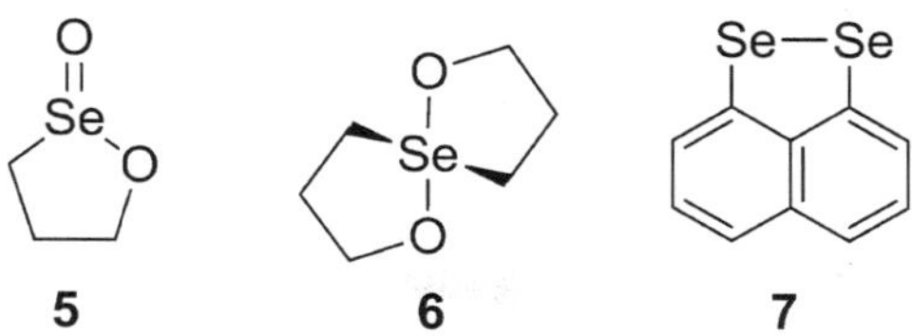

5   **6**   **7**

Figure 2. Structures of prototype GPx mimetics.

Cyclic Seleninate Esters

The oxidation of allyl selenides to selenoxides is typically followed by a rapid [2,3] sigmatropic rearrangement to afford selenenate esters, which can be hydrolyzed to selenenic acids. The latter were expected to provide entry into catalytic cycles analogous to those in Schemes 1 and 3, where EnzSeOH and the selenenic acid **4** had been shown to comprise key intermediates. Furthermore, the beneficial effects of coordinating amino substituents had been observed by previous researchers (*vide supra*) in other types of GPx mimetics and so a variety of allyl selenides containing aminoalkyl substituents was investigated. An example of a GPx mimetic containing a coordinating hydroxyl substituent had been previously reported by Wirth and coworkers (*31*) and several allyl hydroxyalkyl selenides were also prepared for testing in our assay to compare the effects of amino vs. hydroxyl substituents. During these experiments, allyl 3-hydroxypropyl selenide (**8**) exhibited the highest catalytic activity of all of the compounds that we had tested so far and further investigation revealed that it was readily transformed into the novel cyclic seleninate ester **5** by further oxidation with *t*-butyl hydroperoxide (Scheme 4). The seleninate ester was easily isolated and characterized, and could be stored for extended periods in a refrigerator. When **5** was subjected to the assay, the catalytic activity was further improved, compared to that of its precursor selenide **8**. For comparison, the $t_{1/2}$ values of **1**, **8**, **5** and ebselen under the conditions of Scheme 2 with *t*-butyl hydroperoxide as the oxidant were 18 h, 4.8 h, 2.5 h and 42 h, respectively (*51, 52*). Thus, the cyclic seleninate ester **5** displayed a 17-fold improvement over ebselen. Based on a series of control experiments, we proposed the catalytic cycle shown in Scheme 5 for the seleninate ester. However, during the course of this work, it was also discovered that the selenenyl sulfide **9** was generated and that its formation was predominant when high thiol:hydroperoxide ratios were employed. In contrast to the corresponding compound EnzSeSG, which appears to play a key role in the GPx catalytic cycle shown in Scheme 1, the formation of selenenyl sulfide **9** in Scheme 5 resulted in a competing deactivation pathway. This was proved conclusively through the synthesis of authentic **9** and its independent evaluation in the usual assay, where it exhibited a much longer $t_{1/2}$ of 35 h than the seleninate ester **5** (see Figure 3) (*51, 52*). Since thiols are abundant in living cells, the existence of the deactivation pathway casts some doubt about the potential efficacy of cyclic seleninate esters for *in vivo* applications.

147

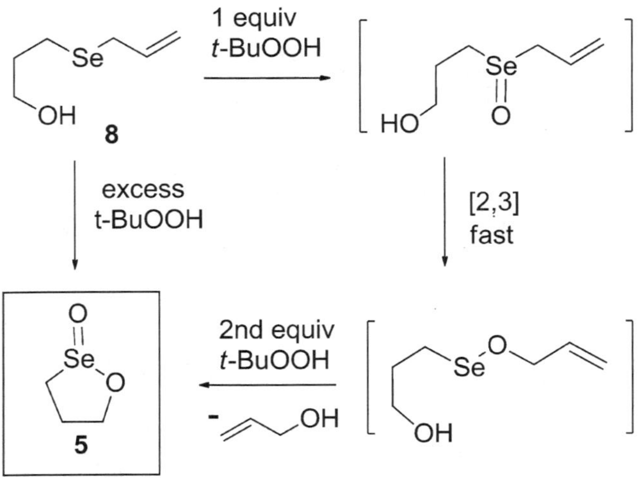

*Scheme 4. Conversion of allylic selenide **8** to cyclic seleninate ester **5***

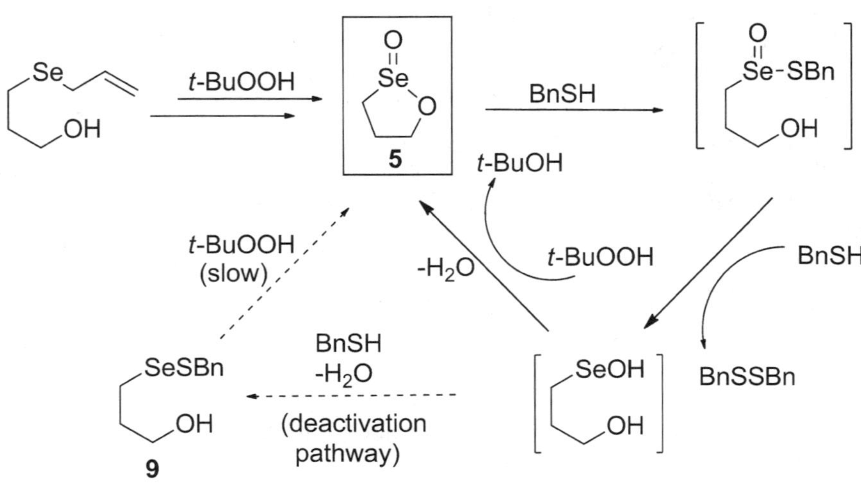

*Scheme 5. Catalytic cycle of cyclic seleninate ester **5***

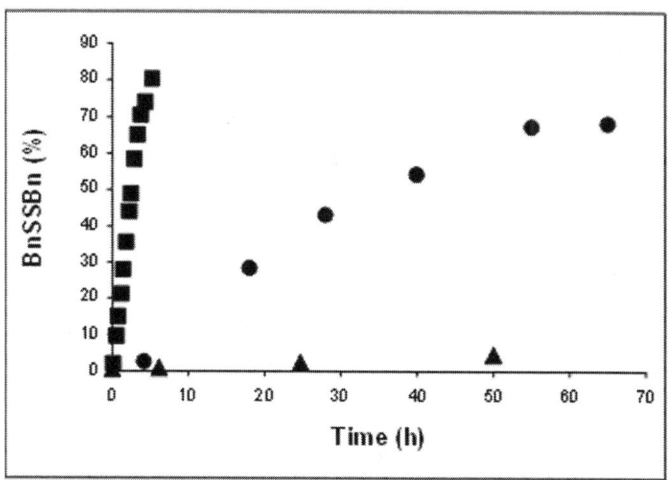

Figure 3. Kinetic plots for BnSSBn formation in the presence of 10 mol % of 5 (■) or 9 (•) and with no catalyst (▲) under the standard assay conditions with t-BuOOH shown in Scheme 2.

Related experiments on several other cyclic seleninate esters and their GPx-like activity have also been reported by H. B. Singh et al. (*53–55*), with special emphasis on the effects of substituents capable of coordinating with the selenium atom, as well as other structural features. Very recently, Bayse and coworkers (*56*) performed detailed computational studies on the model reaction in which reduction of methyl hydroperoxide with methanethiol is catalyzed by **5**. Their results augment the mechanism in Scheme 5 and provide activation energies for the various steps and relative energies for potential intermediates. They also demonstrated that selenurane **10** can be produced in the initial reaction of **5** with methanethiol and that **11** may be a viable intermediate in that step of the catalytic cycle where the seleninate ester **5** is regenerated (Scheme 6).

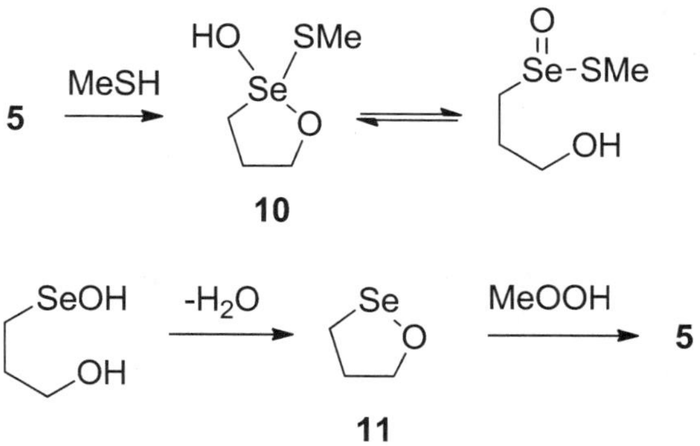

Scheme 6. Other possible intermediates in the catalytic cycle of cyclic seleninate ester 5

Spirodioxyselenuranes

In view of the troublesome deactivation pathway associated with the cyclic seleninate ester **5**, attempts were made to discover an alternative catalytic system that would preclude the formation of the relatively inert selenenyl sulfide **9**. A second set of experiments was therefore performed on a series of bis(hydroxylalkyl) selenides in order to determine whether the pendant hydroxyl groups would affect the catalytic redox properties of the corresponding selenoxides (*57*). In the case of selenide **12**, we observed that upon oxidation with *t*-butyl hydroperoxide, very rapid cyclization to the novel spiroselenurane **6** occurred, preventing isolation of the corresponding selenoxide. The selenurane proved to be a stable and easily handled compound whose structure was confirmed by x-ray crystallography (*57*). Moreover, subsequent treatment of **6** with excess benzyl thiol regenerated the original selenide **12**. When **6** was subjected to the usual assay, it also showed excellent catalytic activity with $t_{1/2}$ = 2.9 h. A plausible mechanism for the process is shown in Scheme 7, where **6** undergoes thiolysis of a Se-O bond and subsequent reaction with a second equivalent of the thiol in an overall reductive elimination of dibenzyl disulfide to produce selenide **12**. Oxidation of **12** then forms the corresponding selenoxide, which cyclizes spontaneously to regenerate **6**. It is therefore possible to introduce the catalyst as either the selenide (reduced form) or the spiroselenurane (oxidized form) when conducting the assay. The hydroxybutyl homolog of **12** behaved similarly, while the corresponding hydroxyethyl derivative failed to cyclize and produced an isolable selenoxide. Both homologs proved inferior catalysts to **12**. As in the case of the cyclic seleninate esters it appears that hydroxypropyl substituents in the starting selenide and five-membered rings in the oxidized product provide the optimum catalytic activity. It is noteworthy that selenenyl sulfides were not produced in experiments conducted with spiroselenuranes, thereby circumventing a potentially limiting deactivation pathway such as was identified with the cyclic seleninate esters.

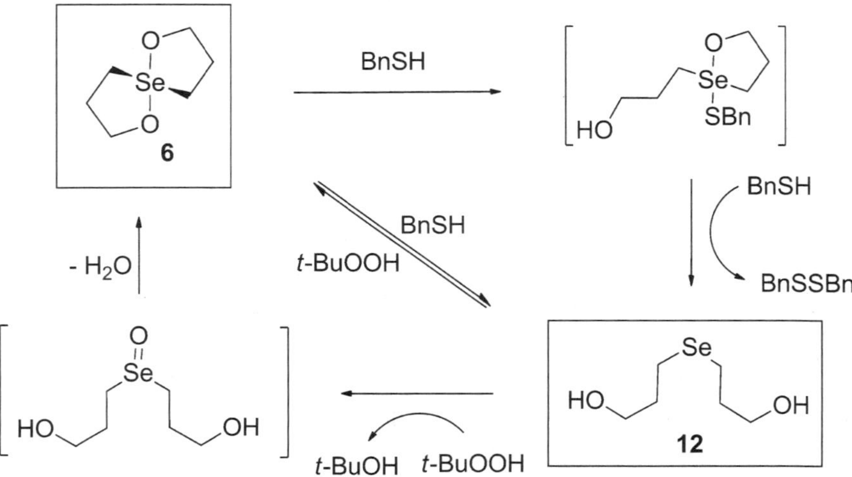

Scheme 7. Catalytic cycle of spirodioxyselenurane 6

The tellurium analogs of **5** and **6** (**13** and **14**, respectively, Figure 4) were also prepared and demonstrated exceptionally high catalytic activity (*58*). In both cases, the precise $t_{1/2}$ was too short to measure effectively by either HPLC or NMR techniques, but was shown to be < 5 min.

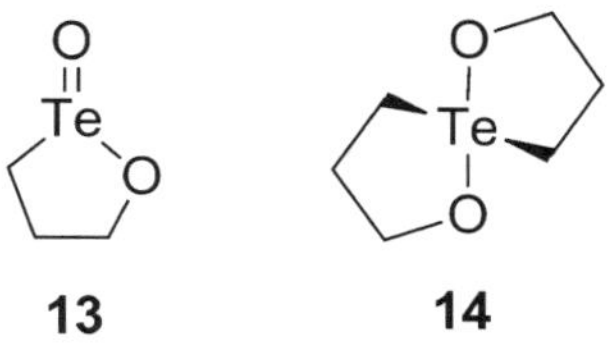

13   **14**

Figure 4. Structures of Te analogs of 5 and 6.

Aromatic Derivatives

While aliphatic cyclic seleninate esters and spirodioxyselenuranes such as **5** and **6**, respectively, proved to be effective catalysts for the reduction of *t*-butyl hydroperoxide with benzyl thiol *in vitro*, there were concerns that the potentially more facile metabolism and higher toxicity (*23*, *59*) of aliphatic selenium compounds, compared to their more stable aromatic counterparts, would render them less suitable for *in vivo* experiments. The stronger $C(sp^2)$-Se bonds of aromatic analogs compared to the weaker $C(sp^3)$-Se bonds of aliphatic ones presumably contributes to the greater stability of the former. Consequently, a series of analogous aromatic compounds was prepared, including compounds **15-18** (Figure 5). Unfortunately, they all displayed poor catalytic activity ($t_{1/2}$ between 62 h and 252 h) with *t*-butyl hydroperoxide as the oxidant. These results proved highly sensitive to the concentration and purity of the hydroperoxide and were challenging to reproduce consistently. We therefore repeated these experiments with hydrogen peroxide, which provided more rapid and more easily reproducible reaction rates, with $t_{1/2}$ between 5.5 h and 73 h (*58*). However, these catalytic activities were still considerably inferior to the prototype aliphatic compounds, especially the spiroselenurane **6**, which gave $t_{1/2}$ = 12 min with hydrogen peroxide (*58*). Thus, the anticipated improvement in metabolic stability and lower toxicity was obtained at the expense of catalytic activity in the aromatic compounds.

One possible approach to improving the reactivity of the aromatic compounds was to exploit inductive and mesomeric effects of substituents conjugated to the selenium atoms. We hoped that this would not only lead to the discovery of improved catalysts, but also provide additional mechanistic insight. We therefore synthesized a series of *para*-substituted seleninates and spiroselenuranes **19** and **20** (Figure 5) in order to compare them in the standard assay with benzyl

151

thiol and hydrogen peroxide (*60*).   These experiments revealed that in both series of compounds, the reaction rates were increased by the installation of electron-donating groups and retarded by electron-withdrawing substituents. Indeed, in the case of the methoxy derivative **20** (R = OMe), the $t_{1/2}$ decreased to only 30 min. Hammett plots of the two series of experiments (Figures 6 and 7) (*60*) both indicated negative slopes, but with significantly different reaction constants, where ρ = -0.43 for the seleninates **19** and ρ = -3.12 for selenuranes **20**. These results demonstrate that while the catalytic activity in both series of compounds is enhanced by electron donation, the substituent effects are much more pronounced in the selenuranes **20** than in the seleninates **19**. Furthermore, the negative reaction constants are consistent with rate-determining steps where there is increasing development of positive charge on the selenium atom that is stabilized by electron-donating substituents. This occurs most notably during the oxidation of Se(II) to Se(IV) and implicates the oxidation of the selenide starting materials (or selenenate intermediates) with the peroxide as the rate-limiting step in both series of compounds. The kinetic plots for the cyclic seleninate esters **15** and **19**, as well as for spirodioxyselenuranes **17** and **20**, indicated a linear rate in the production of dibenzyl disulfide, but with a significant y-intercept at ca. 10-20% yield, as in the example of **17** in Figure 8. This suggests an abnormally fast rate of reaction in the early stages of the process and is consistent with the rapid thiolysis and Se(IV) to Se(II) reduction of the catalyst (10 mol %) with concomitant disulfide formation, followed by regeneration of the catalyst through a slower, rate-limiting oxidation of Se(II) to Se(IV).

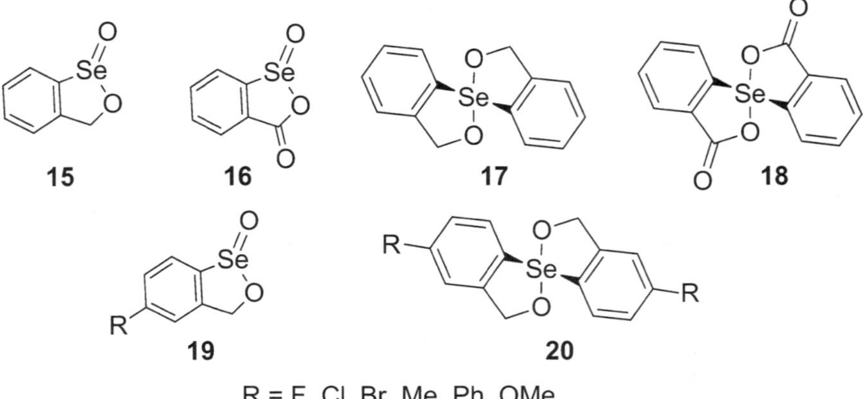

Figure 5. *Structures of representative aromatic GPx mimetics.*

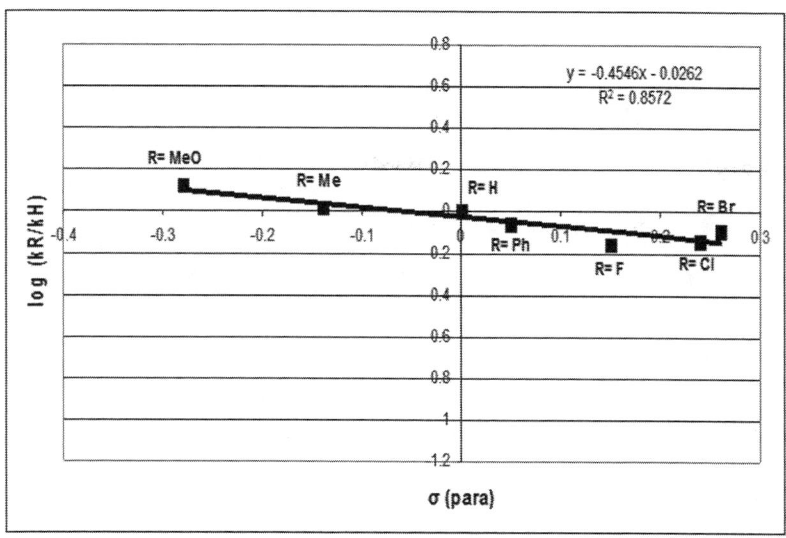

*Figure 6. Hammett plot of cyclic seleninate esters **15** and **19** (60) (ρ = −0.45). Reproduced with permission from reference (60). Copyright 2008 American Chemical Society.*

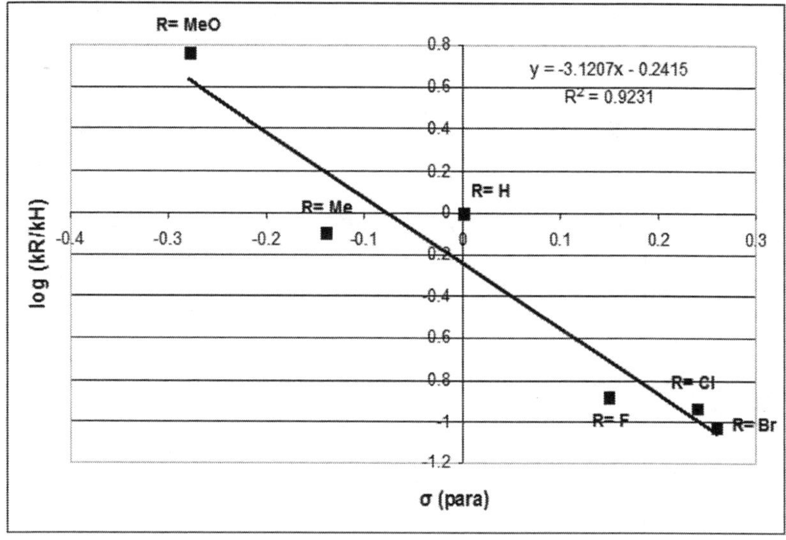

*Figure 7. Hammett plot of spirodioxyselenuranes **17** and **20** (60) (ρ = −3.12). Reproduced with permission from reference (60). Copyright 2008 American Chemical Society.*

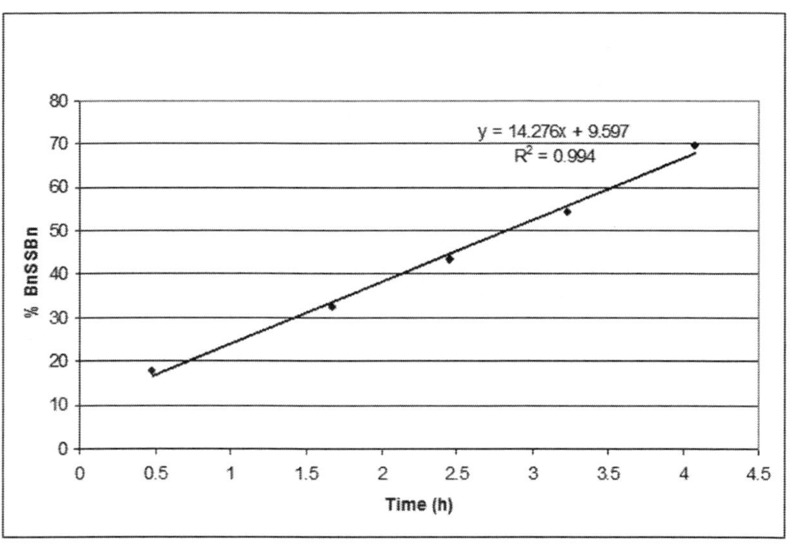

Figure 8. Kinetic plot for BnSSBn formation in the presence of 10 mol % of 17 with hydrogen peroxide under the conditions of ref. (60).

Conformationally Constrained Diselenides

A variety of diselenides have been previously investigated for GPx-like behavior (*28, 29, 31–33, 61, 62*) and even simple diaryl diselenides display a modest level of catalytic activity (*63*). Many years ago, in connection with other endeavors, we prepared the highly hindered diselenide **21** (*64*). Not surprisingly, its X-ray crystal structure revealed that the C-Se-Se-C dihedral angle was unusually large, 112.1° compared with 82.0° for the more typical diphenyl diselenide (*65*) (Figure 9). Furthermore, the UV-visible spectrum of **21** revealed an absorption at 324 nm, with a strong bathochromic shift of 19 nm compared to less hindered dialkyl diselenides. This suggested that the deviation from the ideal, nearly orthogonal geometry had decreased the HOMO-LUMO gap in **21**, presumably through destabilization of the ground state. A lower oxidation potential was also anticipated for such conformationally constrained diselenides, which would improve their GPx-like activity by facilitating the rate-limiting Se(II) to Se(IV) oxidation step. Although the aliphatic diselenide **21** was not considered a promising GPx mimetic for other reasons, it was reasonable to assume that the same effect might be achieved with a diselenide that was confined to a conformation with a decreased dihedral angle. The planar rigid nature of the naphthalene skeleton suggested that the known (*66*) 1,8-*peri*-diselenide **7** would be confined to a C-Se-Se-C dihedral angle of essentially 0° (*67*). Furthermore, the introduction of methoxy groups into the 2- and 7-positions would be expected to provide additional enhancement of catalytic activity, as had been previously demonstrated for cyclic seleninate esters and spirodioxyselenuranes. Indeed, a comparison of the $t_{1/2}$ of diphenyl diselenide with that of di-*o*-methoxyphenyl diselenide showed a modest decrease from 129 h to 90 h. A more dramatic decrease of over an order of magnitude to 9.7 h was observed with the unsubstituted *peri*-diselenide **7** compared to diphenyl diselenide, while a further reduction to 7.4 h was realized with the dimethoxy derivative **22** (Table 1) (*68*). Clearly, these results validate the expectation of improved catalytic activity of conformationally constrained naphthalene *peri*-diselenides such as **7** and **22**. The corresponding ditelluride **23** was also prepared and, as in the case of the tellurium analogs **13** and **14**, proved exceptionally potent as a GPx mimetic.

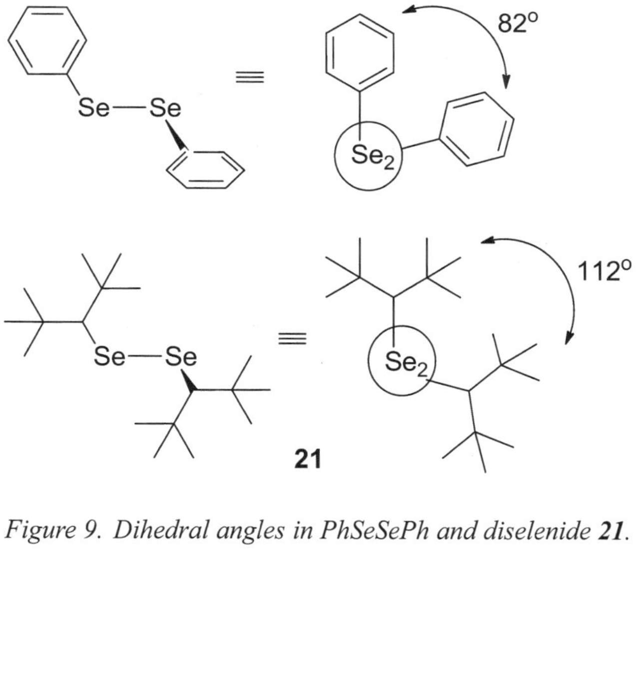

Figure 9. Dihedral angles in PhSeSePh and diselenide 21.

Early work by John Kice (*69*) on diselenide **7** revealed that it underwent facile oxidation with one equivalent of *m*-chloroperoxybenzoic acid (*m*-CPBA) to a mixture of the corresponding selenolseleninate **24** and selenenic anhydride **25**, while 3 equivalents afforded the postulated seleninic anhydride **26**, which proved too insoluble even for NMR analysis (Scheme 8). The behavior of the dimethoxy derivative **22** was investigated through a series of control experiments in order to gain further insight into its catalytic cycle in our standard assay. When **22** was oxidized with 1 equivalent of hydrogen peroxide, it produced solely the selenolseleninate **27**, which could be isolated and fully characterized. The reaction required ca. 30 min to go to completion at room temperature and could be easily monitored by the discharge of the intense purple color of the diselenide and the appearance of the orange selenolseleninate. Further oxidation of **27** with hydrogen peroxide was not observed, even after 24 h. Moreover, there was no significant reaction of diselenide **22** with excess benzyl thiol in the absence of hydrogen peroxide to afford the corresponding selenenyl sulfide, even after 12 h. Finally, treatment of authentic selenolseleninate **27** with excess benzyl thiol resulted in the rapid reappearance of the purple color of the diselenide, which was recovered in 85% yield. These experiments reveal that the first step in the catalytic cycle of **22** is its reaction with hydrogen peroxide to produce **27** and that, once again, oxidation of Se(II) to Se(IV) is the rate-determining step. The subsequent rapid reduction of **27** by the thiol completes the process and regenerates the original diselenide. A plausible mechanism is shown in Scheme 9. The reduction of **27** likely proceeds via the selenenyl sulfide/selenenic acid **28**, which can react with a second equivalent of thiol at the sulfur atom, resulting in cyclization and dehydration of the resulting selenol/selenenic acid to afford the corresponding disulfide and cyclic diselenide directly. Alternatively, reaction with a second equivalent of thiol could produce the bis(selenenyl sulfide) **29** from the selenenic acid moiety of **28**, followed by thiolysis of one selenenyl sulfide moiety and attack by the liberated selenolate upon the neighbouring selenenyl sulfide, again leading to diselenide and disulfide formation (*70*). Finally, treatment of diselenide **22** with excess *m*-CPBA produced a more soluble seleninic anhydride **30** than the anhydride **26** postulated by Kice et al. from the oxidation of diselenide **7**. The product was obtained as a mixture of two diastereomers, as evidenced by its NMR spectra, which as expected, afforded a unique dipotassium salt **31** upon the addition of potassium hydroxide (Scheme 10).

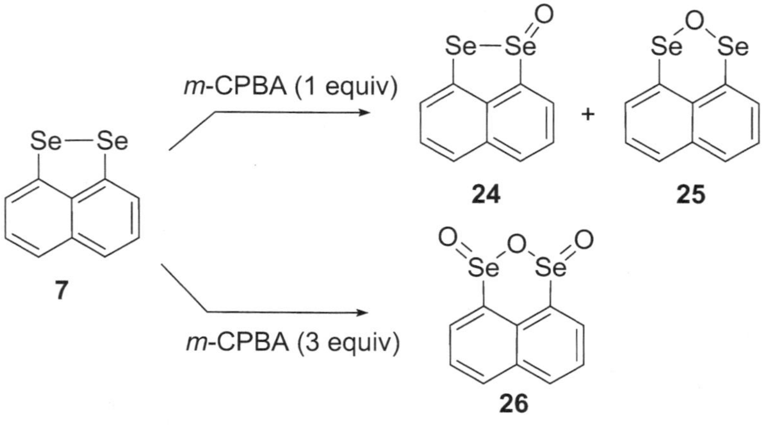

Scheme 8. Oxidation of diselenide 7 with m-CPBA

Table 1. Catalytic activities of diselenides under the standard assay conditions

$$2\ \text{BnSH} + \text{H}_2\text{O}_2 \xrightarrow[\substack{\text{MeOH-CH}_2\text{Cl}_2 \\ (95:5)}]{\text{catalyst (10 mol \%)}} \text{BnSSBn} + 2\text{H}_2\text{O}$$

Catalyst	$t_{1/2}$ (h)
Nil (control)	>300
PhSeSePh	129
(o-MeOPhSe)$_2$	90
7 R = H	9.7
22 R = OMe	7.4
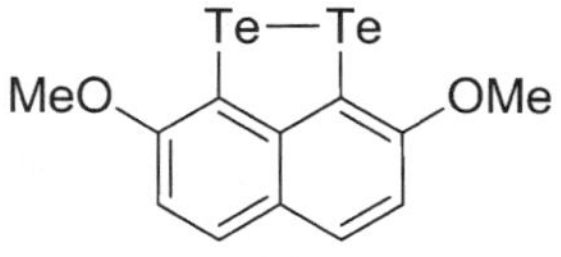 **23**	<0.05

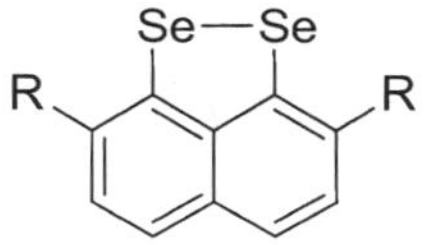

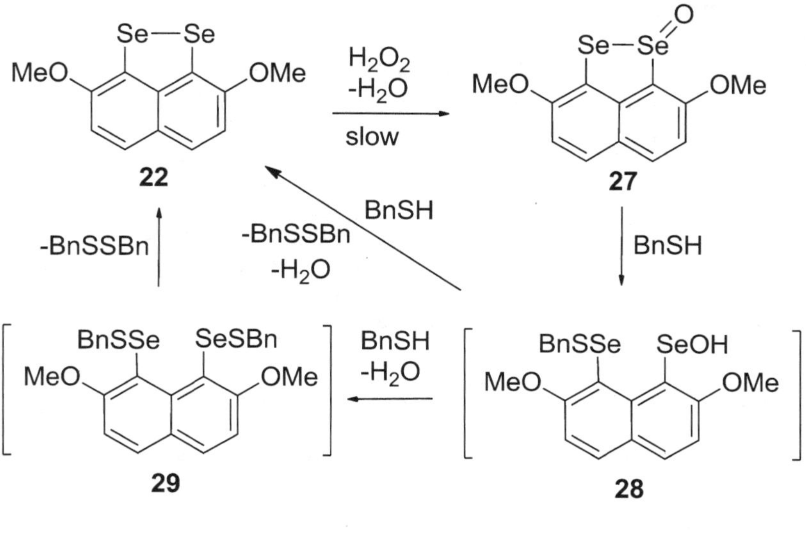

Scheme 9. Catalytic cycle for diselenide **22**

157

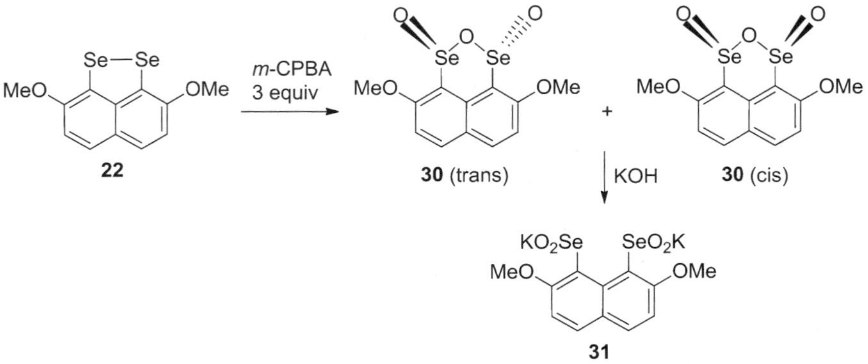

Scheme 10 shows the oxidation of diselenide **22** with *m*-CPBA (3 equiv) to give **30** (trans) and **30** (cis), which upon treatment with KOH gives **31**.

Scheme 10. Oxidation of diselenide **22** *with m-CPBA*

Conclusions

The investigations of the three new classes of highly active GPx mimetics described above have led to several insights into the structural features required for optimum activity, the mechanisms by which these compounds catalyze the reduction of peroxides with thiols and their potential utility in minimizing the deleterious effects of strongly elevated levels of oxidative stress, as occurs in ischemic reperfusion.

These results indicate that a variety of mechanisms significantly different from that of GPx (Scheme 1) can be highly effective in reducing peroxides with thiols. This is illustrated in Schemes 5 (cyclic seleninate esters), 7 (spirodioxyselenuranes) and 9 (naphthalene *peri*-diselenides). The role of selenenyl sulfides in these processes remains contradictory. They are postulated as essential intermediates in the catalytic cycle of GPx itself (Scheme 1) and also appear to play a role in the antioxidant properties of selenenamide **1** and naphthalene *peri*-diselenides. In contrast, their formation results in a deactivation pathway that competes with the primary catalytic cycle in the case of the seleninate esters. While earlier GPx mimetics were often based on N-Se compounds, it is now apparent that O-Se compounds are also highly effective in this regard.

Of the three principal classes investigated, the aliphatic cyclic seleninate ester **5** and spirodialkoxyselenurane **6** displayed exceptionally high catalytic activities, as indicated by their short $t_{1/2}$ values. However, the weaker C-Se bonds present in these compounds, compared to those in aromatic derivatives, raise the expectation of lower stability and more rapid metabolic degradation *in vivo*, along with higher toxicity. On the other hand, the more robust and presumably less toxic aromatic analogs provide lower catalytic activity, which fortunately, can be greatly improved by the introduction of electron-donating substituents *para* to the selenium atom. We postulate that this effect stems from stabilization of the increased positive charge on Se during its oxidation from Se(II) to Se(IV), which comprises the rate-determining step in all three classes of compounds. The deactivation pathway in the case of cyclic seleninate esters (Scheme 5) becomes increasingly dominant as the thiol:peroxide ratio increases. Since thiols are abundant and ubiquitous *in vivo*, this will likely curtail their application as biological antioxidants. In contrast, the spirodioxyselenuranes and conformationally constrained diselenides operate via simpler catalytic cycles that do not appear to become as easily compromised by competing pathways.

In all three classes of selenium compounds studied, the analogous tellurium derivatives proved several orders of magnitude more powerful as catalysts in our standard assay. However, the toxicity of organotellurium compounds has been less studied and may pose serious impediments to the further development of compounds such as **13**, **14** and **23**.

It is interesting to note that direct comparison of catalytic activities of compounds prepared by various groups is limited by the numerous types of assays that have been employed and on the parameters used to define their activities. Our own assay employs an organic solvent mixture that has the advantage of dissolving virtually all of the catalysts investigated in our laboratory to date, as well as the benzyl thiol and *t*-butyl hydroperoxide or hydrogen peroxide employed in our assay. This permits the use of homogeneous solutions during kinetic determinations. However, the advantage of an aqueous system is, of course, that it more closely resembles *in vivo* conditions. Moreover, we have observed nonlinear kinetics, especially when employing catalysts in the Se(IV) state. This is because the initial reaction of the catalyst with the thiol is very rapid until the catalyst is consumed, followed by a slower reaction during the reoxidation of the catalyst from its Se(II) state. Thus, the rate of disulfide production and thiol consumption is anomalously rapid in the initial stages of such processes. On the other hand, the rate of disulfide production also slows considerably at later stages of the reaction in the case of seleninate esters, with the accumulation of the relatively inert selenenyl sulfide. To compensate for these differences, we recommend the $t_{1/2}$ parameter as a useful means of comparing different GPx mimetics.

The poor aqueous solubility of many of the compounds prepared in our laboratory, as well of those reported by other groups, limits their biological applications, notwithstanding the Phase III clinical trials conducted with ebselen, which also has low solubility in water. For testing in animal models, intravenous administration is generally preferable to gauvage. We have prepared water-soluble analogs of several of the compounds described herein and conducted preliminary trials in an ischemic mouse model (with Dr. Paul Kubes, Snyder Institute for Chronic Diseases, University of Calgary). Unfortunately, these compounds had high clearance rates and were ineffective in reducing infarct size. Details of these and related studies will be reported in due course. Several approaches to water-soluble GPx mimetics by other groups have also been reported (*71–73*).

In summary, there has been remarkable progress in numerous laboratories in developing highly catalytically active GPx mimetics of diverse structure and mechanism. However, the translation of these efforts into the design of clinically efficacious drugs remains a considerable challenge for the future.

Acknowledgments

Financial support from the Natural Sciences and Engineering Research Council of Canada is gratefully acknowledged, as well as the participation of many students over the years, whose names are indicated in the cited references

References

1. Arthur, J. R. *Cell. Mol. Life Sci.* **2000**, *57*, 1825–1835.
2. Brigelius-Flohe, R; Maiorino, M. *Biochim. Biophys. Acta* **2013**, *1830*, 3289–3303.
3. Brigelius-Flohe, R.; Kipp, A. P. *Ann. N. Y. Acad. Sci.* **2012**, *1259*, 19–25.
4. Epp, O.; Ladenstein, R.; Wendel, A. *Eur. J. Biochem.* **1983**, *133*, 51–69.
5. Ganther, H. E. *Chem. Scr.* **1975**, *8a*, 79–84.
6. Ganther, H. E.; Kraus, R. J. In *Methods in Enzymology*; Colowick, S. P.;, Kaplan, N. O., Eds.; Academic Press: New York, 1984; Vol. 107, pp 593–602.
7. Stadtman, T. C. *J. Biol. Chem.* **1991**, *266*, 16257–16260.
8. Tappel, A. L. *Curr. Top. Cell. Regul.* **1984**, *24*, 87–97.
9. Flohé, L. *Curr. Top. Cell. Regul.* **1985**, *27*, 473–478.
10. Rotruck, J. T.; Pope, A. L.; Ganther, H. E.; Swanson, A. B.; Hafeman, D. G.; Hoekstra, W. G. *Science* **1973**, *179*, 588–590.
11. Jamier, V.; Ba, L. A.; Jacob, C. *Chem.—Eur. J.* **2010**, *16*, 10920–10928.
12. Jacob, C.; Winyard, P. G. *Redox Signaling and Regulation in Biology and Medicine*; Wiley-VCH: Weinheim, 2009.
13. Schafer, F. Q.; Buettner, G. R. *Free Radicals Biol. Med.* **2001**, *30*, 1191–1212.
14. Crack, P. J.; Taylor, J. M.; de Haan, J. B.; Kola, I.; Hertzog, P.; Iannello, R. C. *J. Cereb. Blood Flow Metab..* **2003**, *23*, 19–22.
15. Wong, C. H. Y.; Bozinovski, S.; Hertzog, P. J.; Hickey, M. J.; Crack, P. J. *J. Neurochem.* **2008**, *107*, 241–252.
16. Dhalla, N. S.; Elmoselhi, A. B.; Hata, T.; Makino, N. *Cardiovasc. Res.* **2000**, *47*, 446–456.
17. Lim, C. C.; Bryan, N. S.; Jain, M; Garcia-Saura, M. F.; Fernandez, B. O.; Sawyer, D. B.; Handy, D. E.; Loscalzo, J.; Feelisch, M.; Liao, R. *Am. J. Physiol.: Heart Circ. Physiol.* **2009**, *297*, H2144–H2153.
18. Maulik, N.; Yoshida, T.; Das, D. K. *Mol. Cell. Biochem.* **1999**, *196*, 13–21.
19. Mužáková, V.; Kandár, R.; Vojtíšek, P.; Skalický, J.; Vaňková, R.; Čegan, A.; Červinková, Z. *Physiol. Res.* **2001**, *50*, 389–396.
20. Moutet, M.; D'Alessio, P.; Malette, P.; Devaux, V.; Chaudière, J. *Free Radical Biol. Med.* **1998**, *25*, 270–281.
21. Day, B. J. *Biochem. Pharmacol.* **2009**, *77*, 285–296.
22. Bhabak, K. P.; Mugesh, G. *Acc. Chem. Res.* **2010**, *43*, 1408–1419.
23. Mugesh, G.; Singh, H. B. *Chem. Soc. Rev.* **2000**, *29*, 347–357.
24. Mugesh, G.; du Mont, W.-W.; Sies, H. *Chem. Rev.* **2001**, *101*, 2125–2179.
25. Bhuyan, B. J.; Mugesh, G. In *Organoselenium Chemistry: Synthesis and Reactions*; Wirth, T., Ed.; Wiley VCH Verlag: Weinheim, 2012; pp 369–379.

26. Reich, H. J.; Jasperse, C. P. *J. Am. Chem. Soc.* **1987**, *109*, 5549–5551.

27. Engman, L.; Andersson, C.; Morgenstern, R.; Cotgreave, I. A.; Andersson, C. M.; Hallberg, A. *Tetrahedron* **1994**, *50*, 2929–2938.

28. Wilson, S. R.; Zucker, P. A.; Huang, R.-R. C.; Spector, A. *J. Am. Chem. Soc.* **1989**, *111*, 5936–5939.

29. Galet, V.; Bernier, J.-L.; Hénichart, J.-P.; Lesieur, D.; Abadie, C.; Rochette, L.; Lindenbaum, A.; Chalas, J.; Renaud de la Faverie, J.-F.; Pfeiffer, B.; Renard, P. *J. Med. Chem.* **1994**, *37*, 2903–2911.

30. Ostrovidov, S.; Franck, P.; Joseph, D.; Martarello, L.; Kirsch, G.; Belleville, F.; Nabet, P.; Dousset, B. *J. Med. Chem.* **2000**, *43*, 1762–1769.

31. Wirth, T. *Molecules* **1998**, *3*, 164–166.

32. Mugesh, G.; Panda, A.; Singh, H. B.; Punekar, N. S.; Butcher, R. J. *J. Chem. Soc., Chem. Commun.* **1998**, 2227–2228.

33. Kumar, S.; Tripathi, S. K.; Singh, H. B.; Wolmershäuser, G. *J. Organomet. Chem.* **2004**, *689*, 3046–3055.

34. You, Y.; Ahsan, K.; Detty, M. R. *J. Am. Chem. Soc.* **2003**, *125*, 4918–4927.

35. Engman, L.; Stern, D.; Cotgreave, I. A.; Andersson, C. M. *J. Am. Chem. Soc.* **1992**, *114*, 9737–9743.

36. Erdelmeier, I.; Tailhan-Lomont, C.; Yadan, J.-C. *J. Org. Chem.* **2000**, *65*, 8152–8157.

37. Jacquemin, P. V.; Christiaens, L. E.; Renson, M. J.; Evers, M. J.; Dereu, N. *Tetrahedron Lett.* **1992**, *33*, 3863–3866.

38. Sarma, B. K.; Mugesh, G. *Chem.—Eur. J.* **2008**, *14*, 10603–10614.

39. Bhabak, K. P.; Mugesh, G. *Chem.—Eur. J.* **2007**, *13*, 4594–4601.

40. Sarma, B. K.; Mugesh, G. *J. Am. Chem. Soc.* **2005**, *127*, 11477–11485.

41. Müller, A.; Cadenas, E.; Graf, P.; Sies, H. *Biochem. Pharmacol.* **1984**, *33*, 3235–3239.

42. Wendel, A.; Fausel, M.; Safayhi, H.; Tiegs, G.; Otter, R. *Biochem. Pharmacol.* **1984**, *33*, 3241–3245.

43. Parnham, M. J.; Kindt, S. *Biochem. Pharmacol.* **1984**, *33*, 3247–3250.

44. Müller, A.; Gabriel, H.; Sies, H. *Biochem. Pharmacol.* **1985**, *34*, 1185–1189.

45. Safayhi, H.; Tiegs, G.; Wendel, A. *Biochem. Pharmacol.* **1985**, *34*, 2691–2694.

46. Wendel, A.; Tiegs, G. *Biochem. Pharmacol.* **1986**, *35*, 2115–2118.

47. Fischer, H.; Dereu, N. *Bull. Soc. Chim. Belg.* **1987**, *96*, 757–768.

48. Haenen, G. R. M. M.; De Rooij, B. M.; Vermeulen, N. P. E.; Bast, A. *Mol. Pharmacol.* **1990**, *37*, 412–422.

49. Glass, R. S.; Farooqui, F.; Sabahi, M.; Ehler, K. W. *J. Org. Chem.* **1989**, *54*, 1092–1097.

50. Back, T. G.; Dyck, B. P. *J. Am. Chem. Soc.* **1997**, *119*, 2079–2083.

51. Back, T. G.; Moussa, Z. *J. Am. Chem. Soc.* **2002**, *124*, 12104–12105.

52. Back, T. G.; Moussa, Z. *J. Am. Chem. Soc.* **2003**, *125*, 13455–13460.

53. Tripathi, S. K.; Patel, U.; Roy, D.; Sunoj, R. B.; Singh, H. B.; Wolmershäuser, G.; Butcher, R. J. *J. Org. Chem.* **2005**, *70*, 9237–9247.

54. Tripathi, S. K.; Sharma, S.; Singh, H. B.; Butcher, R. J. *Org. Biomol. Chem.* **2011**, *9*, 581–587.

55. Singh, V. P.; Singh, H. B.; Butcher, R. J. *Chem. Asian J.* **2011**, *6*, 1431–1442.

56. Bayse, C. A.; Ortwine, K. N. *Eur. J. Inorg. Chem.* **2013**, 3680–3688.

57. Back, T. G.; Moussa, Z.; Parvez, M. *Angew. Chem., Int. Ed.* **2004**, *43*, 1268–1270.

58. Back, T. G.; Kuzma, D.; Parvez, M. *J. Org. Chem.* **2005**, *70*, 9230–9236.

59. Nogueira, C. W.; Zeni, G.; Rocha, J. B. T. *Chem. Rev.* **2004**, *104*, 6255–6286.

60. Press, D. J.; Mercier, E. A.; Kuzma, D.; Back, T. G. *J. Org. Chem.* **2008**, *73*, 4252–4255.

61. Mugesh, G.; Panda, A.; Singh, H. B.; Punekar, N. S.; Butcher, R. J. *J. Am. Chem. Soc.* **2001**, *123*, 839–850.

62. Selvakumar, K.; Shah, P.; Singh, H. B.; Butcher, R. J. *Chem.–Eur. J.* **2011**, *17*, 12741–12755.

63. For example, diphenyl diselenide has been reported to provide GPx-like catalytic activity about twice that of ebselen; see ref. 28.

64. Back, T. G.; Codding, P. W. *Can. J. Chem.* **1983**, *61*, 2749–2752.

65. Marsh, R. E. *Acta Cryst.* **1952**, *5*, 458–462.

66. Meinwald, J.; Dauplaise, D.; Wudl, F.; Hauser, J. J. *J. Am. Chem. Soc.* **1977**, *99*, 255–257.

67. Aucott, S. M.; Milton, H. L.; Robertson, S. D.; Slawin, A. M. Z.; Woollins, J. D. *Heteroat. Chem.* **2004**, *15*, 530–542.

68. Press, D. J.; Back, T. G. *Org. Lett.* **2011**, *13*, 4104–4107.

69. Kice, J. L.; Kang, Y.-H.; Manek, M. B. *J. Org. Chem.* **1988**, *53*, 2435–2439.

70. For a discussion of the reactions of thiols with selenenyl sulfides in GPx mimetics, see refs. 25 and 40, and references cited therein.

71. Kumakura, F.; Mishra, B.; Priyadarsini, K. I.; Iwaoka, M. *Eur. J. Org. Chem.* **2010**, 440–445.

72. McNaughton, M.; Engman, L.; Birmingham, A.; Powis, G.; Cotgreave, I. A. *J. Med. Chem.* **2004**, *47*, 233–239.

73. Dong, Z.-Y.; Huang, X.; Mao, S.-Z.; Liang, K.; Liu, J.-Q.; Luo, G.-M.; Shen, J.-C. *Chem.—Eur. J.* **2006**, *12*, 3575–3579.

Chapter 8

Emulating Antioxidative Functions of Glutathione Peroxidase Using Selenopeptides

Michio Iwoaka*

Department of Chemistry, School of Science, Tokai University, Kitakaname, Hiratsuka-shi, Kanagawa 259-1292, Japan
***E-mail: miwaoka@tokai.ac.jp.**

The antioxidative function of the selenoenzyme glutathione peroxidase (GPx) can be imitated by using a wide range of small organoselenium compounds. However, the GPx-like antioxidant activity of selenocysteine (Sec) derivatives as well as Sec-containing peptides has seldom been investigated in the literature, despite the expectation that they would serve as useful probes or models for understanding the molecular mechanism of GPx. In this chapter, applications of various Sec derivatives, including selenopeptides, to emulating the antioxidative functions of GPx are focused. Selenoglutathione (GSeH) has an unusually high antioxidant capacity mainly due to its structural similarity to glutathione (GSH). On the other hand, shorter selenopeptides, which have the same primary amino acid sequence as the GPx active site or the designed sequence mimicking the catalytic triad structure, would be useful to get insight into the structure-activity relationship of GPx.

1. Oxidative Stress and Glutathione Peroxidase (GPx)

Oxidative stress (*1–3*) a well known cause of various syndromes and diseases related to excess production of reactive oxygen species, such as superoxide ($O_2{}^{\cdot-}$), hydroxyl radial ($HO\cdot$), and hydroperoxides (H_2O_2, $ROOH$). These species react with proteins, DNA, and lipids to seriously damage the cell (*4*). In our body, there are several antioxidant systems to prevent these harmful processes. Antioxidants are commonly involved in such systems to decrease the concentration of reactive oxygen species. There are two types of antioxidants (*5*): one-electron radical scavengers like tocopherols (vitamin E) and two-electron redox catalysts like glutathione peroxidase (GPx) (*6–9*) and peroxiredoxin (*10, 11*). GPx is a family of antioxidant enzymes that catalyze the reduction of hydroperoxides using glutathione (GSH) as a reducing cofactor. This is a unique enzyme because it usually has selenocysteine (Sec; U), so-called the 21st amino acid, at the active site.

The catalytic cycle of GPx consists of three steps (Figure 1). The first step is the oxidation of the selenol form (E-SeH) of the Sec residue with hydroperoxides (for example, H_2O_2) to generate a highly reactive selenenic acid intermediate (E-SeOH). The second step is the reduction of E-SeOH with GSH to produce a selenenyl sulfide intermediate (E-SeSG), and the third step is the reduction of E-SeSG with another GSH to reproduce E-SeH. Small organic selenium compounds have long attracted the interest of researchers attempting to mimic the antioxidative functions of this enzyme (*12*).

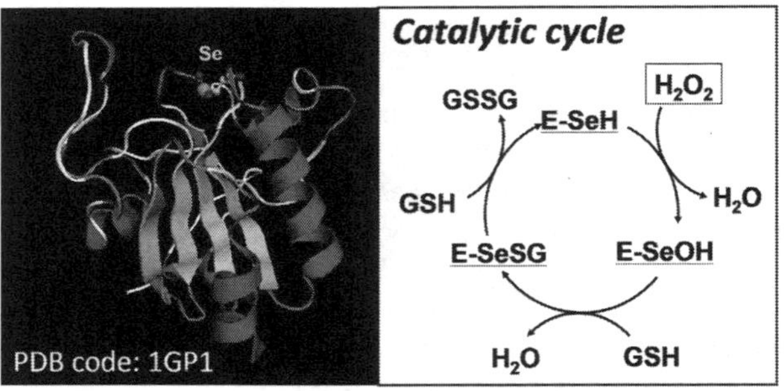

Figure 1. Structure of glutathione peroxidase (GPx) and the catalytic cycle.
(see color insert)

The active site of GPx contains three amino acid residues (i.e., U, Q, and W) (Figure 2) that make up the catalytic triad for antioxidant behavior (*13*). The X-ray structural analysis of bovine GPx in the stable seleninic acid form (E-SeO$_2$H) (*14*) revealed that the nitrogen atoms of Q80 and W158 are located in close proximity to the selenium atom of U45 (3.27 and 3.56 Å, respectively) suggesting that these N atoms form hydrogen bonds to the Se atom in the E-SeH form. The presence of the catalytic triad was supported by the phylogenetic analysis, which showed the conservation of the three amino acid residues over various living species (*13, 15*). On the other hand, mutation of U to C resulted in a significant loss of the enzymatic activity (*16, 17*). The activity was further decreased by additional mutations of the Q and W residues (*18–20*). The observations indirectly demonstrated the importance of Se···Ninteractions between residues of the triad to the antioxidant function of GPx.

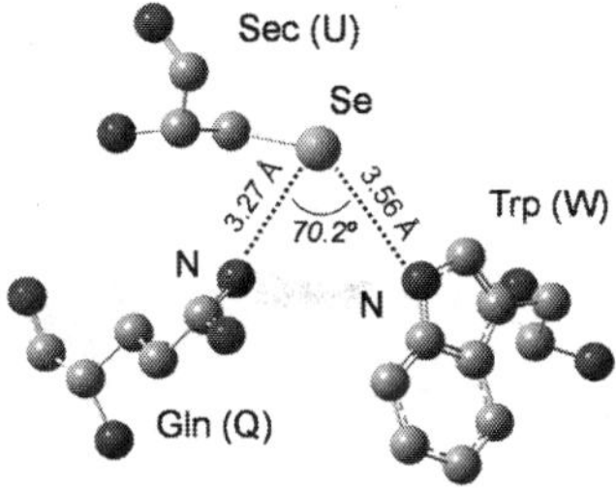

Figure 2. The active site of GPx revealed by X-ray analysis. The two oxygen atoms attached to the Se atom are omitted. (see color insert)

In this chapter, we focus on applications of various Sec derivatives, including selenopeptides to emulate the antioxidant function of GPx. After briefly reviewing typical GPx mimics of small organoselenium compounds, the catalytic cycle of Sec and the different forms of Sec derivatives as a possible precatalyst are discussed. Recent studies of the GPx-like antioxidant capacity of selenopeptide models are presented afterwards. Since the site-directed point mutation of selenoproteins is technically difficult at this moment, the approach using designed selenopeptides to the structure-activity relationship would provide invaluable information about the molecular mechanism of selenoenzyme functions.

2. GPx Mimics by Using Small Organoselenium Compounds

Many types of organoselenium compounds have been tested to date as possible GPx mimics. These mimics have already been summarized in several reviews (*21–23*) and a book chapter (*24*). Here, mimics are categorized in three classes: aromatic, aliphatic, and novel conformationally restricted diselenide compounds. The representative examples of these selenium compounds are shown in Figure 3.

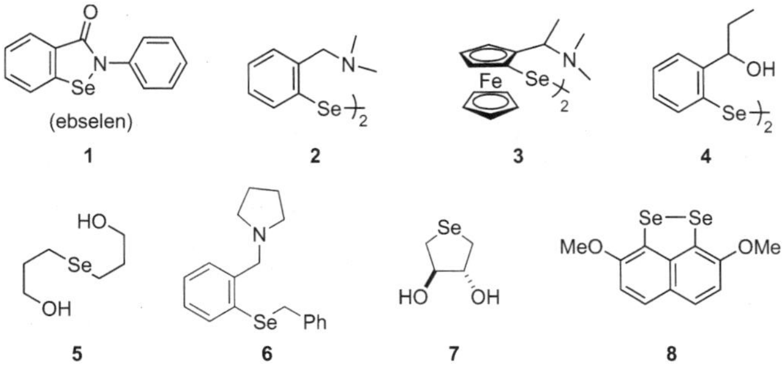

Figure 3. Representative GPx mimics of small organic selenium compounds.

Most GPx mimics are aromatic selenides or diselenides and their analogs. Ebselen (**1**) (*25–29*), with a five-membered isoselenazol ring fused to benzene, is the first lead of the GPx mimics. The recent study by Mugesh (*30, 31*) on the reaction mechanism revealed that precatalyst **1** is activated by the reaction with thiol (RSH) to produce a selenenyl sulfide intermediate. The resulting open-chain intermediate participates in the GPx-like catalytic cycle as shown in Figure 4. When RSH is benzenethiol (PhSH), the selenenyl sulfide intermediate undergoes disproportionation to produce the diselenide, but is converted to the selenol intermediate when a modified PhSH with an oxazoline ring at the ortho position is applied. Thus, the catalytic cycle of **1** depends on the nature of the thiol substrate. In the absence of enough thiol, the selenenic acid intermediate easily cyclizes to precatalyst **1** (*30, 31*). The reaction mechanism of **1** was also studied in details by application of sophisticated theoretical calculations (*32*).

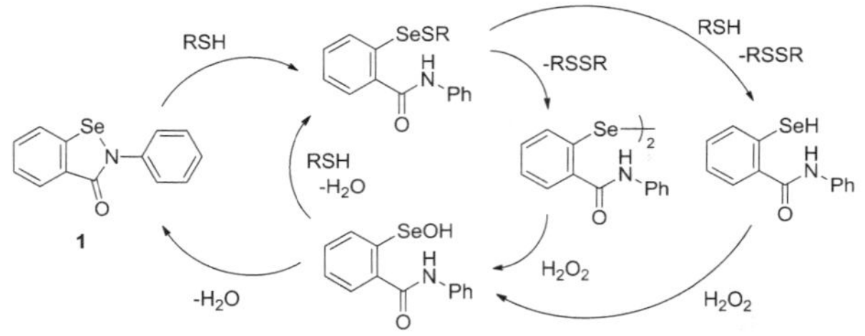

*Figure 4. The catalytic cycle of ebselen (**1**).*

The second generation of GPx mimics consists of the diselenide with an intramolecular coordinating heteroatom, such as **2-4**. Aromatic diselenide **2**, which has a tertiary amino group at the ortho position of the benzene ring, was designed by Wilson (*33*). Similar diselenides, such as **3** and **4**, were later synthesized by Mugesh (*34, 35*) and Wirth (*36, 37*), respectively. In these diselenides, the roles of intramolecular Se···N(*38*) and Se···O(*39*) nonbonding interactions were remarkable. Such weak interactions may exist at the active site of GPx to regulate the reactivity of the active intermediates in the catalytic cycle. Interestingly, diselenides similar to **2-4** can be utilized not only as GPx mimics but also as organoselenium catalysts in the transformation of alkenes to allylic alcohols or ethers (*40–42*).

Aliphatic selenium compounds have less frequently been examined as possible GPx mimics despite Sec itself being an aliphatic selenium compound. This is because the Se–C(sp^3) bond of aliphatic compounds is more easily cleaved than the Se–C(sp^2) bond, resulting in the precipitation of red or black selenium particles. Moreover, the difference in the hybridization of the C atom attached to the Se atom should cause a significant change in the reactivity of the Se atom. Therefore, more attention should be paid to aliphatic selenium compounds as GPx mimics. Indeed, Back (*43*) has developed several aliphatic selenides, such as **5**, as good GPx mimics. When oxidized, selenide **5** can be converted to the spirodioxaselenanonane, which is responsible for the observed high GPx activity. On the other hand, Braga (*44*) reported a high GPx-like activity of benzyl selenide **6**, which can be converted to the highly active hydroxy perhydroxy selenane intermediate (RR′Se(OH)OOH) by the reaction with excess H_2O_2. Our group (*45*) recently found that water-soluble cyclic selenide **7** can be utilized as a mimic of GPx. In the catalytic cycle, the selenide is oxidized with H_2O_2 to the corresponding selenoxide, which is stable but is able to oxidize thiols to disulfides selectively (*46*). Nevertheless, the catalytic cycles proposed for these aliphatic selenide GPx mimics are significantly different from the one determined for GPx enzymes (Figure 1). Therefore, they are not useful for elucidation of structure-activity relationships at the active site of GPx. Thus, the synthesis of various selenocysteine derivatives and investigation of their GPx-like antioxidant activity in relation to their molecular structure are desirable.

An example of a the third type of GPx mimic, the conformationally restricted naphthalene diselenide **8**, was recently investigated by Back (*47*). Upon oxidation with H_2O_2, **8** is converted to the selenolseleninate (Se–Se(O)), which is then reduced with thiol back to **8**. This type of constrained diselenide has also been reported to be a useful mimic of iodothyronine deiodinase (*48*), another representative selenoenzyme that controls the activation and deactivation of thyroid grand hormone thyroxine (*49–51*).

3. GPx Cycle of Selenocysteine

In the GPx cycle, the Sec residue is repeatedly oxidized and reduced by hydroperoxides and GSH, respectively. Singh (*52*) investigated in detail the GPx-like catalytic cycle of selenocystine (Sec$_2$), which is an oxidized dimer of Sec having a diselenide (SeSe) bond. By employing a standard method of GPx assay (Figure 5), the GPx-like antioxidant activity of Sec$_2$ was measured using three different hydroperoxides (H_2O_2, $PhC(CH_3)_2OOH$, and tBuOOH). In this assay, the produced glutathione disulfide (GSSG) is reduced back to GSH with NADPH by glutathione reductase (GR). Therefore, the progress of the reaction can be monitored by the decrease of NADPH, which absorbs at 340 nm. The kinetics of hydroperoxide reduction was analyzed by the Dalziel equation (*53, 54*) applying double-reciprocal plots between the initial velocity of the ROOH reduction (v_0) and the concentration of ROOH, i.e., $1/v_0$ vs. $1/[ROOH]$. The results of this kinetic analysis showed that the active intermediates derived from Sec$_2$ have more affinity toward ROOH than GSH.

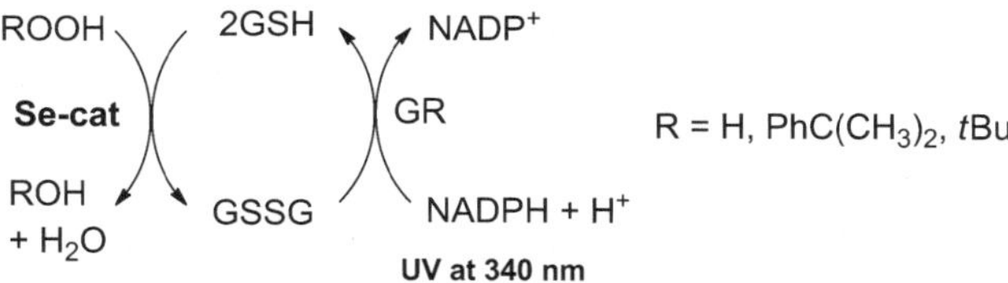

Figure 5. The reaction scheme of the standard assay for the GPx-like antioxidant activity of a selenium catalyst.

An NMR investigation on the reaction mechanism (*52*) revealed that the reaction of Sec$_2$ with dithiothreitol (DTT) is slow in water at pH 7. On the other hand, the reaction of Sec$_2$ with an equimolar amount of H$_2$O$_2$ afforded the seleninic acid (SecO$_2$H; δ_{Se} 1183 ppm) in ca. 33 % yield, suggesting the transient formation of the selenenic acid (SecOH), which is rapidly oxidized to SecO$_2$H in the reaction mixture. When two more equivalents of H$_2$O$_2$ were added, most of Sec$_2$ (88 %) was converted to SecO$_2$H. The reaction mixture was then treated with 6 equivalents of cysteine (Cys) as a thiol substrate to yield the selenenyl sulfide species (SecCys; δ_{Se} 349 ppm) having a mixed SeS bond.

From the NMR experiments as well as the kinetics, the catalytic cycle of Sec$_2$ was proposed as shown in Figure 6 (*52*). Sec$_2$ is preferentially oxidized to SecOH and SecO$_2$H. The subsequent reactions of the oxidized products with GSH (k_1 and k_2) produce SecSG, which would further react with GSH (k_3) to produce the selenol intermediate (SecH). Since SecCys was observed in the reaction of SecO$_2$H and Cys, it can be assumed that the k_1 and k_2 processes are much faster than the k_3 process. This means that the reaction process from the selenenyl sulfide to the selenol (k_3) is the rate-determining step as usually observed in the catalytic cycles of other GPx mimics (*55*).

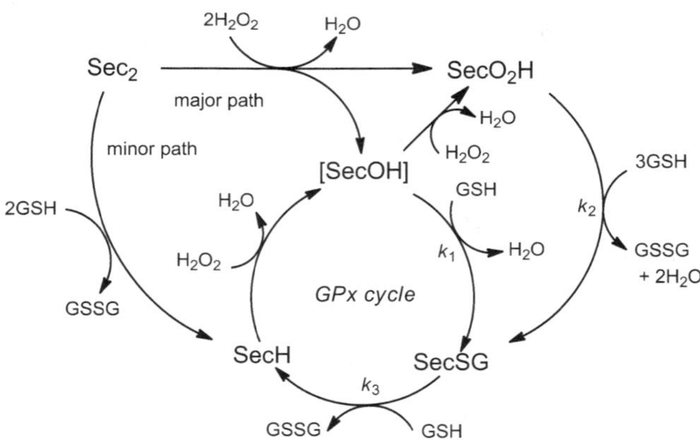

Figure 6. The reaction mechanism for the GPx-like antioxidant activity of selenocystine (Sec₂).

4. Selenocysteine Derivatives as Possible GPx Mimics

In the GPx cycle (Figure 1), the selenol form (E-SeH) and the selenenic acid form (E-SeOH) are interconverted rapidly to promote the redox reaction between H$_2$O$_2$ and GSH. Since these intermediates are highly reactive, it is worth seeking a useful form of Sec derivatives that will be sufficiently stable and reactive. In this section, four types of Sec derivatives are considered as possible GPx precatalysts: diselenides (**A**), selenides (**B**), isoselenazolines (**C**), and selenazolines (**D**) (Figure 7), although the other compounds may also be possible.

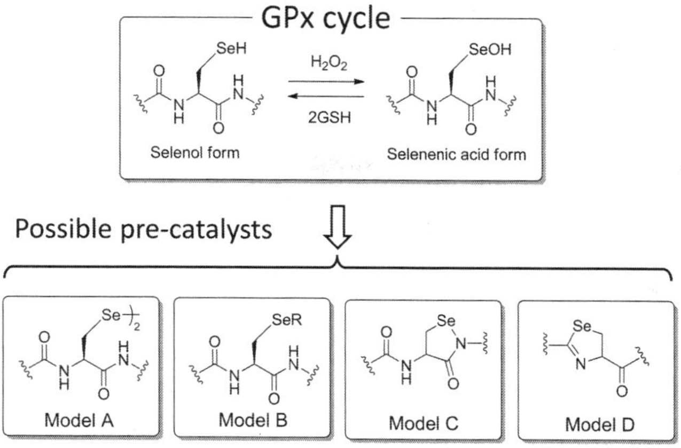

Figure 7. Four models of selenocysteine derivatives as possible GPx precatalysts.

4.1. Diselenide Precatalyst

The diselenide form (**A**) would be the most useful Sec derivative as a GPx precatalyst. Indeed, Sec_2 is a stable material, and the SeSe bond can be activated either by the oxidation with hydroperoxides or by the reduction with GSH to produce the selenenic acid form or the selenol form, respectively. As seen in Figure 6, the former oxidation path is predominant in the case of simple Sec_2 (*52*). But the mechanism could change depending on modifications to Sec_2. Currently, only diselenide precatalysts have been applied as practical selenopeptide models of GPx as described in the next section.

4.2. Selenide Precatalyst

A selenide model (**B**) may be applied as a GPx precatalyst because particular alkyl groups (R) on the Se atom of Sec can be removed under specific conditions. When R is a *p*-methoxyphenylmethyl (MPM or PMB) or *p*-methylbenzyl (MBzl) group, R can be removed from the Se atom under acidic conditions (*56–62*). However, severe conditions not applicable to a GPx assay, such as the use of strong acids at high temperature, are required. Alternatively, the oxidation of the selenide moiety with hydroperoxides would produce the corresponding selenoxide, which can be subsequently converted via *β*-elimination to the selenenic acid and an alkene (*63*). If this elimination occurs selectively for R attached to the Se atom, the selenide model (**B**) would function as a GPx precatalyst. However, the elimination can occur also in the Sec moiety to afford dehydroalanine (Figure 8) (*64*), probably because of the stability of the double bond due to conjugation of the π electrons. In our preliminary experiments, significant formation of dehydroalanine was observed when R was a simple alkyl group like Et and CH_2CH_2Ph. So, in order to design a good GPx model using **B**, this undesirable *β*-elimination of the selenoxide moiety must be suppressed. This will be an interesting issue in future work.

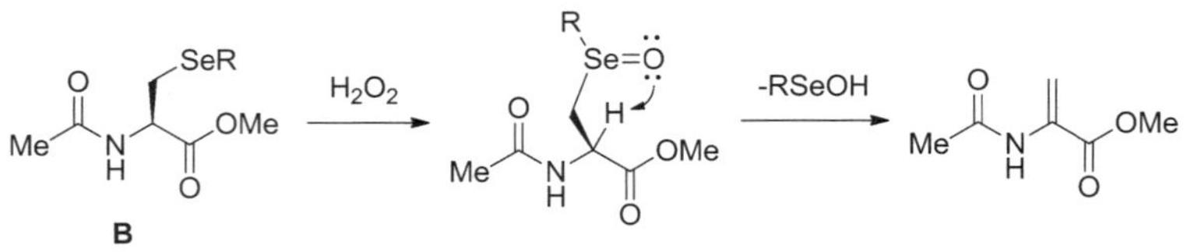

Figure 8. *Formation of the Sec selenoxide by the oxidation of selenide model **B*** *and the rearrangement to dehydroalanine.*

4.3. Isoselenazoline Precatalyst

The next possible form of Sec derivatives as a GPx precatalyst is an isoselenazoline ring model (**C**), which may be derived from the unstable selenenic acid intermediate through the dehydration process as shown in Figure 9. Actually this cyclic model is already found in ebselen (**1**) (*25–29*) and can readily be converted to the selenenyl sulfide form by a ring-opening reaction with thiol (Figure 4). However, without the fused benzene ring, this type of isoselenazoline compounds has never been isolated in the literature, except for 2,4,4-trimethyl-3-isoselenazolidinone (**9**) (*65*). Therefore, the application of the isoselenazoline model (**C**) as a GPx precatalyst will likely be difficult in strong contrast to the case of cysteine, for which the corresponding isothiazoline ring has been characterized as an important intermediate of protein tyrosine phosphatase 1B (PTP1B) (*66, 67*).

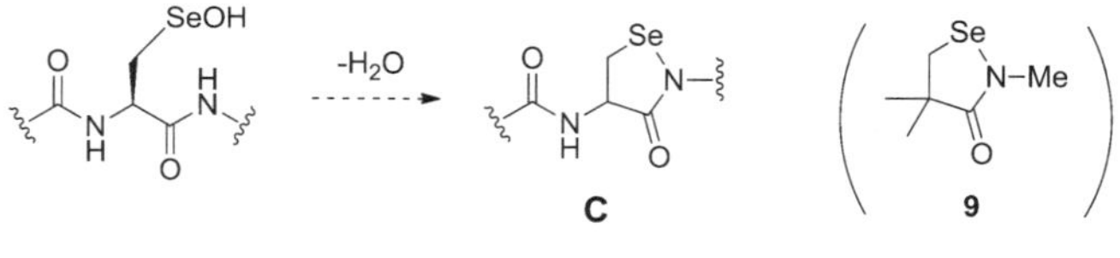

Figure 9. *Hypothetical isoselenazoline model **C**.*

4.4. Selenazoline Precatalyst

Another possible form is a selenazoline model (**D**). The selenazoline would be derived from the selenol intermediate through the nucleophilic cyclization to the backbone carbonyl group of the adjacent amino acid followed by dehydration of the produced selenazolidine intermediate (**10**). However, no selenazoline derivative of selenocysteine is known, expect for the case of α-methyl selenocysteine (*68*). Instability of the selenazoline form is supported by the observation that in the selenocysteine-mediated native chemical ligation, the saturated selenazolidine intermediate (**10**) is preferentially transformed to selenocysteine (**11**), and the selenazoline (**D**) is not obtained (Figure 10) (*69–71*).

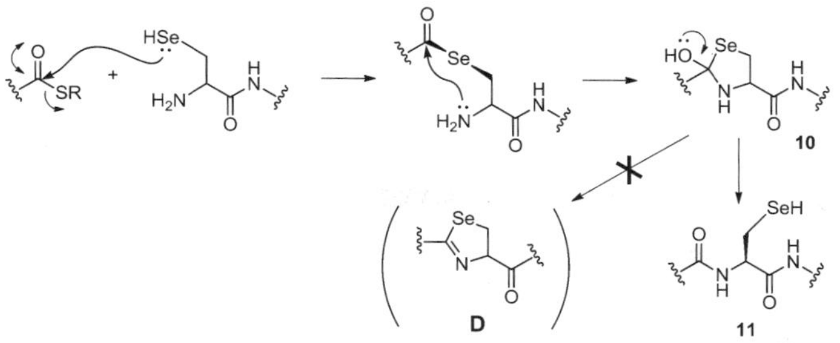

*Figure 10. The mechanism of Sec-mediated native chemical ligation. Selenazoline model **D** is not produced.*

5. Selenopeptide Models

Because the diselenide is the only form of Sec derivatives useable as a GPx precatalyst, several selenopeptides have been explored in their diselenide form as GPx mimics (*72–74*). To obtain these selenopeptides, the solid-phase peptide synthesis (SPPS) method (*72*), chemical mutation of Ser to Sec (*73*) as well as Sec-mediated native chemical ligation (*74*) methods, were employed. Both the Fmoc and Boc protocols are available in the SPPS, in which the MPM and MBzl groups, respectively, are utilized as protecting groups of the Se atom (*75*).

5.1. Selenoglutathione

The GPx-like activity of selenoglutathione (GSeH) was investigated by Yoshida et al (*72*) who showed that the oxidized diselenide form of GSeH (GSeSeG) exhibits a higher GPx-like antioxidant activity than Sec$_2$ and the short selenopeptides discussed in the next subsection. The observed high activity was attributed to the structural similarity to GSSG. When GSeSeG was reacted with NADPH in the presence of GR, generation of the selenol intermediate (GSeH) was observed. Since the reaction did not proceed in the absence of GR, the result clearly indicated that GSeSeG can be a substrate of GR. This would be possible because GSSG and GSeSeG are structurally similar to each other. Thus, GR can accept not only GSSG but also GSeSeG in its reaction pocket (*76, 77*). The subsequent addition of H$_2$O$_2$ produced a mixture of GSeSG and GSSG. The proposed catalytic cycle of GSeSeG is shown in Figure 11. GSeSeG is reduced with NADPH by the function of GR. The produced selenol participates in the main GPx cycle through selenenic acid and selenenyl sulfide intermediates. In addition to this main cycle, there are two possible bypass routes. One involves conversion from the selenenic acid to the diselenide, and the other involves direct conversion from the selenenyl sulfide (GSeSG) to the selenol with NADPH by the function of GR. These bypass routes would significantly enhance the GPx-like activity of GSeSeG.

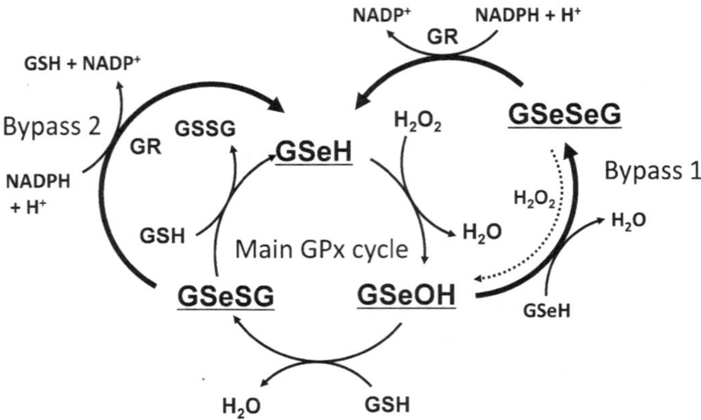

Figure 11. A GPx-like catalytic cycle of selenoglutathione diselenide (GSeSeG).

The presence of these bypass routes would be the main reason for the observed high GPx activity of GSeSeG. However, when the GPx activity of GSeSeG was tested by using dithiothreitol (DTT) instead of GSH in the absence of NADPH and GR, it was still found that GSeSeG exhibits a little higher GPx-like activity than Sec$_2$ does. The result suggested that the carboxylic group of the γ-glutamic acid of GSeH may have an effect on the activity (*72*), although the details are not yet well understood.

5.2. Selenopeptides with the Same Primary Amino Acid Sequence as the GPx Active Site

To investigate the roles of the primary amino acid sequence on the GPx activity, two short selenopeptides in the diselenide form with a different length of the amino acid sequence were investigated (*72*). The first target was a tripeptide with the sequence LUG. The second target was a pentapeptide (SLUGT) that has two more amino acids, one on both sides of the tripeptide. The amino acid sequences of these selenopeptides are the same as the GPx active site (Figure 12).

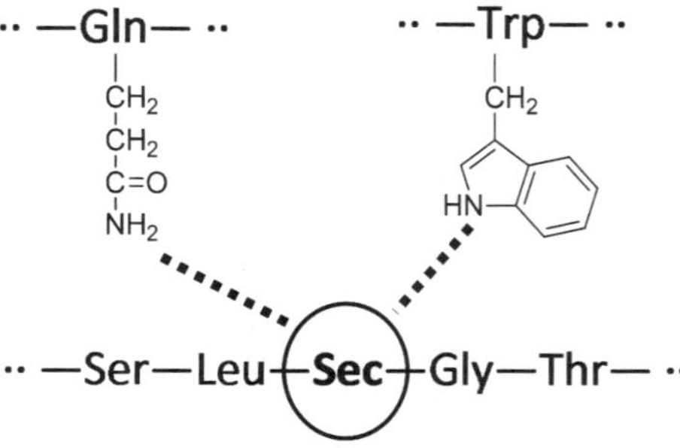

Figure 12. A cartoon diagram of the GPx active site.

172

The GPx-like activity measurement was carried out for the tripeptide and pentapeptide by using the standard assay system (Figure 5). However, the GPx activity of these selenopeptides was found to be lower than that of Sec$_2$. Moreover, the activity gradually decreased with increasing length of the amino acid sequence. These results suggested that the primary amino acid sequence around the GPx active site does not play an important role in the GPx activity when it is isolated from the reaction pocket of the enzyme.

5.3. Selenopeptides Modeling the Catalytic Triad

Since the primary amino acid sequence around the Sec residue of the GPx active site exhibited no significant effect on antioxidant activity (*72*), the next target of the GPx model is a selenopeptide emulating the catalytic triad structure. One simple model would be a short selenopeptide containing Sec (U), Gln (Q), and Trp (W) residues along the amino acid sequence. However, there are challenges to the design of such a target. First, it is not easy to find an amino acid sequence for which U, Q, and W can interact as in GPx. Theoretical calculation of possible conformers of the peptide will be helpful to identify potential peptides. Second, the synthesis of a selenopeptide with a longer amino acid sequence is not easy due to a significant decrease in the yield of SPPS, especially after the introduction of U to the peptide on the resin. Needless to say, the shorter the sequence, the easier the synthesis of the selenopeptide would be. But a short peptide sequence would result in significant structural hindrance, preventing interactions between the U, Q, and W residues. In order to obtain suitable conformational flexibility, these residues must be separated from each other by adequate spacers.

These selenopeptides modeling the catalytic triad of GPx are now being developed in our laboratory. The GPx-like activity measurement of such targets will provide valuable information about the importance of the triad structure in enzymatic function.

5.4. Miscellaneous

An extraordinary high GPx-like antioxidant activity was reported for the 15-mer selenopeptide (WPFLRHNVYGRPRAU)$_2$ (*73*). This selenopeptide was designed by combination of a phage-display peptide library followed by computer-aided molecular modeling and chemical mutation so that it mimics the catalytic center and the substrate binding site of GPx. However, these results were later questioned by Casi and Hilvert (*74*), who suggested that accurate placement of functional groups within a precisely defined three-dimensional structure is important to achieve efficient catalysis.

6. Summary and Perspectives

Although many organic selenium compounds have been examined as GPx mimics (*12, 21–24*), the use of selenocysteine derivatives or selenopeptides has only recently been explored (*52, 72–74*). There are four forms of selenocysteine derivatives applicable as possible precatalysts; the diselenide, selenide, isoselenazoline, and selenazoline. Among these, only the diselenide model is currently a useful precatalyst. Indeed, some selenopeptides in the diselenide form have been targeted as active site models of GPx (*72*). From GPx activity measurements, it was revealed that selenoglutathione diselenide (GSeSeG) is a unique selenopeptide with high GPx-like antioxidant capacity, mainly due to its structural similarity to GSSG (*72*). It was also found that the primary amino acid sequence around the GPx active site, when isolated, does not play a significant role in GPx activity (*72*).

In the future work in this field, it will be interesting to explore new types of selenocysteine precatalysts other than the diselenide model. The selenide model is probably one of the promising candidates. On the other hand, the structure-activity relationship at the GPx active site is still ambiguous. Recently, the presence of the catalytic tetrad, in which a proximate Asn residue is involved as the fourth member of the active site, has been suggested (*78*). Selenopeptides emulating not only the GPx catalytic triad but also the tetrad will also be an interesting target to be explored. Once the reasonable mimics are obtained, we will be able to utilize them as a probe to investigate the mechanism of the enzymatic function based on the structure-activity relationship. Since the site-directed point mutation of selenoproteins is still difficult, a similar approach utilizing designed selenopeptides also would be useful for other selenoenzymes, for which the versatile enzymatic functions have not yet been well elucidated in terms of active site structure.

References

1. Sies, H. *Exper. Physiol.* **1997**, *82*, 291–295.
2. Arteel, G. E.; Sies, H. *Environ. Toxicol. Pharmacol.* **2001**, *10*, 153–158.
3. Santos, O. S.; Saraiva, D. F.; Costa, D. C. F.; Scofano, H. M.; Alves, P. C. C. *Exp. Brain Res.* **2007**, *177*, 347–357.
4. Ren, X.; Yang, L.; Liu, J.; Su, D.; You, D.; Liu, C.; Zhang, K.; Luo, G.; Mu, Y.; Yan, G.; Shen, J. *Arch. Biochem. Biophys.* **2001**, *387*, 250–256.
5. *Redox Chemistry*; Banerjee, R., Ed.; John Wiley & Sons, Inc.: Hoboken, NJ, 2008.
6. Flohé, L.; Gunzler, W. A.; Schock, H. H. *FEBS Lett.* **1973**, *32*, 132–134.
7. Savaskan, N. E.; Ufer, C.; Kühn, H.; Borchert, A. *Biol. Chem.* **2007**, *388*, 1007–1017.
8. Conrad, M.; Schneider, M.; Seiler, A.; Bornkamm, G. W. *Biol. Chem.* **2007**, *388*, 1019–1025.
9. Toppo, S.; Flohé, L.; Ursini, F.; Vanin, S.; Maiorino, M. *Biochim. Biophys. Acta.* **2009**, *1790*, 1486–1500.
10. Hofmann, B.; Hecht, H.-J.; Flohé, L. *Biol. Chem.* **2002**, *383*, 347–364.
11. Wood, Z. A.; Schröder, E.; Harris, J. R.; Poole, L. B. *Trends Biochem. Sci.* **2003**, *28*, 32–40.
12. Mugesh, G.; du Mont, W.-W. *Chem.—Eur. J.* **2001**, *7*, 1365–1370.

13. Ursini, F.; Maiorino, M.; Brigelius-Flohé, R.; Aumann, K. D.; Roveri, A.; Schomburg, D.; Flohé, L. *Method Enzymol.* **1995**, *252*, 38–53.

14. Epp, O.; Ladenstein, R.; Wendel, A. *Eur. J. Biochem.* **1983**, *133*, 51–69.

15. Dimastrogiovanni, D.; Anselmi, M.; Miele, A. E.; Boumis, G.; Petersson, L.; Angelucci, F.; Di Nola, A.; Brunori, M.; Bellelli, A. *Proteins Struct. Funct. Bioinf.* **2010**, *78*, 259–270.

16. Rocher, C.; Lalanne, J.-L.; Chaudière, J. *Eur. J. Biochem.* **1992**, *205*, 955–960.

17. Schnurr, K.; Borchert, A.; Gerth, C.; Anton, M.; Kuhn, H. *Protein Expres. Purif.* **2000**, *19*, 403–410.

18. Maiorino, M.; Aumann, K. D.; Brigelius-Flohé, R.; Doria, D.; van den Heuvel, J.; McCarthy, J.; Roveri, A.; Ursini, F.; Flohé, L. *Biol. Chem. Hoppe-Seyler* **1995**, *376*, 651–660.

19. Scheerer, P.; Borchert, A.; Krauss, N.; Wessner, H.; Gerth, C.; Höhne, W.; Kuhn, H. *Biochemistry* **2007**, *46*, 9041–9049.

20. Mannes, A. M.; Seiler, A.; Bosello, V.; Maiorino, M.; Conrad, M. *FASEB J.* **2011**, *25*, 2135–2144.

21. Mugesh, G.; Singh, H. B. *Chem. Soc. Rev.* **2000**, *29*, 347–357.

22. Mugesh, G.; Du Mont, W. -W.; Sies, H. *Chem. Rev.* **2001**, *101*, 2125–2179.

23. Huang, X.; Liu, X.; Luo, Q.; Liu, J.; Shen, J. *Chem. Soc. Rev.* **2011**, *40*, 1171–1184.

24. Bhuyan, B. J.; Mugesh, G. In *Organoselenium Chemistry*; Wirth, T., Ed.; Wiley-VCH Verlag GmbH & Co. KGaA: Weinheim, Germany, 2012; pp 361–396.

25. Müller, A.; Cadenas, E.; Graf, P.; Sies, H. *Biochem. Pharmacol.* **1984**, *33*, 3235–3239.

26. Wendel, A.; Fausel, M.; Safayhi, H.; Tiegs, G.; Otter, R. *Biochem. Pharmacol.* **1984**, *33*, 3241–3245.

27. Tabuchi, Y.; Sugiyama, N.; Horiuchi, T.; Furusawa, M.; Furuhama, K. *Eur. J. Pharmacol.* **1995**, *272*, 195–201.

28. Imai, H.; Graham, D. I.; Masayasu, H.; Macrae, I. M. *Free Radic. Biol. Med.* **2003**, *34*, 56–63.

29. Ardais, A. P.; Santos, F. W.; Nogueira, C. W. *J. Appl. Toxicol.* **2008**, *28*, 322–328.

30. Sarma, B. K.; Mugesh, G. *Chem.—Eur. J.* **2008**, *14*, 10603–10614.

31. Bhabak, K. P.; Mugesh, G. *Acc. Chem. Res.* **2010**, *43*, 1408–1419.

32. Antony, S.; Bayse, C. A. *Inorg. Chem.* **2011**, *50*, 12075–12084.

33. Wilson, S. R.; Zucker, P. A.; Huang, R. -R. C.; Spector, A. *J. Am. Chem. Soc.* **1989**, *111*, 5936–5939.

34. Mugesh, G.; Panda, A.; Singh, H. B.; Punekar, N. S.; Butcher, R. J. *Chem. Commun.* **1998**, 2227–2228.

35. Mugesh, G.; Panda, A.; Singh, H. B.; Punekar, N. S.; Butcher, R. J. *J. Am. Chem. Soc.* **2001**, *123*, 839–850.

36. Wirth, T. *Molecules* **1998**, *3*, 164–166.

37. Jauslin, M. L.; Wirth, T.; Meier, T.; Schoumacher, F. *Hum. Mol. Genet.* **2002**, *11*, 3055–3063.

38. Iwaoka, M.; Tomoda, S. *J. Am. Chem. Soc.* **1996**, *118*, 8077–8084.

39. Iwaoka, M.; Komtasu, H.; Katsuda, T.; Tomoda, S. *J. Am. Chem. Soc.* **2004**, *126*, 5309–5317.

40. Fujita, K.; Iwaoka, M.; Tomoda, S. *Chem. Lett.* **1994**, 923–926.

41. Fukuzawa, S.; Takahashi, K.; Kato, H.; Yamazaki, H. *J. Org. Chem.* **1997**, *62*, 7711–7716.

42. Wirth, T.; Häuptli, S.; Leuenberger, M. *Tetrahedron Asymmetry* **1998**, *9*, 547–550.

43. Back, T. G.; Moussa, Z.; Parvez, M. *Angew. Chem., Int. Ed.* **2004**, *43*, 1268–1270.

44. Nascimento, V.; Alberto, E. E.; Tondo, D. W.; Dambrowski, D.; Detty, M. R.; Nome, F.; Braga, A. L. *J. Am. Chem. Soc.* **2012**, *134*, 138–141.

45. Kumakura, F.; Mishra, B.; Priyadarsini, K. I. *Eur. J. Org. Chem.* **2010**, 440–445.

46. Araia, K.; Noguchia, M.; Singh, B. G.; Priyadarsini, K. I.; Fujio, K.; Kubo, Y.; Takayama, K.; Ando, S.; Iwaoka, M. *FEBS Open Bio* **2013**, *3*, 55–64.

47. Press, D. J.; Back, T. G. *Org. Lett.* **2011**, *13*, 4104–4107.

48. Manna, D.; Mugesh, G. *J. Am. Chem. Soc.* **2011**, *133*, 9980–9983.

49. Williams, G. R.; Bassett, J. H. D. *J. Endocrinol.* **2011**, *209*, 261–272.

50. Maia, A. L.; Goemann, I. M.; Meyer, E. L. S.; Wajner, S. M. *J. Endocrinol.* **2011**, *209*, 283–297.

51. Dentice, M.; Ambrosio, R.; Salvatore, D. *Expert Opin. Ther. Targets* **2009**, *13*, 1363–1373.

52. Singh, B. G.; Bag, P. P.; Kumakura, F.; Iwaoka, M.; Priyadarsini, K. I. *Bull. Chem. Soc. Jpn.* **2010**, *83*, 703–708.

53. Dalziel, K. *Acta Chem. Scand.* **1957**, *11*, 1706–1723.

54. Sztajer, H.; Gamain, B.; Aumann, K. -D.; Slomianny, C.; Becker, K.; Brigelius-Flohé, R.; Flohé, L. *J. Biol. Chem.* **2001**, *276*, 7397–7403.

55. Sarma, B. K.; Mugesh, G. *J. Am. Chem. Soc.* **2005**, *127*, 11477–11485.

56. Koide, T.; Itoh, H.; Otaka, A.; Yasui, H.; Kuroda, M.; Esaki, N.; Soda, K.; Fujii, N. *Chem. Pharm. Bull.* **1993**, *41*, 502–506.

57. Moroder, L.; Besse, D.; Musiol, H. J.; Rudolph-Bohner, S.; Siedler, F. *Biopolymers* **1996**, *40*, 207–234.

58. Besse, D.; Moroder, L. *J. Pept. Sci.* **1997**, *3*, 442–453.

59. Hondal, R. J. *Protein Pept. Lett.* **2005**, *12*, 757–764.

60. Harris, K. M.; Flemer, S., Jr.; Hondal, R. J. *J. Pept. Sci.* **2007**, *13*, 81–93.

61. Oikawa, T.; Esaki, N.; Tanaka, H.; Soda, K. *Proc. Natl. Acad. Sci. U.S.A.* **1991**, *88*, 3057–3059.

62. Rajarathnam, K.; Sykes, B. D.; Dewald, B.; Baggiolini, M.; Clark-Lewis, I. *Biochemistry* **1999**, *38*, 7653–7658.

63. Nishibayashi, Y.; Uemura, S. In *Organoselenium Chemistry*; Wirth, T., Ed.; Wiley-VCH Verlag GmbH & Co. KGaA: Weinheim, Germany, 2012; pp 287–320.

64. Okeley, N. M.; Zhu, Y.; van der Donk, W. A. *Org. Lett.* **2000**, *2*, 3603–3606.

65. Reich, H. J.; Jasperse, C. P. *J. Am. Chem. Soc.* **1987**, *109*, 5549–5551.

66. Salmeen, A.; Andersen, J. N.; Myers, M. P.; Meng, T.-C.; Hinks, J. A.; Tonks, N. K.; Barford, D. *Nature* **2003**, *423*, 769–773.

67. van Montfort, R. L. M.; Congreve, M.; Tisi, D.; Carr, R.; Jhoti, H. *Nature* **2003**, *423*, 773–777.

68. Makiyama, A.; Komatsu, I.; Iwaoka, M.; Yatagai, M. *Phosphorus, Sulfur Silicon Relat. Elem.* **2011**, *186*, 125–133.

69. Hondal, R. J.; Nilsson, B. L.; Raines, R. T. *J. Am. Chem. Soc.* **2001**, *123*, 5140–5141.

70. Gieselman, M. D.; Xie, L.; van der Donk, W. A. *Org. Lett.* **2001**, *3*, 1331–1334.

71. Quaderer, R.; Sewing, A.; Hilvert, D. *Helv. Chim. Acta* **2001**, *84*, 1197–1206.

72. Yoshida, S.; Kumakura, F.; Komatsu, I.; Arai, K.; Onuma, Y.; Hojo, H.; Singh, B. G.; Priyadarsini, K. I.; Iwaoka, M. *Angew. Chem., Int. Ed.* **2011**, *50*, 2125–2128.

73. Sun, Y.; Li, T.; Chen, H.; Zhang, K.; Zheng, K.; Mu, Y.; Yan, G.; Li, W.; Shen, J.; Luo, G. *J. Biol. Chem.* **2004**, *279*, 37235–37240.

74. Casi, G.; Hilvert, D. *J. Biol. Chem.* **2007**, *282*, 30518–30522.

75. Muttenthaler, M.; Alewood, P. F. *J. Pept. Sci.* **2008**, *14*, 1223–1239.

76. Beld, J.; Woycechowsky, K. J.; Hilvert, D. *Biochemistry* **2007**, *46*, 5382–5390.

77. Beld, J.; Woycechowsky, K. J.; Hilvert, D. *Biochemistry* **2008**, *47*, 6985–6987.

78. Tosatto, S. C. E.; Bosello, V.; Fogolari, F.; Mauri, P.; Roveri, A.; Toppo, S.; Flohé, L.; Ursini, F.; Maiorino, M. *Antioxid. Redox Signaling* **2008**, *10*, 1515–1525.

Chapter 9

Modeling of Mechanisms of Selenium Bioactivity Using Density Functional Theory

Craig A. Bayse[*]

Department of Chemistry and Biochemistry, Old Dominion University, Hampton Boulevard, Norfolk, Virginia 23529
*E-mail: cbayse@odu.edu.

Selenium plays an important role in normal biological function through its incorporation into processes that protect against oxidative stress and promote thyroid function. The complex nature of the mechanisms involved has led to extensive study through computational modeling. However, density functional theory (DFT) modeling of the redox mechanisms of selenium is complicated by proton exchange processes catalyzed by the bulk solvent. To address this issue, we have used a method of microsolvation which we have dubbed solvent-assisted proton exchange (SAPE). These models include a few strategically placed water molecules to act as a proton shuttle and are able to obtain activation barriers that are in good agreement with experimental data and trends. Models that do not use these networks necessarily have higher barriers because they force the proton exchange through an 'unnatural' pathway that is not representative of the reaction as it occurs in solution. As a first approximation to solution-phase reactivity, SAPE models are more appropriate for drawing conclusions about the trends and energetics of these complex mechanisms. From these models, we have gained significant insight into the biological pathways of these compounds. In this chapter, we discuss our implementation of SAPE models for redox scavenging by various biorelevant organoselenium compounds, specifically arylselenols, seleninic acids and ebselen, and the related mechanism of cysteine. Finally, we discuss approaches to modeling additional important biochalcogen processes in the thyroid and in zinc finger proteins.

Introduction

The redox chemistry of selenium is important to biological control of oxidative stress and biochemical signaling related to reactive oxygen species (ROS) (*1–4*). Selenoenzymes based on the 21st amino acid selenocysteine (SeCys) help prevent ROS-induced cellular damage and may reduce risk of a variety of diseases such as cancer (*5, 6*) and cardiovascular disease, including stroke and atherosclerosis (*7–9*). In addition to several classes of antioxidant proteins such as glutathione peroxidase (GPx) and thioredoxin reductase (*10*), selenium plays an important role in the thyroid. Studies have also shown that dietary intake of selenium correlates with reduced risk of cancer even at levels above optimal protein function (*11*). As a result, natural and synthetic selenium compounds such as selenomethionine and ebselen have been recommended as supplements to the antioxidant properties of selenoproteins. The synthetic compound ebselen has even been subjected to clinical trials for stroke prevention (*12*). GPx has been a key target of these studies with large numbers of compounds investigated for their ability to catalyze the same overall reaction (GPx mimics) (*13–15*). Selenium compounds have also been shown to inhibit tumor growth (*16, 17*). However, the extensive SELECT trial (*18*) that examined the effect of selenomethionine and vitamin E on the incidence of prostate cancer demonstrated no benefit from either supplement (*19*). For selenium, this result was surprising in terms of the earlier NPC study which suggested potential broad effect against non-cutaneous cancers (*20*). One potential conclusion from the SELECT and NPC studies may be that beneficial effects of selenium may be limited to populations with low nutritional Se intake, whereas those with normal intake may see a negative effect (*19*). These results show a need for a greater understanding of the biochemical mechanisms of selenium compounds.

The narrow range over which selenium supplementation is beneficial is also a roadblock to therapeutic applications. Above dosages of 800 µg/day, symptoms of selenosis including fatigue, hair loss, bad breath, etc. begin to appear. Selenium toxicity may result from exhaustion of the glutathione pool (imbalance in the GSH:GSSG ratio), oxidation or selenenization of protein thiols, or other mechanisms. Reducible organoselenium compounds have been shown to inhibit zinc-finger transcription factors by oxidizing the Cys residues and releasing Zn^{2+} (*21–23*). While this mechanism may be related to selenium activity against some tumors, disruption of transcription and DNA repair in normal cells could result in long-term damage to genomic stability (*22*). Understanding the role of selenium in these important biological processes is important to harnessing these compounds to treat various illnesses.

Over the past several years, our group has worked to unravel these mechanisms through a series of DFT calculations on redox properties of biochalcogen compounds using a technique of microsolvation that we call solvent-assisted proton exchange (SAPE) (*24–29*). SAPE models incorporate a network of explicit solvent molecules, in our case water, into the gas-phase DFT model in order to provide a pathway for proton transfer. Including a few solvent molecules allows a model to be designed that is consistent with the expected changes in the heavy-atom bonding along reaction pathway while approximating the role that the bulk solvent plays in shuttling protons. Gas-phase DFT models that do not include such networks of explicit water molecules tend to have unrealistically high activation barriers because they force the proton exchange through highly strained transition states that are not representative of how the reaction occurs in solution. In this chapter, we review our experiences applying these models to the ROS-scavenging mechanisms of several representative biologically important organoselenium compounds and the related redox processes of Cys. We also discuss preliminary results on aspects of release of Zn^{2+} from zinc finger proteins and our progress in the modeling of the role of selenium in thyroid activity.

Models of the Redox Mechanisms of GPx Mimics

The antioxidant selenoprotein GPx has been proposed to scavenge ROS by a three-step mechanism. From the resting state selenol E-SeH, GPx consumes one equivalent of ROS and is oxidized to the selenenic acid E-SeOH. Two equivalents of thiol are required to reduce E-SeOH back to the resting state through a selenenyl sulfide E-SeSG intermediate (Scheme 1). Various organoselenium compounds including selenols, diselenides, selenides, selenenamides, selenenates and seleninates catalyze the same overall reaction as GPx (eq 1) (*13, 14*).

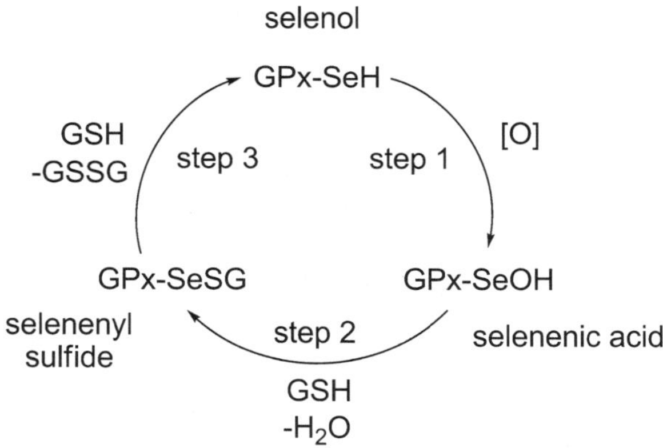

Scheme 1. Mechanism for redox scavenging by GPx and other selenols

$$ROOH + 2\ R'SH \rightarrow ROH + R'SSR' + H_2O \qquad (eq\ 1)$$

Many of these GPx mimics were originally designed to include either a Se-N bond or a close Se⋯N interaction due to the conservation of two nitrogen-containing amino acids (Trp and Gln) in the GPx active site (*30*). Together with SeCys, these residues have been referred to as a "catalytic triad" for GPx activity. Wirth later showed that Se···O interactions also enhanced GPx activity in a series of diaryldiselenides with oxygen-containing pendants (*31*). Because the selenol functionality most directly related to GPx is easily oxidized, researchers have typically synthesized diselenides as a precursor. Diselenides are likely to be activated by thiol cleavage of the Se-Se bond to form equivalents of selenol and selenenyl sulfide which enter the GPx-like cycle. Although diselenides have also been proposed as a key intermediate in ROS scavenging (*15, 32–34*), we do not believe that they contribute significantly because diselenide formation is second order in the concentration of selenium and the disproportionation from selenenyl sulfide is slow. Ortho substitution of the N- or O-containing Lewis basic donor in a diaryl diselenide has been shown to modify GPx-like activity (*13, 14, 30*). Mugesh and du Mont (*30*) described the role of the pendant group as (a) activating the diselenide bond for cleavage by the thiol, (b) stabilizing the selenenic acid against over-oxidation to the seleninic acid, and (c) increasing the electrophilicity of the sulfur atom of the selenenyl sulfide. Se···N,O interactions in particular have been used to stabilize reactive selenium functionalities as well as tune the redox activity of GPx mimics with a large body of experimental literature. Diselenides with strong Se···N,O interactions tend to be inactive (*35*), but overall trends in the reactivity were difficult to anticipate (*36*). In addition to the diselenide-based mimics, a number of selenides and heterocyclic organoselenium compounds containing an Se-N or Se-O bond have been reported (Scheme 2). The selenenamide ebselen is probably the most studied GPx mimic despite its low overall activity. Back and Singh reported heterocyclic GPx mimics based upon selenuranes and selene(i)nate esters (*37–45*). While the mechanism of most diselenides is assumed to proceed through a GPx-like mechanism of monoselenium compounds, these heterocyclic compounds have much more complex mechanisms due the possible pathways resulting from the cyclic Se-N and Se-O bonds. Experimental study of the mechanisms of these compounds is made difficult by the instability of key intermediates and the complex nature of the mechanisms.

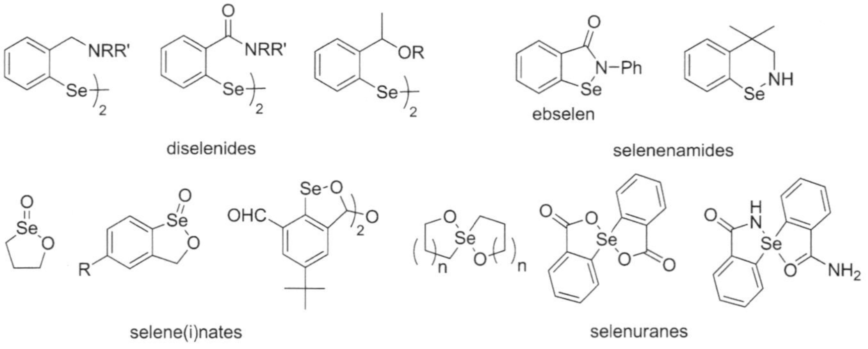

diselenides   ebselen   selenenamides

selene(i)nates   selenuranes

Scheme 2. Representative examples of various classes of GPx mimics.

$$\text{L: Se–X} \leftrightarrow \text{L–Se}^+ \text{X}^- \qquad (eq\ 2)$$

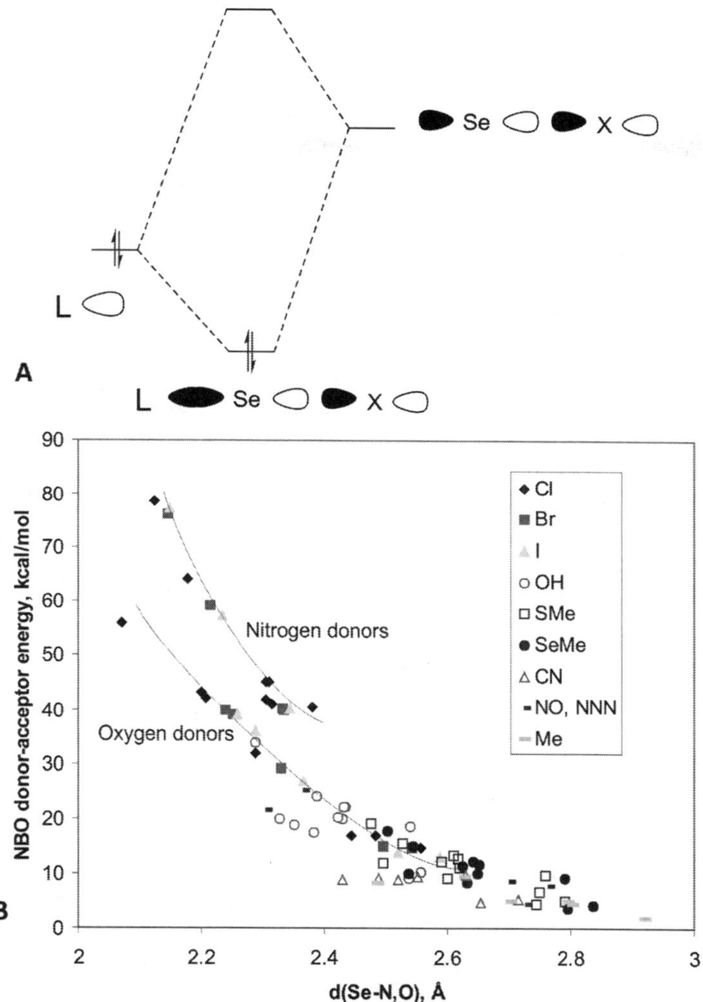

Figure 1. (A) MO diagram for the Se···N,O donor-acceptor interaction. (B) Trends in Se···N,O donor-acceptor strength ($\Delta E_{d \to a}$) versus the Se-N,O bond distance from reference (28). Reproduced by permission of The Royal Society of Chemistry.

Early on, we sought to relate the GPx activity of known mimics to the strength of their Se···N,O interactions through a series of DFT calculations (*46*). In terms of valence bond theory (*47*), the resonance structures for the donation of a pair of electrons from L to an Se-X bond can be written for the interaction similar to a three-center-four-electron hypervalent bond (eq 2). The strength of the interaction depends upon both the Lewis basicity of L and the strength of the Se-X bond. The admixture of the right-hand resonance structure will be increased with both strong Lewis bases and weak Se-X bonds. In the molecular orbital (MO) picture, the occupied lone pair MO of the donor mixes with the Se-X antibonding MO to form a bonding MO (Figure 1A). Stronger Lewis basic donors destabilize the lone pair MO and weaker Se-X bonds stabilize $\varphi_{Se-X}*$. The donor-acceptor energy $\Delta E_{d \to a}$ of the interaction can be calculated from through the localization scheme in a Natural Bond Orbital (NBO) analysis (*48*). Various groups have also used NBO analysis to explore the donor-acceptor interaction between selenium and nitrogen-, oxygen- and halogen-containing pendant groups (*49–58*). Trends in the $\Delta E_{d \to a}$ values and the Se$\cdots$N,Obond distance for a large series of divalent ortho-substituted organoselenium compounds with a variety of Se-X bonds (Figure 1B) are consistent with the relative strengths of both the donor and the Se-X acceptor (*28*). The strongest interactions in terms of both of these properties are found for nitrogen donors and highly polarized selenium-halogen bonds and the weakest for oxygen donors and highly covalent selenium-carbon bonds. Although nitrogen donors are generally stronger than oxygen donors, the functionality of each is important. In amides, the carbonyl oxygen is more Lewis basic than the nitrogen due to the resonance structures. Based upon our calculations, we predicted that the Gln residue in the catalytic triad of GPx would form Se$\cdots$Ointeractions with the Se-OH and Se-SG intermediates rather than the Se$\cdots$Ninteraction expected from earlier discussions (*46*). However, Morokuma later showed that neither interaction is favored in their DFT models of the GPx active site (*59*).

Arylselenols

Although these DFT calculations of isolated molecules are useful in understanding the trends in the strengths of the Se$\cdots$N,Ointeractions themselves, they do not give much information on the effect of these interactions on the activity of the GPx mimics. Each step in the GPx-like mechanism requires the exchange of a proton from one heavy atom to another over the course of the reaction. For example, in the second step (eq 3), the thiol proton is transferred to the leaving water with the formation of the Se-S bond in the overall reaction.

$$E\text{-SeOH} + RS\underline{H} \rightarrow E\text{-SeSR} + \underline{H}OH \qquad\qquad (eq\ 3)$$

A "traditional" gas-phase theoretical model would require the proton to be passed directly from the sulfur to the oxygen. This constraint forces the model to adopt a four-centered transition state that is necessarily high in energy as well as unrealistic from a solution-phase chemistry point of view (Scheme 3A). The reaction should occur through a backside attack of the thiol on the −SeOH group to displace water, yet the purely gas-phase model cannot accommodate both the Se-S bond formation and the proton exchange simultaneously. Our solution, one adopted from other computational models that have faced similar problems (*60, 61*), was to orient the reacting molecules for the backside attack and add water molecules to serve as a proton shuttle pathway (Scheme 3B). This type of microsolvation model, which we refer to as solvent-assisted proton exchange or

SAPE, can be considered a first approximation to role of the solvent in proton transfer processes. To test the ability of these SAPE models to effectively represent the redox chemistry of organoselenium compounds, we selected the GPx-like mechanism of PhSeH because the mechanism was simple and the results could be related to experimental and theoretical kinetics data of GPx. As discussed for step two of the GPx-like mechanism, the SAPE models of each step in Scheme 1 were designed by considering how these reactions would be likely to occur in solution. The reacting molecules were then positioned to allow the expected approach of the heavy atoms and water molecules were added to bridge the site of protonation in the reactant with the site in the product (as in Scheme 3). We use the smallest network required to bridge these sites in order to avoid computationally expensive conformation searches (generally two to four waters). Lundin et al. showed that adding more water molecules does not significantly reduce the barrier for the oxidation of alkenes to epoxides (*62*). Additionally, two- and three-water SAPE networks produced similar activation barriers for oxygen atom transfer reactions to organoselenium compounds (*25*). For the process in Scheme 3, at least two water molecules are required to maintain the hydrogen bonding network between the leaving group and a thiol approaching from the backside. However, examination of the dependence of the size of the SAPE network in step 2 showed that proton transfer between opposite corners of a square four-water network produced a lower barrier than a two-water network (*24*). Strain in the hydrogen bonding network of the smaller SAPE model led to a less than optimal approach of the thiol nucleophile and a higher activation barrier. Similarly, a one-water network would force the thiol to approach from the side and also lead to a strained transition state. Therefore, care must be taken in the design of the overall model to ensure that the reactant molecules are oriented for a proper approach after the water molecules are included and the system relaxed.

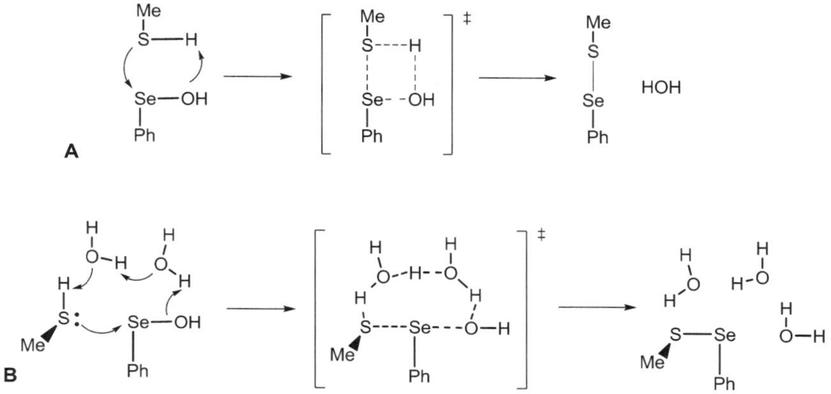

Scheme 3. Comparison of the modeling approach to step 2 of the GPx-like mechanism using (A) direct proton transfer through a strained four-centered TS and (B) indirect transfer through a network of explicit water molecules as in SAPE.

DFT geometry optimizations of the SAPE reaction pathway were conducted using a TZVP+-quality basis set on the atoms involved in the reaction. Energies were corrected for solvation in bulk water. The type of xc functional is important to obtaining reasonable activation barriers for SAPE models (*63*). In test calculations of step 1 for PhSeH, pure functionals such as BLYP produced activation barriers that were unrealistically low. Hybrid functionals, such as mPW1PW91 (used in all of our SAPE studies), generally provided the best comparison with benchmark MP2 calculations, but activation barriers tended to be dependent upon the admixture of Hartree-Fock exchange in individual functionals. Functionals with large contributions from the HF exchange (i.e., 50% in BHandHLYP) significantly over-estimated the activation barrier. Testing of additional functionals is currently being explored.

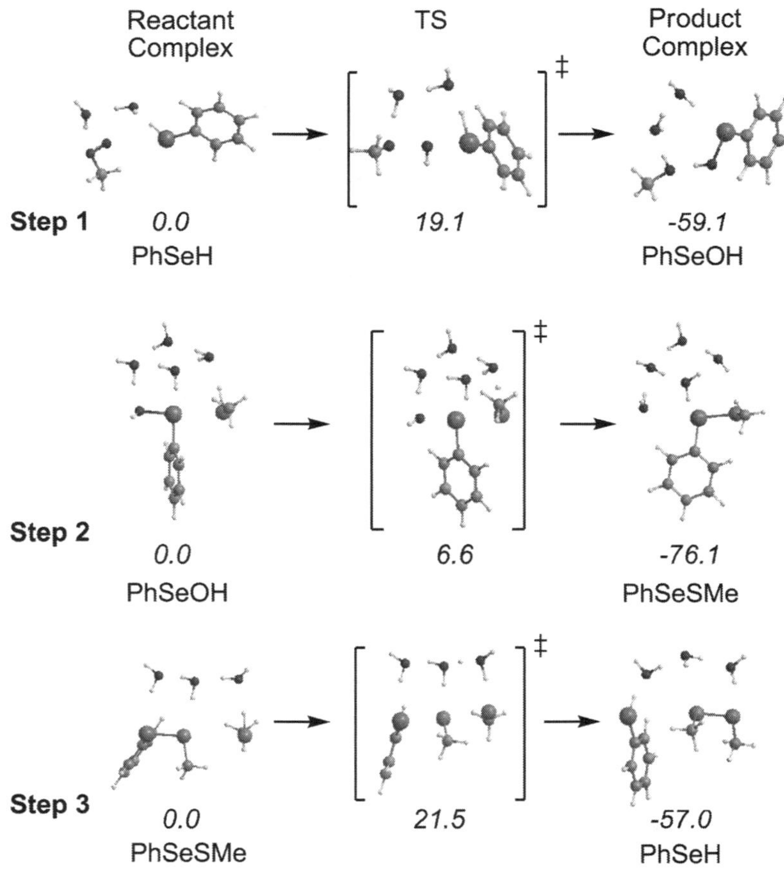

Figure 2. Structures and activation barriers the GPx-like cycle of PhSeH determined by SAPE modeling (24).

The DFT-SAPE activation barriers we obtained for the three steps of the GPx-like mechanism were consistent with the known kinetics of GPx (Figure 2) (*24*). The activation barrier for the oxidation of PhSeH (step 1) was in agreement with both experimental rate constants and Morokuma's results for the DFT transition state of a model of the GPx active site (*59*). The low barrier (6.6 kcal/mol) for the reduction of the selenenic acid (step 2) is consistent with the reactivity of that functional group (*64*). The highest barrier was found for the regeneration of the selenol (step 3, 21.5 kcal/mol), which is known to be the rate-determining step. This qualitative agreement between the results of this study and experimental and theoretical results was promising for study of more complex mechanisms of other organoselenium compounds.

The importance of providing a path for proton exchange in these steps is shown through a comparison of models of the GPx-like mechanism through direct proton transfer (*65, 66*). In these models, strained, high-energy transition states were obtained by the requirement that the protons be transferred directly to the acceptor atom. The transition states were not only strained, but they did not allow the heavy atoms to approach one another naturally. The inclusion of explicit solvent molecules and a path for proton exchange that allows the heavy atoms to approach in the expected fashion provides a more realistic model for the reaction. Further improvements to SAPE models would include an additional shell of water molecules handled with either full QM or a hybrid QM/MM theory. However, more explicit solvent molecules will require molecular dynamics to account for the large number of conformations available to the system.

Following the promising results of the PhSeH SAPE study, we used similar models to test the effect of Se$\cdots$N,O interactions on the three steps of the GPx-like mechanism. SAPE studies were conducted on a series of calculations on aryl selenols ortho-substituted with pendant donor groups of varying strength. The resulting activation barriers were compared to PhSeH and p-MeOC$_6$H$_4$SeH as a contrast to unsubstituted selenols and substitutions unable to interact with the selenium center. Pendant groups were selected based upon their strength –CH$_2$OH (weak interaction), –CH$_2$NR$_2$ (strong interaction) and –NO$_2$ (strong interaction with aromatic stabilization). In the latter, the Se$\cdots$O interaction is further strengthened by the unsaturated nature of the nitro group. Donation of the lone-pair from oxygen to the selenium forms a five-membered ring with the correct number of electrons to satisfy the Hückel rule for aromaticity (*67, 68*) such that the resulting interaction is stronger than one without the additional stabilization. For example, comparing ortho-substituted areneselenenyl chlorides, the Se$\cdots$O interaction with the nitro group is twice that of the benzylic alcohol ($\Delta E_{d\rightarrow a}$ = 32.0 vs 16.8 kcal/mol). Consideration of a range of pendant groups was important to understanding the varied effects of Se$\cdots$N,O interactions upon the activity of GPx mimics.

In the SAPE studies of the GPx-like cycle for these substituted aryl selenols, the effect of the pendant groups on step 1 is an overall lowering of the barrier to selenol oxidation. However, the Se$\cdots$N,O interaction actually increases the uncorrected activation barriers by 2-6 kcal/mol relative to PhSeH. It is the increased solvation of the transition state that leads to the overall decrease in activation barrier. For step 2, the pendant groups increase the activation barrier because they must be displaced from their donor-acceptor interaction with the selenenic acid in order for the thiol to approach the Se center from the backside of the –OH group. Although the activation barrier from the thiol-coordinated intermediate is comparable to the reduction of PhSeOH to PhSeSMe, the overall barrier is increased by the penalty for replacing the pendant group with the thiol (+5-15 kcal/mol). This penalty correlates well with the strength of the Se$\cdots$N,O interaction with the –SeOH group. The increase in the barrier as a result of the Se$\cdots$N,O interaction is consistent with the use of internal coordination as a means to protect reactive selenium functionalities such as selenenyl halides. For step 3, Se$\cdots$N,O interactions with the –SeSR group increase the partial negative charge of the sulfur at which the attack must occur for disulfide bond formation. Mugesh has shown that these interactions decrease the GPx-like activity of diselenides by favoring thiol exchange at the selenium center (*56*). Our calculations support these results as those compounds with strong interactions tend to have higher barriers to regeneration of the selenol. However, if the Se$\cdots$N,O interaction is displaced, the partial charge at sulfur will decrease and the center will be more susceptible to nucleophilic attack. Although Se$\cdots$N,O interactions to the selenenyl sulfide group are generally weak, the strong donation of the tertiary amine and nitro groups are difficult to displace leading to higher overall barriers. This weaker interaction, and potentially the greater solubility in water, of the alcohol and primary amine donor lead to barriers comparable to the unsubstituted PhSeH or *p*-MeOC$_6$H$_4$SeH. Therefore, weak Se$\cdots$N,O interactions may be key to cycling through the GPx-like mechanism because they increase the rate of oxidation and allow a low barrier to regenerate the selenol. Comparison of the SAPE results to experimental trends are difficult because various conditions are used by different researcher in their kinetics studies. However, our conclusions appear to be in agreement with a series of paper on amino-substituted diselenides (*69, 70*). Interesting, and a problem for future experimental studies, is the possibility that alternate pathways for catalytic redox scavenging open due to the high barrier to step 3 of the GPx cycle.

PhSeO$_2$H

Although GPx mimics have been considered as treatments or preventatives for a number of ROS-related diseases, the low threshold for selenium toxicity is a major concern for these applications. Seleninic acids and other reducible selenium (rSe) compounds react with a range of biological thiols including the Cys ligands of ZF proteins. Seleninic acids (RSeO$_2$H) release Zn^{2+} from zinc-finger proteins (*21, 22*), regulate the activity of tumor-suppressor protein p53 (*71*) and inhibit thioredoxin reductase (*5, 72*). This over-oxidized form of selenium is found in commercial selenium supplements (*73*), as an oxidative product of selenoamino acids (*74*), and as a intermediate in a proposed GPx-like mechanism of ebselen (*75*). The X-ray structures of GPx and the semisynthetic selenosubtilisin show that the active site SeCys has been air-oxidized to the seleninic acid form (*76, 77*). Two equivalents of thiol are required to reduce RSeO$_2$H to RSeOH and a third gives the selenenyl sulfide form. The role that thiol reduction plays in the biological activity of this oxidized form of selenium is important to understanding its toxicity.

step 1 (k_a) : $RSeO_2H + RSH \rightarrow$ intermediate (λ_{max} = 265 nm)
step 2 (k_b):   intermediate + RSH → RSeOH   *slow*   (eq 4)
$RSeOH + RSH \rightarrow RSeSR$ *fast*

Experimental studies by Kice et al. (*78*) suggested that the reduction of a seleninic acid to the selenenyl sulfide was a two step process through an observable intermediate (eq 4). They identified the intermediate as either a seleninyl sulfide (**A**) or a hypervalent selenurane (**B**) resulting from reaction of the first equivalent of thiol with seleninic acid (Scheme 4). Selenuranes such as **B** have been proposed as intermediates in ROS scavenging by several organoselenium compounds (*79–81*). The second step consisted of the conversion of the intermediate to the selenenic acid, which is rapidly converted to the selenenyl sulfide. SAPE models of this mechanism (*26*) considered pathways for the conversion of $RSeO_2H$ to RSeOH through both **A** and **B**. Activation barriers for the competing pathways to **A** or **B** suggested that the direct reduction of $RSeO_2H$ to **A** was unlikely due to its high barrier (R = Ph: 18.1 kcal/mol) relative to the addition of thiol to form **B** (7.8 kcal/mol). However, the conversion of **B** to **A** by loss of water was shown to have a low barrier (4.5 kcal/mol). Therefore, the intermediate observed by Kice and Lee is **A** formed in a stepwise fashion through **B** (Scheme 4). Although the subsequent activation barriers for reduction to RSeOH were lower from the selenurane **B** (14.1 kcal/mol vs 16.9 kcal/mol from **A**), reaction through **B** may be possible under more acidic conditions. The final reaction from RSeOH to RSeSR′ is a low barrier process (*vide supra*) consistent with the assignment of this step as fast (eq 4). The results of this modeling study served as an archetype for key steps in the mechanisms of ebselen and cyclic seleninates.

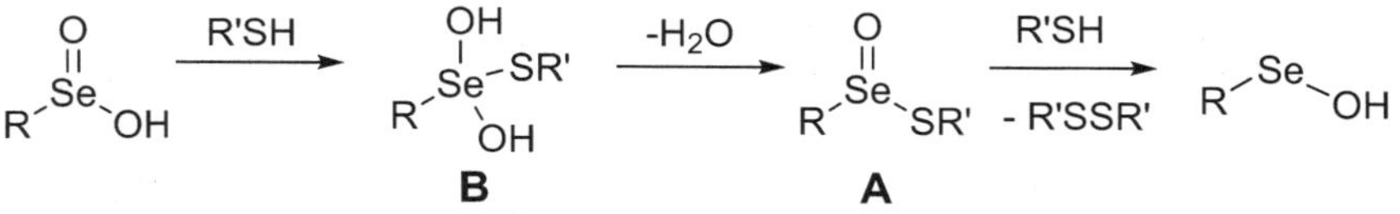

Scheme 4. DFT-SAPE reaction pathway for reduction of seleninic acid groups. Adapted from reference (26).

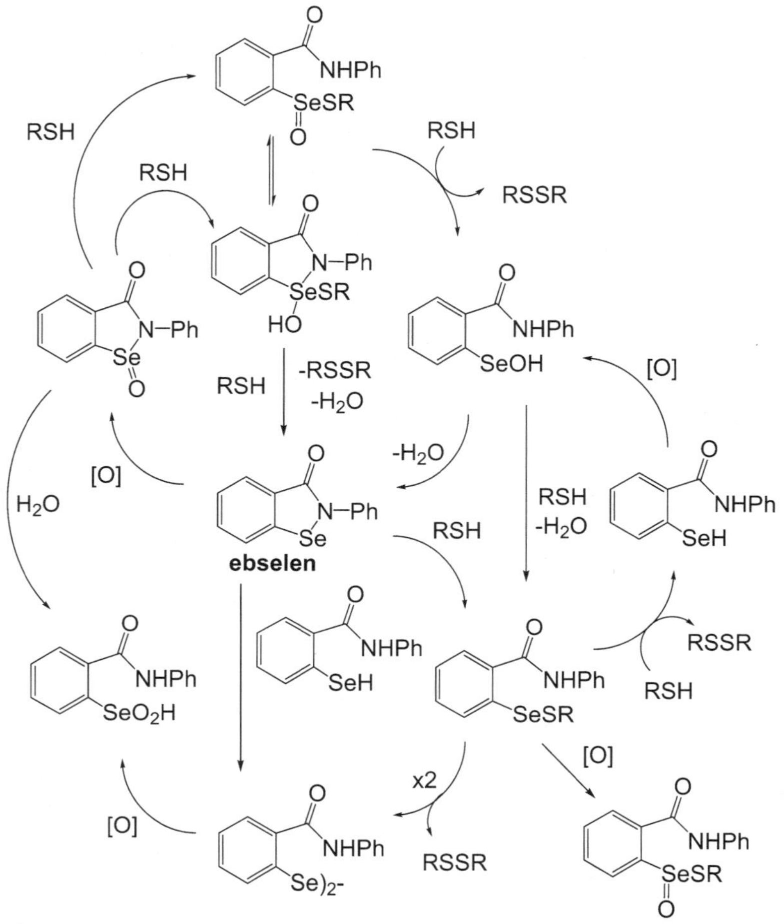

*Scheme 5. Composite of mechanistic steps proposed for ebselen. Adapted from reference (29). Reprinted with permission from Inorg. Chem. **2011**, 50, 12075-12084. Copyright 2011 American Chemical Society.*

Ebselen

Our experiences with the simpler mechanisms of aryl selenols and seleninic acids allowed us to tackle the much more complex mechanism of the most well-known GPx mimic, ebselen. Although this compound has been subjected to many studies, its mechanism of activity remains an open question. Ebselen's cyclic selenenamide bond (Se-N) suggests two alternate cycles for initial reaction: ring opening by thiol reduction or oxidation of the selenium center to the selenoxide. Experimental studies have suggested that either of these steps may be prevalent under various conditions. Fischer and Dereu showed that ebselen could be oxidized to a selenoxide that could be reduced back to ebselen with two equivalents of thiol (*32*). However these experiments were conducted under oxidizing conditions that may not be representative of biological systems. Most studies favor ring opening via thiol addition as the initial step, since ebselen is highly reactive with glutathione and other thiols (*32, 33, 82*) and binds to the Cys thiols of albumin when administered intraveneously (*83*). A composite of the proposed reaction pathways from these two initial reactions (Scheme 5) demonstrates the challenge we faced in computational modeling of ROS scavenging by ebselen (*29*). SAPE models of these steps indicate that ring opening is roughly 8 kcal/mol more favorable than ebselen oxidation, in agreement with experimental results showing that ebselen favors initial reduction. However, a shift to the oxidation pathway under conditions of oxidative stress as observed by Fischer and Dereu is also consistent with the DFT-SAPE results. The selenenyl sulfide pathway is energetically preferred as expected from experimental results, but the reduction to the selenol is a key bottleneck and SAPE calculations predict a significant barrier (32 kcal/mol) (*29*). However, in our SAPE modeling of possible mechanisms for redox cycling for ebselen (Scheme 5) and, more recently, a cyclic seleninate (*84*), all low barrier pathways appear to converge on the selenenyl sulfide as an intermediate. To bypass this problem, we have proposed that ebselen and other GPx mimics avoid the bottleneck of selenol regeneration by oxidizing the selenenyl sulfide to the seleninyl sulfide (Scheme 6). Other authors have proposed that the selenenyl sulfide disproportionates to the diselenide, which is then assigned as the intermediate that reacts with ROS (*15*), but, as discussed above, the slow rate of disproportionation to the diselenide is too low to sustain catalysis.

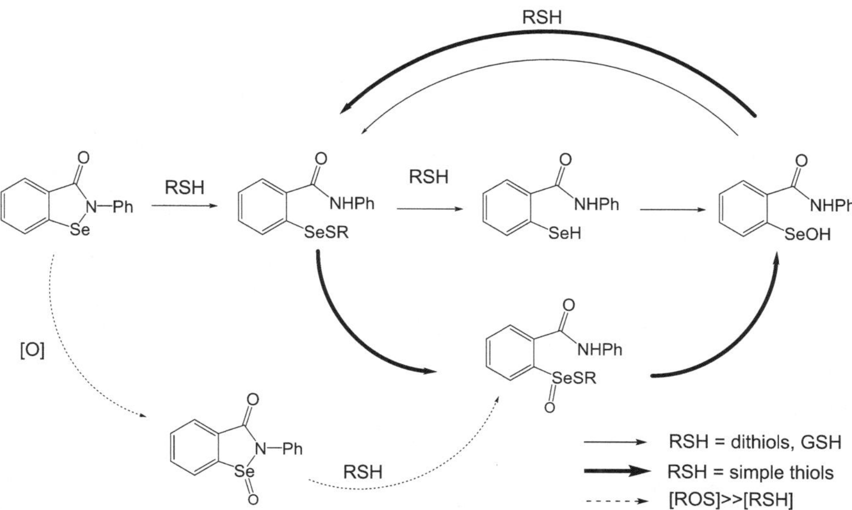

Scheme 6. Pathways for ROS scavenging by ebselen recommended from the results of SAPE-DFT modeling. Adapted from reference (29).

As a side note, the choice of the thiol reductant appears to be important to the overall mechanism observed experimentally. Selenol formation is observed for certain thiols, especially dithiols and GSH, but not PhSH and BzSH, which we have proposed to cycle through the seleninyl sulfide. For dithiols, the unimolecular rate law for selenol regeneration likely accounts for the improved cycling. Mugesh et al. showed that thiols with nearby donor groups may interact with the selenenyl sulfide to improve the kinetics of GPx-like activity (56). The effect is likely to be similar to the Se···N,O interactions, which shift electron density to the sulfur center that is the site of attack for the regeneration of the selenol and lead to a preference for thiol exchange. For these thiols, an S···N,O interaction could instead shift the partial negative charge to Se with an increase in the electrophilicity of S. Biological thiols such as GSH may operate similarly, with the S···O interaction forming from the backbone carbonyls. Simple thiols enjoy neither of these advantages and appear to cycle by a different mechanism than the thiols that are likely to be important biologically (Scheme 6). This dependence of the redox mechanism on the thiol may lead to misleading kinetics in laboratory testing.

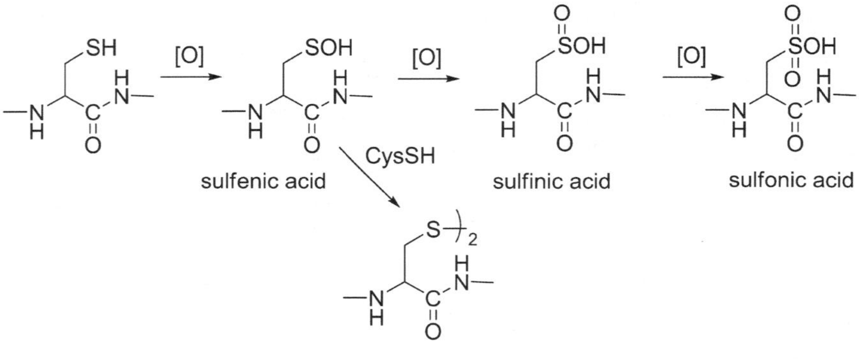

Scheme 7. Selected steps in the oxidation and disulfide coupling of Cys

Models of the Cysteine Redox Mechanism

Although most of our modeling studies have focused on selenium, we have also reported SAPE activation barriers for the progressive oxidation of Cys thiol to the sulfonic acid and the reductive coupling of Cys-SH and Cys-SOH to cystine (Scheme 7) (27). Thiol oxidation tends to correlate to the pK_a of the sulfhydryl with greater reactivity for those compounds that prefer the thiolate form (85). Cys residues are often located in polar active sites and/or with acidic residues which stabilize the thiolate (86). To account for the pH-dependence of thiol oxidation, we examined the two extremes of the neutral Cys-SH and basic Cys-S-. The thiolate model required a larger SAPE network to stabilize the anion in addition to representing the proton exchange. The SAPE activation barrier for oxidation of Cys-S- is 6 kcal/mol lower than Cys-SH, consistent with the greater nucleophilicity of the thiolate and the 100-fold difference in experimental rate at low and high pH (*87*). Further oxidation of the sulfenic acid Cys-SOH to the sulfinic (Cys-SO$_2$H) and sulfonic acids (Cys-SO$_3$H) resulted in activation barriers that were highly dependent upon the solvent model (cyclohexene > chlorobenzene > water). The results were consistent with enhancement of over-oxidation at solvent-exposed and/or highly polar sites (*86*).

In addition to the oxidation of Cys, we calculated the barrier for disulfide bond formation via coupling of Cys-SOH and Cys-SH. Compared to the reduction of PhSeOH to PhSeSMe, the lower affinity of the sulfenic acid for the thiol leads to a higher, but still relatively low, activation barrier (12.5 kcal/mol vs 6.6 kcal/mol). Disulfide bond formation competes with over-oxidation to Cys-SO$_2$H and the relative SAPE activation barriers of these competing reactions is in agreement with the branching ratios obtained in kinetics studies by Luo et al. (*87*). These experimental studies showed that disulfide formation is dominant at [H$_2$O$_2$]:[Cys] ratios below 10:1, but over-oxidation is important during conditions of extreme oxidative stress. With the success of DFT-SAPE calculations in reproducing experimental trends for Cys redox processes, we will be expanding these calculations to protein-sized systems to directly model the effect of the residues in the active site on the reactivity of Cys.

Models of the Reduction of Zinc-Sulfur Centers by Organoselenium Compounds

Reducible selenium compounds such as ebselen, benzeneseleninic acid, benzeneselenenyl chloride and the anti-tumor agent 1,4-phenylenebis(methylene)selenocyanate (p-XSC) have been shown to release Zn^{2+} from zinc-finger (ZF) proteins (*21, 22, 88–90*). Based upon experimental observations, the mechanism of the Zn^{2+} release may occur through two related pathways: (a) oxidation of the Cys thiolates to disulfides or (b) per-selenenylation of the Cys residues (Figure 3A) (*23*). The former pathway appears to be prevalent with the less nucleophilic Cys$_2$His$_2$ ZFs while the latter is preferred for Cys$_4$ (experimental data is not available for Cys$_3$His-type ZFs). In either case, formation of a covalent bond between Cys and either sulfur or selenium lowers the electron density and charge of the thiolate sulfur to weaken the Zn-S interaction. Competition with surrounding water molecules for coordination allows the Zn^{2+} ion to be released from the protein. Fully reduced selenium compounds, such as selenols or selenides, cannot react with Cys and do not release Zn^{2+}. However, a telluride was shown to catalytically release Zn^{2+} by cycling through a telluride oxide (*91*).

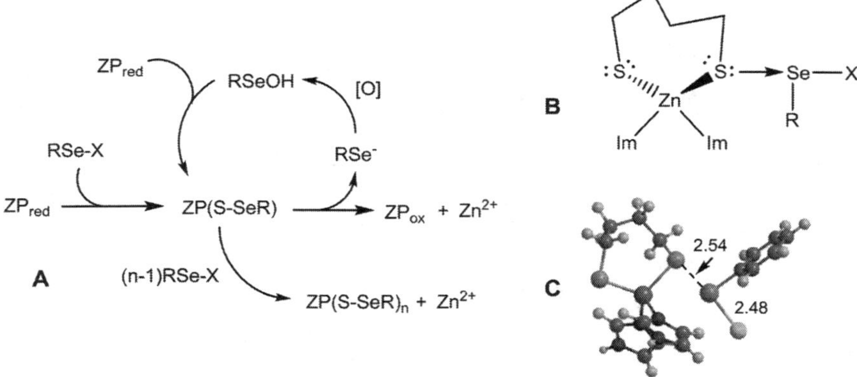

Figure 3. (A) Composite mechanism for Zn^{2+} release from zinc-sulfur proteins (ZP) by rSe compounds. (B) Se···S interaction between a general rSe compound RSe-X and a simplified model of a Cys_2His_2-type ZF (23). (C) Bond distances (Å) for sample RSe-ZF model complex with PhSeCl (23).

As a prelude to modeling Zn^{2+} release by rSe compounds, we have explored the strength of an Se···S donor-acceptor interaction between the electrophilic rSe compound and the thiolate of a simplified model of the Cys_2His_2-type of ZFs (Figure 3B) (*23*). In these models, the Se···S interaction is related to Se···N,O interactions in GPx mimics and is dependent upon the nature of both the strength of the Se-X bond of the rSe electrophile and the nucleophilicity of the ZF thiolate. Increasing the number of Cys ligands on Zn^{2+} increases the overall negative charge of the ZF center and the Lewis basicity of the Cys sulfur lone pairs (Cys_4 > Cys_3His > Cys_2His_2). This trend suggests a stronger Se···S interaction in our hypothetical model complex as well as an increasing reactivity toward electrophiles. In our preliminary study (*23*), we focused on Se···S interactions with a model of the least nucleophilic Cys_2His_2 ZF to compare the effect of the Se-X bond on the interaction. The rSe compounds with the most polarized bonds (e.g., Se-Cl) tended to be those that formed the greatest interaction with the ZF model (Figure 3C). The complex with PhSeCl forms a very strong interaction ($\Delta E_{d \to a}$ = 94 kcal/mol) with distances around Se suggestive of hypervalent bonding. The increase in Se-X and Zn-S bond distances are consistent with an activation of the rSe compound toward loss of X^- and weakening of the coordinate covalent bond of the thiolate to Zn^{2+}. The weak Se···S interactions found for disulfides and diselenides (~5 kcal/mol) may explain their overall lower reactivity toward ZFs, but environmental factors such as pH and bulky groups surrounding the zinc-sulfur center will also affect the ability of rSe compounds to release Zn^{2+}. In addition, test calculations with the more Lewis basic Cys_3His and Cys_4 ZFs show that the some Se-X bonds are displaced by Se-S bond formation with the ZF suggesting that the assumed Se···S intermediate may not be a valid assumption for these more reactive ZFs. Further calculations on Se···S interactions of rSe compounds with Cys_3His and Cys_4 models and a SAPE model of Zn^{2+} release are being pursued by our group.

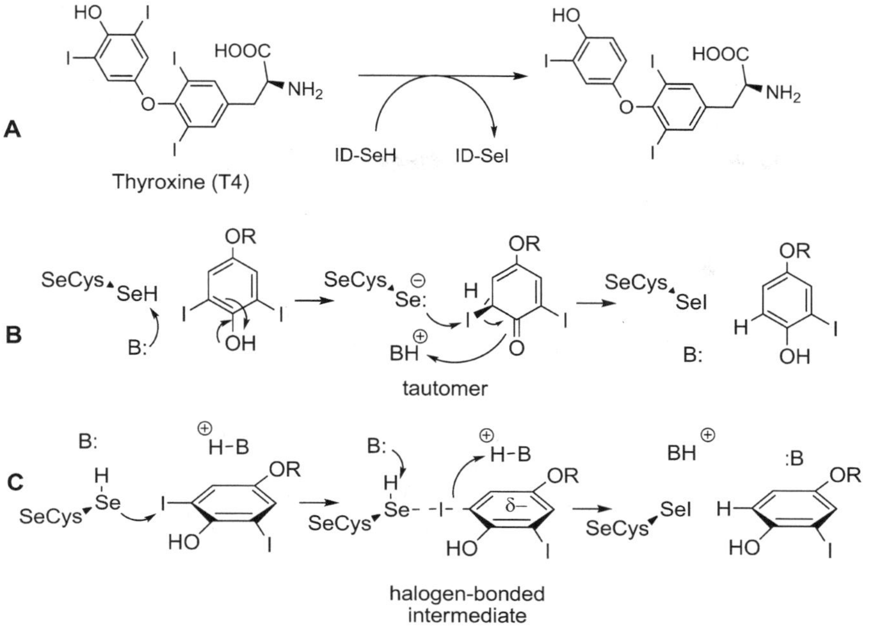

Scheme 8. *Mechanisms for deiodination of thyroxine by iodothyronine deiodinase ID: (A) general 'ping-pong' mechanism. (B) Deiodination through a tautomer intermediate. (C) Deiodination through a XB intermediate.*

Models of Iodothyronine Deiodinase

In addition to the ROS scavenging mechanisms of sulfur and selenium, we have reported models of mechanism of the thyroid enzyme iodothyronine deiodinase (ID) from the perspective of halogen bonding (XB) (*92*). The ID selenoproteins activate the pro-hormone thyroxine by deiodinating either the inner or outer ring (Scheme 8A), but the difficulty in isolating the protein has prevented detailed mechanistic studies. Literature mechanisms of the deiodination have suggested that the phenol tautomerizes to the keto form to activate the C-I bond for nucleophilic attack (Scheme 8B) (*93, 94*). However, ID-X and model compounds can deiodinate the inner ring for which a tautomer is not accessible. Intermolecular XB interactions between iodine centers of the thyroid hormones are important to transport of these highly insoluble compounds (e.g., thyroxine binds to transthyretin through an I⋯O interaction) (*95*). Se⋯I interactions have been studied widely, including in context of antithyroid drugs (*96*), but no connection between the deiodination of thyroxine and other derivatives had yet been made in the literature. In examining the strength of the Se···I interaction in model complexes (*92*), we noted its similarity with the Se···N,O interaction common to GPx mimics and wrote comparable resonance structures (eq 5).

$$R\text{-}I \quad X \leftrightarrow R^- \quad I\text{-}X \qquad\qquad (eq\ 5)$$

From a XB intermediate, if the right hand side could be stabilized by protonation of the carbanion, the bonding could be shifted to the right for nucleophilic substitution/deiodination (Scheme 8C). DFT studies of a model reaction using 1,4-dihydroxy-2,6-diiodobenzene as model of thyroxine with methylselenolate as the nucleophile and imidazoles as proton donor/acceptors were used to model the reaction through a XB intermediate (*92*). The activation barrier from the XB complex (17.6 kcal/mol) is reasonable for a biological reaction. The protein would be expected to further stabilize this reaction. The higher barrier found for a thiolate nucleophile (19.8 kcal/mol) is in agreement with the 100-fold decrease in activity for the sulfur version of ID-1. Therefore, the selection of selenium and iodine for this important biological process can be attributed to nucleophilicity of selenium and the polarizability of iodine. Subsequent to our study, Mugesh reported that XB was indeed critical to ID activity and further suggested that Se···S interactions may also be important (*97, 98*).

Conclusions and Future Directions

Understanding the mechanisms underlying the role of selenium in biological systems is important to applying its compounds for chemoprevention without toxic side-effects. DFT modeling is an important tool for unraveling some of the unanswered questions in selenium redox chemistry. In particular, the use of microsolvation techniques such as SAPE has been integral to the modeling of organoselenium-catalyzed redox scavenging by providing a pathway for proton exchange that is representative of the reaction in solution phase. As a result, our DFT-SAPE models have provided realistic transition states and activation barriers which have allowed greater insight into the reaction pathways favored by these compounds. A further step to improve the results of SAPE studies would be to encase the DFT-SAPE models in an explicit solvation shell and explore the reactions through QM/MM molecular dynamics methods. Given the success of applying SAPE modeling to the mechanisms of selenium compounds as well as Cys, we intend to extend these models to explore the effect of the more complex environments in proteins. These models will allow us to include the effects of nearby residues as well as surrounding solvent to understand how the activity of selenium and sulfur are modified in protein environments.

Acknowledgments

We thank the National Science Foundation (CHE-0750413) and the Thomas F. Jeffress and Kate Miller Jeffress Memorial Trust for support of our research.

References

1. Tapiero, H.; Townsend, D.; Tew, K. *Biomed. Pharmacother.* **2003**, *57*, 134–144.
2. Hatfield, D. L.; Berry, M. L.; Gladyshev, V. N. *Selenium. Its Molecular Biology and Role in Human Health*, 2nd ed.; Springer: New York, 2006.
3. Rayman, M. P. *Lancet* **2000**, *356*, 233–241.
4. Forman, H. J.; Fukuto, J. M.; Torres, M. *Signal Transduction by Reactive Oxygen and Nitrogen Species*; Kluwer: Dordrecht, 2003.
5. Ganther, H. E. *Carcinogenesis* **1999**, *20*, 1657–1666.
6. Davis, C. D.; Tsuji, P. A.; Milner, J. A. *Annu. Rev. Nutr.* **2012**, *32*, 73–95.
7. Cherubini, A.; Ruggiero, C.; Polidori, M. C.; Mecocci, P. *Free Radicals Biol. Med.* **2005**, *39*, 841–852.
8. Crack, P. J.; Taylor, J. M. *Free Radicals Biol. Med.* **2005**, *38*, 1433–1444.
9. Madamanchi, N. R.; Vendrov, A.; Runge, M. S. *Arterioscler. Thromb. Vasc. Biol.* **2005**, *25*, 29–38.
10. Gromer, S.; Eubel, J. K.; Lee, B. L.; Jacob, J. *Cell. Mol. Life Sci.* **2005**, *62*, 2414–2437.
11. Ganther, H. E.; Lawrence, J. R. *Tetrahedron* **1997**, *53*, 12299–12310.
12. Lapchak, P. A.; Zivin, J. A. *Stroke* **2003**, *34*, 2013–2018.
13. Mugesh, G.; Singh, H. B. *Chem. Soc. Rev.* **2000**, *29*, 347–357.
14. Mugesh, G.; du Mont, W.-W.; Sies, H. *Chem. Rev.* **2001**, *101*, 2125–2180.
15. Bhabak, K. P.; Mugesh, G. *Acc. Chem. Res.* **2010**, *43*, 1408–1419.
16. Yang, Y.; Huang, F.; Ren, Y.; Xing, L.; Wu, Y.; Li, Z.; Pan, H.; Xu, C. *Oncol. Res.* **2009**, *18*, 1–8.
17. Cherukuri, D. P.; Goulet, A.-C. *Cancer Biol. Ther.* **2005**, *4*, 175–180.
18. Lippman, S. M.; Goodman, P. J.; Klein, E. A.; Parnes, H. L.; Thompson, I. M.; Kristal, A. R.; Santella, R. M.; Probstfield, J. L.; Moinpour, C. M.; Albanes, D.; Taylor, P. R.; Minasian, L. M.; Hoque, A.; Thomas, S. M.; Crowley, J. J.; Gaziano, J. M.; Stanford, J. L.; Cook, E. D.; Fleshner, N. E.; Lieber, M. M.; Walther, P. J.; Khuri, F. R.; Karp, D. D.; Schwartz, G. G.; Ford, L. G.; Coltman, C. A. *J. Natl. Cancer Inst.* **2005**, *97*, 94–102.
19. Nicastro, H.; Dunn, B. *Nutrients* **2013**, *5*, 1122–1148.
20. Clark, L. C. *J. Am. Med. Assoc.* **1996**, *276*, 1957.
21. Jacob, C.; Maret, W.; Vallee, B. L. *Proc. Natl. Acad. Sci. U.S.A.* **1999**, *96*, 1910–1914.
22. Blessing, H.; Kraus, S.; Heindl, P.; Bal, W.; Hartwig, A. *Eur. J. Biochem.* **2004**, *271*, 3190–3199.
23. Bayse, C. A.; Whitty, S. M.; Antony, S. *Curr. Chem. Biol.* **2013**, *7*, 57–64.
24. Bayse, C. A. *J. Phys. Chem. A* **2007**, *111*, 9070–9075.
25. Bayse, C. A.; Antony, S. *J. Phys. Chem. A* **2009**, *113*, 5780–5785.
26. Bayse, C. A. *J. Inorg. Biochem.* **2010**, *104*, 1–8.

27. Bayse, C. A. *Org. Biomol. Chem.* **2011**, *9*, 4748–4751.

28. Bayse, C. A.; Pavlou, A. *Org. Biomol. Chem.* **2011**, *9*, 8006–8015.

29. Antony, S.; Bayse, C. A. *Inorg. Chem.* **2011**, *50*, 12075–12084.

30. Mugesh, G.; du Mont, W.-W. *Chem.—Eur. J.* **2001**, *7*, 1365–1370.

31. Wirth, T. *Molecules* **1998**, *3*, 164–166.

32. Fischer, H.; Dereu, N. *Bull. Soc. Chim. Belg.* **1987**, *96*, 757.

33. Haenen, G. R.; De Rooij, B. M.; Vermeulen, N. P.; Bast, A. *Mol. Pharmacol.* **1990**, *37*, 412–422.

34. Maiorino, M.; Roveri, A.; Coassin, M.; Ursini, F. *Biochem. Pharmacol.* **1988**, *37*, 2267–2271.

35. Mugesh, G.; Panda, A.; Singh, H. B.; Butcher, R. J. *Chem.—Eur. J.* **1999**, *5*, 1411–1421.

36. Mugesh, G.; Panda, A.; Singh, H. B.; Punekar, N. S.; Butcher, R. J. *J. Am. Chem. Soc.* **2001**, *123*, 839–850.

37. Back, T. G.; Moussa, Z. *J. Am. Chem. Soc.* **2002**, *124*, 12104–12105.

38. Back, T. G.; Moussa, Z. *J. Am. Chem. Soc.* **2003**, *125*, 13455–13460.

39. Back, T. G.; Moussa, Z.; Parvez, M. *Angew. Chem., Int. Ed.* **2004**, *43*, 1268–1270.

40. Kuzma, D.; Parvez, M.; Back, T. G. *Org. Biomol. Chem.* **2007**, *5*, 3213–3217.

41. Press, D. J.; Mercier, E. A.; Kuzma, D.; Back, T. G. *J. Org. Chem.* **2008**, *73*, 4252–4255.

42. Mercier, E. A.; Smith, C. D.; Parvez, M.; Back, T. G. *J. Org. Chem.* **2012**, *77*, 3508–3517.

43. Tripathi, S. K.; Patel, U.; Roy, D.; Sunoj, R. B.; Singh, H. B.; Wolmershäuser, G.; Butcher, R. J. *J. Org. Chem.* **2005**, *70*, 9237–9247.

44. Singh, V. P.; Singh, H. B.; Butcher, R. J. *Indian J. Chem.* **2011**, *50A*, 1263–1272.

45. Tripathi, S. K.; Sharma, S.; Singh, H. B.; Butcher, R. J. *Org. Biomol. Chem.* **2010**, *9*, 581–587.

46. Bayse, C. A.; Baker, R. A.; Ortwine, K. N. *Inorg. Chim. Acta* **2005**, *358*, 3849–3854.

47. Barton, D. H. R.; Hall, M. B.; Lin, Z.; Parekh, S. I.; Reibenspies, J. *J. Am. Chem. Soc.* **1993**, *115*, 5056–5059.

48. Reed, A. E.; Curtiss, L. A.; Weinhold, F. *Chem. Rev.* **1988**, *88*, 899–926.

49. Iwaoka, M.; Tomoda, S. *J. Am. Chem. Soc.* **1994**, *116*, 2557–2561.

50. Iwaoka, M.; Tomoda, S. *J. Org. Chem.* **1995**, *60*, 5299–5302.

51. Iwaoka, M.; Tomoda, S. *J. Am. Chem. Soc.* **1996**, *118*, 8077–8084.

52. Iwaoka, M.; Komatsu, H.; Katsuda, T.; Tomoda, S. *J. Am. Chem. Soc.* **2004**, *126*, 5309–5317.

53. Iwaoka, M.; Tomoda, S. *Phosphorus, Sulfur Silicon Relat. Elem.* **2005**, *180*, 755–766.

54. Iwaoka, M.; Komatsu, H.; Katsuda, T.; Tomoda, S. *J. Am. Chem. Soc.* **2002**, *124*, 1902–1909.

55. Iwaoka, M.; Katsuda, T.; Komatsu, H.; Tomoda, S. *J. Org. Chem.* **2005**, *70*, 321–327.

56. Sarma, B. K.; Mugesh, G. *J. Am. Chem. Soc.* **2005**, *127*, 11477–11485.

57. Sarma, B. K.; Mugesh, G. *ChemPhysChem* **2009**, *10*, 3013–3020.
58. Behera, R. N.; Panda, A. *RSC Adv.* **2012**, *2*, 6948–6956.
59. Prabhakar, R.; Vreven, T.; Morokuma, K.; Musaev, D. G. *Biochemistry* **2005**, *44*, 11864–11871.
60. Kallies, B.; Mitzner, R. *J. Mol. Model.* **1998**, *4*, 183–196.
61. Wu, Z.; Ban, F.; Boyd, R. J. *J. Am. Chem. Soc.* **2003**, *125*, 6994–7000.
62. Lundin, A.; Panas, I.; Ahlberg, E. *J. Phys. Chem. A* **2007**, *111*, 9080–9086.
63. Bayse, C. A.; Antony, S. *Main Group Chem.* **2007**, *6*, 185–200.
64. Goto, K.; Nagahama, M.; Mizushima, T.; Shimada, K.; Kawashima, T.; Okazaki, R. *Org. Lett.* **2001**, *3*, 3569–3572.
65. Benkova, Z.; Kóňa, J.; Gann, G.; Fabian, W. M. F. *Int. J. Quant. Chem.* **2002**, *90*, 555–565.
66. Pearson, J. K.; Boyd, R. J. *J. Phys. Chem. A* **2006**, *110*, 8979–8985.
67. Minyaev, R. M.; Minkin, V. I. *Can. J. Chem.* **1998**, *76*, 776–788.
68. Minyaev, R. M.; Minkin, V. I. *Mendeleev Commun.* **2000**, *10*, 173–174.
69. Bhabak, K. P.; Mugesh, G. *Chem.—Eur. J.* **2009**, *15*, 9846–9854.
70. Bhabak, K. P.; Mugesh, G. *Chem.—Eur. J.* **2008**, *14*, 8640–8651.
71. Smith, M. L.; Lancia, J. K.; Mercer, T. I.; Ip, C. *Anticancer Res.* **2004**, *24*, 1401–1408.
72. Gromer, S.; Gross, J. H. *J. Biol. Chem.* **2002**, *277*, 9701–9706.
73. Ayouni, L.; Barbier, F.; Imbert, J.-L.; Lantéri, P.; Grenier-Loustalot, M.-F. *J. Toxicol. Environ. Health, Part A* **2007**, *70*, 735–741.
74. Gammelgaard, B.; Cornett, C.; Olsen, J.; Bendahl, L.; Hansen, S. H. *Talanta* **2003**, *59*, 1165–1171.
75. Sarma, B. K.; Mugesh, G. *Chem.—Eur. J.* **2008**, *14*, 10603–10614.
76. Ren, B.; Huang, W.; Åkesson, B.; Ladenstein, R. *J. Mol. Biol.* **1997**, *268*, 869–885.
77. Syed, R.; Wu, Z. P.; Hogle, J. M.; Hilvert, D. *Biochemistry* **1993**, *32*, 6157–6164.
78. Kice, J. L.; Lee, T. W. S. *J. Am. Chem. Soc.* **1978**, *100*, 5094–5102.
79. Cowan, E. A.; Oldham, C. D.; May, S. W. *Arch. Biochem. Biophys.* **2011**, *506*, 201–207.
80. Arai, K.; Dedachi, K.; Iwaoka, M. *Chem.—Eur. J.* **2011**, *17*, 481–485.
81. Ritchey, J. A.; Davis, B. M.; Pleban, P. A.; Bayse, C. A. *Org. Biomol. Chem.* **2005**, *3*, 4337–4342.
82. Glass, R. S.; Farooqui, F.; Sabahi, M.; Ehler, K. W. *J. Org. Chem.* **1989**, *54*, 1092–1097.
83. Wagner, G.; Schuch, G.; Akerboom, T. P. M.; Sies, H. *Biochem. Pharmacol.* **1994**, *48*, 1137–1144.
84. Bayse, C. A.; Ortwine, K. N. *Eur. J. Inorg. Chem.* **2013**, *2013*, 3680–3688.
85. Winterbourn, C. C.; Metodiewa, D. *Free Radicals. Biol. Med.* **1999**, *27*, 322–328.
86. Claiborne, A.; Miller, H.; Parsonage, D.; Ross, R. P. *FASEB J* **1993**, *7*, 1483–1490.
87. Luo, D.; Smith, S. W.; Anderson, B. D. *J. Pharm. Sci.* **2005**, *94*, 304–316.
88. Jacob, C.; Maret, W.; Vallee, B. L. *Biochem. Biophys. Res. Comm.* **1998**, *248*, 569–573.

89. El-Bayoumy, K.; Das, A.; Narayanan, B.; Narayanan, N.; Fiala, E. S.; Desai, D.; Rao, C. V.; Amin, S.; Sinha, R. *Carcinogenesis* **2006**, *27*, 1369–1376.

90. Larabee, J. L.; Hocker, J. R.; Hanas, J. S. *J. Inorg. Biochem.* **2009**, *103*, 419–426.

91. Giles, N. M.; Gutowski, N. J.; Giles, G. I.; Jacob, C. *FEBS Lett.* **2003**, *535*, 179–182.

92. Bayse, C. A.; Rafferty, E. R. *Inorg. Chem.* **2010**, *49*, 5365–5367.

93. Vasilev, A. A.; Engman, L. *J. Org. Chem.* **1998**, *63*, 3911–3917.

94. Kunishima, M.; Friedman, J. E.; Rokita, S. E. *J. Am. Chem. Soc.* **1999**, *121*, 4722–4723.

95. Wojtczak, A.; Luft, J.; Cody, V. *J. Biol. Chem.* **1992**, *267*, 353–357.

96. Du Mont, W.-W.; Martens-von Salzen, A.; Ruthe, F.; Seppälä, E.; Mugesh, G.; Devillanova, F. A.; Lippolis, V.; Kuhn, N. *J. Organomet. Chem.* **2001**, *623*, 14–28.

97. Manna, D.; Mugesh, G. *Angew. Chem., Int. Ed.* **2010**, *122*, 9432–9435.

98. Manna, D.; Mugesh, G. *J. Am. Chem. Soc.* **2012**, *134*, 4269–4279.

Chapter 10

Selenotrisulfide as a Metabolic Intermediate in Biological Systems

Mamoru Haratake,*[1,2] Katsuyoshi Fujimoto,[1] Hongoh Masafumi,[1] Sakura Yoshida,[1] Takeshi Fuchigami,[1] and Morio Nakayama[1]

[1]Graduate School of Biomedical Sciences, Nagasaki University, 1-14 Bunkyo-machi, Nagasaki 852-8521, Japan
[2]Faculty of Pharmaceutical Sciences, Sojo University, 4-22-1 Ikeda, Kumamoto 860-0082, Japan
*E-mail: haratake@ph.sojo-u.ac.jp.

In 1941, Painter showed that the reaction of selenious acid with low mass thiols (RSH) produced selenotrisulfide *in vitro*. Later, the reactivity of glutathione selenotrisulfide (GSSeSG) with thiols of pancreatic ribonuclease in an acidic medium was characterized by Ganther. Recently, several RSSeSR species were actually identified in biological systems by modern mass spectrometry. Due to the reversibility of the thiol exchange reaction, RSSeSR′ is generated from the reaction of RSSeSR with R′SH, when increasing the molar ratio of R′SH to RSSeSR. On the basis of these facts, we investigated the molecular events of the biogenic thiol-associated selenium transfer in the bloodstream, with particular attention to selenotrisulfide. We demonstrated that the selenium transfer from red blood cells to hepatocytes involves a relay mechanism of thiol exchange that occurs between the selenotrisulfide and thiols (RSSeSR + R′SH → R′SSeSR + R″SH → R′SSeSR″).

Background

Selenium is a metalloid element in group 16 of the periodic table that shares similar chemical properties especially with sulfur and, to a lesser extent, with tellurium. Selenium is a nutritionally essential trace element in most animal species and broadly distributed over the entire body (*1, 2*). This element is taken up in inorganic or organic compounds from the food supply, and these selenium source compounds are enzymatically and/or non-enzymatically metabolized in the biological environment (*3*). Selenium is finally incorporated into the selenoproteins (*e.g.*, glutathione peroxidases and thioredoxin reductase) (*4*). Inorganic selenious acid is rare as a chemical form of the food-source compounds, but it is a highly effective source compound most frequently used in the selenium supplementation for medical treatments. Plant sources of selenium contain mostly selenomethionine, but selenium-enriched yeast (a common form of selenium supplement) is found to contain significant amounts of selenite (*5, 6*), and selenite is also used to supplement infant formula and other products (*7, 8*). Thus, a better understanding of the systemic delivery mechanisms of selenium from selenite is of significance from the viewpoints of medical treatments and toxicology. From the interest of selenium bioinorganic chemistry, the mechanism of metabolism and incorporation of selenious acid are also desirable.

The reaction of selenious acid with low mass thiols such as cysteine (Cys) and glutathione (GSH) to form bis(alkylthio)selenide, selenotrisulfide (STS) is one of the principal pathways by which inorganic selenium is initially incorporated into biological systems. Painter first described this reaction pathway in 1941 (*9*). In this reaction, the usual ratio of thiol to selenious acid is 4 to 1 and the normal stoichiometry is expressed in the following equation: $H_2SeO_3 + 4RSH \rightarrow$ RSSeSR + RSSR + $3H_2O$ (Painter reaction) (*10*). Later, the reactivity of GSH STS (GSSeSG) with thiols of ribonuclease in an acidic medium was characterized by Ganther (*11*). In 1980, a possible formation mechanism of STS from selenious acid and thiols was investigated in aqueous dioxane acid solutions (*12*). Thirty years later, GSSeSG was detected in a biological system, a selenium-enriched yeast sample, using modern mass spectrometric techniques (*13*).

Synthesis of Penicillamine-Based Selenotrisulfide Compounds as Metabolic Intermediates of Selenious Acid

It has been presumed that the Painter reaction is involved in the reduction of selenious acid *in vivo*. Ganther found GSSeSG to be relatively stable in an acid solution (*10*). However, once formed in acidic solution, many STS rapidly break down at physiological pH to form a red elemental selenium species. As a result, is quite difficult to obtain a stable RSSeSR′ with endogenous low molecular weight thiols such as GSH and coenzyme A at physiological pH *in vitro* since GSSeSG is fairly unstable and glutathione selenopersulfide (GSSe⁻) is generated by the subsequent GSH reduction. Furthermore, GSSe⁻ can easily decompose to produce elemental selenium (Se^0) as a terminal product or be reduced further to hydrogen selenide (H_2Se) in the presence of a large excess of GSH. Since details of the reduction mechanism from GSSeSG to H_2Se are not elucidated *in vivo*, the actual behavior of these reactive metabolic intermediates in a biological environment is still unclear.

We prepared a STS compound using L-penicillamine (Pen) as a thiol instead of Cys (*14*). Pen is structurally similar to Cys, but it has two methyl groups at the β carbon atom of Cys and is capable of generating a chemically stable STS species (PenSSeSPen, Figure 1) that can be easily isolated from the reaction mixture. Isolated PenSSeSPen was analyzed by X-ray photoelectron spectroscopy. The selenium 3d spectrum of PenSSeSPen has an absorption peak at 56.30 eV (Figure 2), significantly different from that of selenious acid (60.50 eV) but similar to that of elemental selenium (55.65 eV). The chemical stability of PenSSeSPen in phosphate buffer at pH 7.4 and 37 °C was followed by reverse-phase liquid chromatography. No remarkable degradation of PenSSeSPen was observed during 24 h incubation under these physiological conditions. Actually, the PenSSeSPen solution was colorless and clear even after 24 h incubation. We also prepared a GSSeSG mimic, Pen-substituted GSSeSG (PenGSSeSGPen, Figure 3) (*15*). This compound is also chemically stable under these physiological conditions.

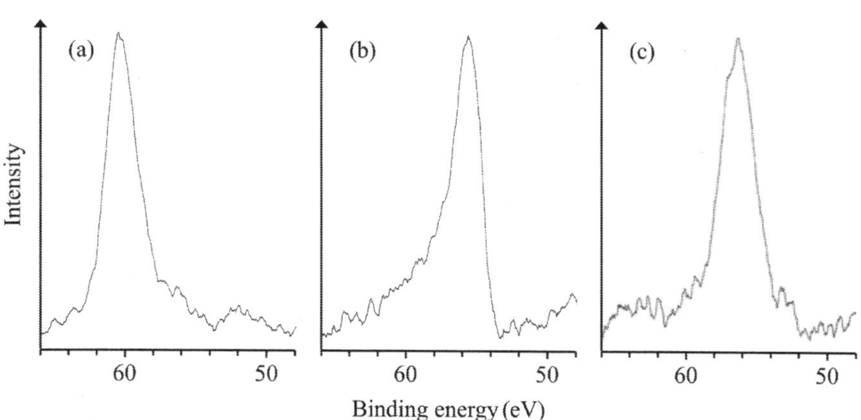

Figure 1. Chemical structure of PenSSeSPen.

Figure 2. X-ray photoelectron spectroscopy of the selenium 3d spectrum of PenSSeSPen. Instrument: Shimadzu/Kratos AXIS-ULTRA, X-ray source: monochromated aluminum Ka line (wavelength: 0.8339 nm, 1.486 keV), Ground state electronic configuration of selenium: [Ar] $3d^{10}\ 4s^2\ 4p^4$.

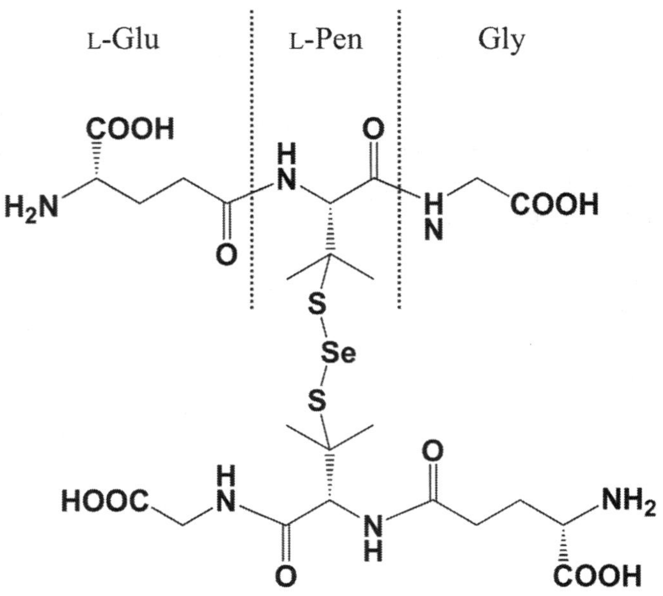

Figure 3. Chemical structure of penicillamine-substituted glutathione selenotrisulfide, PenGSSeSGPen. L-Glu: L-glutamic acid, L-Pen: L-penicillamine, Gly: glycine.

Interactions of Pen-Based STS Compounds with Hemoglobin

In the bloodstream, selenious acid is reported to be immediately taken up into the red blood cell (RBC) through the anion exchanger 1 (AE1) protein and then reappears in the plasma after reductive metabolism in the RBC (*16*). The reappearing selenium species is thought to bind to albumin and subsequently be transferred to the peripheral organs and tissues. By using Pen-based STS compounds, we studied the relevance of STS species in the metabolic pathway of selenious acid in the bloodstream.

After the uptake of selenious acid into RBCs, the RBC hemolysate was separated into three fractions: less than 30 kDa, more than 30 kDa, and the plasma membrane. The selenium content of each fraction was determined by the fluorescence method after acid digestion (*17*). Selenium in the RBC hemolysate was mostly found in the high-mass fraction (> 30 kDa) containing abundant Hb. These selenium distributions remained the same up to 3 h after the uptake of selenious acid, and most of the selenium seemed to be stably bound to Hb (*15*).

RBCs contain GSH in the reduced form at approximately 2 mM (1.77–2.39 mM) (*18*), and Hb tetramer is present as a 34 wt% solution (5.27 mM) (*19*). Its concentration is nearly three times higher than the GSH concentration. The Hb tetramer consists of four polypeptide chains, α, α', β and β', and has one reactive Cys residue on the respective β and β' chains, Cysβ93 (*16*). Cysβ93 is most likely to react with STS and other metabolites.

First, we examined the reactivity of selenious acid with Hb in phosphate buffer at pH 7.4 and 37 °C. Selenious acid did not directly react with Hb at all. When GSH was added to the mixture of selenious acid and Hb, selenium binding to Hb was observed in the presence of increasing GSH concentration, and the selenious acid concentration in the mixture correspondingly decreased, indicating that a selenium species reduced by GSH can bind to Hb. Similarly, after combining PenSSeSPen and Hb, free PenSSeSPen was monitored by reverse-phase liquid chromatography, and PenSSeSPen rapidly bound to Hb. After a 15-minute incubation of PenSSeSPen with Hb, Hb was removed from the mixture by ultrafiltration, and the thiol content in the Hb-free filtrate was measured by using 5,5′-dithiobis(2-nitrobenzoic acid). The thiol exchange, in general, proceeds without changing the thiol content in the reaction mixture. If this reaction involves thiol exchange (RSSeSR + R′SH → R′SSeSR + RSH), free Pen should be released from PenSSeSPen. In fact, the thiol content in the filtrate was nearly the same as that from the initially added Hb for the reaction, suggesting that thiol exchange is occurring during PenSSeSPen binding to Hb.

This selenium binding to Hb was characterized using the Langmuir type binding equation ($1 / r = 1 / nK \times 1 / [\text{selenium}]_{\text{free}} + 1 / n$; n: apparent number of selenium binding sites on Hb; $[\text{selenium}]_{\text{free}}$: free selenium concentration; K: binding constant, $= [\text{Hb} \cdot \text{selenium}] / [\text{selenium}]_{\text{free}} \times [\text{Hb}]_{\text{free}}$; r: amount of selenium bound per Hb tetramer) (*20*). The calculated n value of 1.588 was nearly equal to the number of Hb thiols (1.45). Also, a large binding constant K value (4.8×10^8 M^{-1}) would suggests a specific interaction between the PenSSeSPen and Hb tetramer.

To explore the participation of Hb thiols in selenium binding, its thiols were covalently blocked by iodoacetamide (IAA) treatment and then the reactivity of this treated Hb and PenSSeSPen was examined. Selenium binding to Hb was almost completely inhibited by the IAA blockade of thiols, showing that PenSSeSPen can react with Hb through the thiols of its Cys residues. The reaction product of PenSSeSPen and Hb was directly characterized by matrix-assisted laser desorption ionization-time of flight (MALDI-TOF) mass spectrometric analysis (*21*). A newly generated peak 226 mass units higher than the free β chain was detected with increasing concentrations of PenSSeSPen. This observed increase in mass number is due to the binding of the PenSSe moiety to the β chains. In contrast, the mass number of the α chains did not change at all, even at a one to ten concentration ratio of Hb and PenSSeSPen. Consequently, the reaction of PenSSeSPen with Hb involves thiol exchange between Pen and Hb Cysβ93.

To explore whether selenium bound to Hb is stable or not in the presence of GSH, the Hb-Se conjugate was incubated with GSH, and the amount of released selenium from the conjugate was determined using fluorescence methods (*17*). PenSSeSPen was completely degraded within 2 min after GSH addition. No selenium elimination was observed from the Hb-Se conjugate. Thus, GSSeSG formed in the RBC could possibly be reactive to Hb thiols and stably bound to Hb, and Hb could participate in the metabolism of selenious acid after the formation of GSSeSG (*21*).

Finally, Hb β chain in selenious-acid-treated RBCs was analyzed by MALDI-TOF mass spectrometry. The β chain gave a peak at molecular mass 16,254, resulting in an increase by 386 compared to that of the non-treated Hb chain. Such an increase in the molecular mass of the β chain corresponds to the selenenyl-GSH moiety, suggesting the formation of Hb-Cysβ93-SSeSG species in selenious acid-treated RBCs (*22*).

Selenium Transfer from Hemoglobin to Red Blood Cell Membrane

AE1 is the most abundant integral membrane protein in the RBC, nearly one-third by protein weight, and catalyzes the exchange of chloride and bicarbonate anions across the plasma membrane (Figure 4) (*23, 24*). The amino-terminal cytoplasmic domain of AE1 (N-CPD) offers binding sites for Hb (*25*). The free Cys residues of N-CPD, Cys201 and Cys317, are known to form a disulfide linkage with Cysβ93 under catalytic oxidative conditions (*26, 27*).

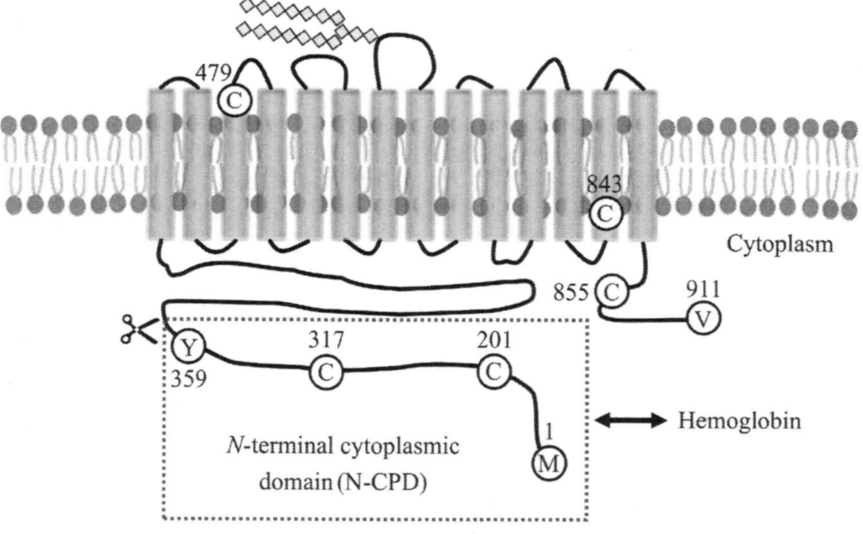

Figure 4. Structural model of anion exchanger 1, a major binding site of hemoglobin on the red blood cell membrane. N-CPD has intrinsic binding affinity for hemoglobin tetramer. ✂ *: a cleaving point by α-chymotrypsin digestion.*

To study the interaction of Hb-Se conjugate with the RBC membrane, PenGSSeSGPen was combined with Hb in phosphate buffer at pH 7.4. This mixture was incubated for 10 min at 37 °C, and then the unreacted PenGSSeSGPen was removed by gel filtration. The purified Hb-Se conjugate contained 1.6 selenium atoms per Hb tetramer (*21*).

We also examined whether the Hb-Se conjugate would bind to the inner surface of the RBC membrane, as the non-treated Hb does, using the RBC inside-out vesicle, IOV. When the binding constants of the Hb-Se conjugate and Hb interacting with IOV ($K = $ [Hb-Se $\cdot$ IOV] / [Hb-Se]$_\text{free}$ [IOV]$_\text{free}$) were estimated using the Langmuir-type binding equation, the values for the Hb-Se conjugate and Hb were 2.10 ± 0.43 and 1.86 ± 0.26 μM^{-1} (mean $\pm$ SEM, $P = 0.64$), respectively.

After incubation of the Hb-Se conjugate, selenious acid, and PenGSSeSGPen with the IOV, the amounts of selenium bound to the membranes were compared. The amount of selenium from the Hb-Se conjugate was much higher than those from selenious acid and PenGSSeSGPen; thus, the Hb-Se conjugate has a binding affinity for RBC membrane comparable to free Hb, but selenious acid and PenGSSeSGPen do not.

We further studied the selenium transfer from the Hb-Se conjugate to IOVs. Thiols in the IOVs were alkylated with IAA before combining them with Hb-Se conjugate. Thiol-modification inhibited the selenium transfer from the Hb-Se conjugate to the IOV in a concentration-dependent manner. The treatments with IAA showed no significant changes in the binding affinity of the Hb-Se conjugate for IOV (binding constant $K = 1.64$ μM^{-1}).

After incubation with the Hb-Se conjugate, IOV membranes were washed with a pH 8 buffer solution to remove Hb, and then the amount of selenium in IOVs was determined. Even after the Hb-Se washout, nearly 50% of the selenium remained in the IOV, suggesting that selenium in the Hb-Se conjugate can be transferred to IOV components due to direct interactions between the two.

After incubation with the Hb-Se conjugate, IOVs were first washed with Hb removal buffer and then selectively digested with α-chymotrypsin (α-Chy) to cleave the N-CPD of AE1. The enzyme digestion released approximately 50% of the selenium from the IOVs, which likely comes from N-CPD-bound selenium.

Separately, AE1 N-CPD in the selenious acid-treated RBC membrane was analyzed by MALDI-TOF mass spectrometry (*22*). A Cys317-containing fragment of the N-CPD was detected at mass number 3,635 and identified by theoretical mass calculation using the Protein Prospector program with a molecular mass gain of 125 after N-ethylmaleimide (NEM) thiol modification. The membrane sample from selenious-acid-treated RBCs provided a peak at 4,021, resulting in an increase of 386 compared to that of the non-treated RBC membranes. The observed increase in the molecular mass of the Cys317-containing fragment corresponds to addition of the GSSe moiety, similar to that observed for the Cysβ93 of Hb. Overall, we determined a metabolic pathway for selenious acid in the RBC; selenious acid is first non-enzymatically metabolized to GSSeSG, GSSeSG can interact with Hb to form a Hb-Se conjugate by thiol exchange, and selenium then can be transferred from the Hb-Se conjugate to the N-CPD of AE1 by a similar transfer mechanism (Figure 5).

The thiol dependency of the subsequent RBC membrane transport of selenium from AE1 was examined using membrane permeable thiol reagents, *i.e.*, NEM and tetrathionate (TTN). Treatment of RBCs with NEM, a thiol-alkylating reagent, resulted in modification of the thiol groups in the N-CPD of AE1, but not those in the membrane domain. This NEM treatment resulted in a marked inhibition of the selenium export from the RBC to the blood plasma. In addition, the treatment with TTN, a thiol-oxidizing reagent that forms intermolecular disulfide bonds, appeared to oxidize thiol groups in both the N-CPD and the membrane domain of AE1, resulting in complete inhibition of selenium export even during the initial period in which the export had a maximum velocity without TTN treatment. Such complete inhibition of selenium export from the TTN-treated RBCs appears to be due to the oligomerized AE1 proteins resulting from intermolecularly formed disulfide bonds. These inhibitory effects observed using NEM and TTN demonstrate that thiol groups in the integral protein AE1 play essential roles in the membrane transport of selenium from the RBCs to the blood plasma. (*28*).

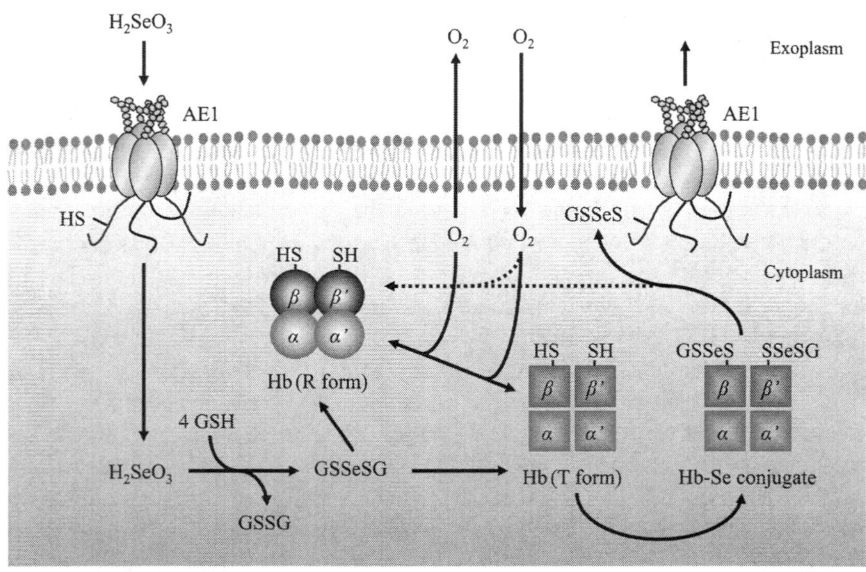

Figure 5. A metabolic pathway of selenious acid in a red blood cell. Hb (T form): Hb in a taut (tense) form; Hb (R form): Hb in a relaxed form.

Selenium Transfer from Red Blood Cell to Human Serum Albumin to Hepatocyte

We also investigated selenium transfer from the RBC to human serum albumin (HSA) (*29*). We demonstrated an HSA-mediated selenium transfer; the selenium exported from RBCs was bound to HSA through the selenotrisulfide, and then transferred into the hepatocyte. After treatment of the RBCs with selenite, the selenium efflux from the RBCs occurred in a HSA-concentration-dependent manner. Pretreatment of HSA with IAA almost completely inhibited the selenium efflux from the RBC to the HSA solution. The selenium efflux experiment was carried out in a HSA solution (45 mg/mL), and subsequently this HSA solution was subjected to gel permeation chromatographic separation. The peak fraction with the selenium content was consistent with that of HSA. The selenium bound to HSA in this solution was completely eliminated by treatment with Pen, resulting in the generation of PenSSeSPen. Selenium efflux from the RBCs also occurred in a Pen solution, and PenSSeSPen was observed in the resulting Pen solution. Thus, selenium exported from the RBC is thought to bind to the HSA via a STS linkage with its single free thiol.

A model of the selenium-bound HSA was prepared by treating HSA with PenSSeSPen. The selenium from PenSSeSPen binds to HSA by thiol exchange between Pen and the free thiol of HSA, producing the selenotrisulfide-containing HSA (HSA-SSeSPen). When HSA-SSeSPen was incubated with isolated rat hepatocytes, the selenium content in the hepatocytes increased together with its decrease in the incubation medium. To verify the results from these model experiments using HSA-SSeSPen, we conducted a HSA-mediated selenium transfer experiment with selenite-treated RBC in the presence of hepatocytes. The STS-containing HSA was able to transport the selenium into the hepatocyte. Overall, selenium transfer from the RBC to the hepatocytes involves a relay mechanism of thiol exchange that occurs between the STS and thiol compounds (STS relay mechanism: RSSeSR + HSA-SH → HSA-SSeSR + R′SH → RSSeSR′; Figure 6).

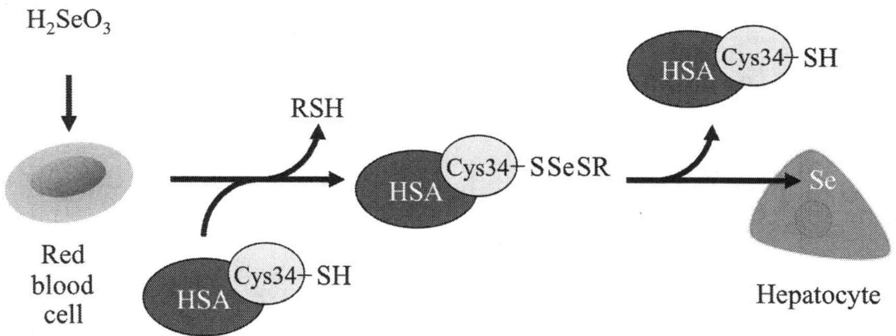

Figure 6. Albumin-mediated selenium transfer from red blood cell to hepatocytes.

Conclusions

We demonstrated that a STS can bind to Hb through its thiols with high affinity. GSSeSG generated during the course of selenious acid metabolism binds Hb, and then Hb mediates selenium transfer to the Cys thiols in the N-CPD of AE1 in the RBC membrane. We also demonstrated HSA-mediated selenium transfer; selenium exported from the RBC binds to the HSA Cys34 thiol, and then is transferred into hepatocytes. This selenium transfer involves a relay mechanism of thiol exchange that occurs between the STS and cysteine thiols. STS reactivity with HSA is probably dependent on the high concentration of HAS in plasma, not the specificity of the HSA thiol for STS, because STS reacts with other thiols, such as Pen, depending on their concentrations. Such selenium transfer processes in the bloodstream (distribution) could possibly involve a relay mechanism of thiol exchange that occurs between the STS species and protein thiols (STS relay mechanism; Scheme 1). Based on the success of this series of experiments, Pen-based STS compounds are useful probes for exploring the metabolic pathways of selenious acid in biological systems. Further investigations using these compounds as models for naturally occurring STSs may uncover the variety of biologically relevant thiols involved in selenium metabolism *in vivo*.

$$H_2SeO_3 + 4\,RSH \rightarrow RSSeSR + RSSR + 3\,H_2O \cdots\cdots\cdots (1)$$

$$RSSeSR + R'SH \rightarrow RSSeSR' + R''SH \rightarrow RSSeSR'' \cdots\cdots (2)$$

Scheme 1. Formation of selenotrisulfide from selenious acid by the Painter reaction (equation 1), and subsequent selenium transfer by a selenotrisulfide relay mechanism (equation 2). RSH: low mass thiol, R'SH and R''SH: protein thiols.

References

1. Schwarz, K.; Foltz, M. C. *J. Am. Chem. Soc.* **1957**, *79*, 3292.
2. Rayman, M. P. *Lancet* **2000**, *356*, 233.
3. Navarro-Alarcon, M.; Cabrera-Vique, C. *Sci. Total Environ.* **2008**, *400*, 115.
4. Kryukov, V. G.; Castellrano, S.; Novoselov, S. V.; Labanov, A. V.; Zehtab, O.; Guigo, R.; Gladyshev, V. *Science* **2003**, *300*, 1439.
5. Infante, H. G.; O'Connor, G.; Rayman, M.; Wahlen, R.; Entwisle, J.; Norris, P.; Hearna, R.; Cattericka, T. *J. Anal. At. Spectrom.* **2004**, *19*, 1529.
6. Ayouni, L.; Barbier, F.; Imbert, J. L.; Gauvrit, J. Y.; Lantéri, P.; Grenier-Loustalot, M. F. *Anal. Bioanal. Chem.* **2006**, *385*, 1504.
7. Dael, P. V.; Davidsson, L.; Muñoz-Box, R.; Fay, L. B.; Barclay, D. *Br. J. Nutr.* **2001**, *85*, 157.
8. Carver, J. D. *Am. J. Clin. Nutr.* **2003**, *77*, 1550S.
9. Painter, E. P. *Chem. Rev.* **1941**, *28*, 179.
10. Ganther, H. E. *Biochemistry* **1968**, *7*, 2898.
11. Ganther, H. E.; Corcoran, C. *Biochemistry* **1969**, *8*, 2557.
12. Kice, J. L.; Lee, T. W. S.; Pan, S.-T. *J. Am. Chem. Soc.* **1980**, *102*, 4448.
13. Lindemann, T.; Hintelmann, H. *Anal. Chem.* **2002**, *74*, 4602.
14. Haratake, M.; Ono, M.; Nakayama, M. *J. Health Sci.* **2004**, *50*, 366.
15. Haratake, M.; Fujimoto, K.; Hirakawa, R.; Ono, M.; Nakayama, M. *J. Biol. Inorg. Chem.* **2008**, *13*, 471.
16. Suzuki, K. T.; Shiobara, Y.; Itoh, M.; Ohmichi, M. *Analyst* **1998**, *123*, 63.
17. Watkinson, J. H. *Anal. Chem.* **1966**, *38*, 92.
18. Hirono, A.; Iyori, H.; Sekine, I.; Ueyama, J.; Chiba, H.; Kanno, H.; Fujii, H.; Miwa, S. *Blood* **1996**, *87*, 2071.
19. Brewer, M.; Scott, T. In *Concise Encyclopedia of Biochemistry*; Walter de Gruyter: New York, 1983; p 201.

20. Langmuir, I. *J. Am. Chem. Soc.* **1917**, *39*, 1848.

21. Haratake, M.; Katsuyoshi, F.; Ono, M.; Nakayama, M. *Biochim. Biophys. Acta* **2005**, *1723*, 215.

22. Hongoh, M.; Haratake, M.; Fuchigami, T.; Nakayama, M. *Dalton Trans.* **2012**, *41*, 7340.

23. Kopito, R. R.; Lodish, H. F. *Nature* **1985**, *316*, 234.

24. Popov, M.; Tam, L. Y.; Li, J; Reithmeier, R. A. F. *J. Biol. Chem.* **1997**, *272*, 18325.

25. Walder, J. A.; Chatterjee, R.; Steck, T. L.; Low, P. S.; Musso, G. F.; Kaiser, E. T.; Rogers, P. H.; Arnone, A. *J. Biol. Chem.* **1984**, *259*, 10238.

26. Alper, S. L. *Exp. Physiol.* **2006**, *91*, 153.

27. Reithmeier, R. A.; Rao, A. *J. Biol. Chem.* **1979**, *254*, 6151.

28. Haratake, M.; Hongoh, M.; Ono, M.; Nakayama, M. *Inorg. Chem.* **2009**, *49*, 7805.

29. Haratake, M.; Hongoh, M.; Miyauchi, M.; Hirakawa, R.; Ono, M.; Nakayama, M. *Inorg. Chem.* **2008**, *47*, 6273.

Editors' Biographies

Craig A. Bayse

Craig A. Bayse received his B.S. in chemistry from Roanoke College in 1994 and his Ph.D. at Texas A&M University in 1998, working with Michael B. Hall on theoretical studies of the structure and bonding of transition metal hydride complexes. As a graduate student, he attended the European Summer School in Quantum Chemistry in Sweden and the Sostrup Summer School on Quantum Chemistry and Molecular Properties in Denmark. He took a brief hiatus from computational chemistry during a Kodak-funded postdoctoral appointment at Cornell University with Barry K. Carpenter, synthesizing novel azoalkanes. More recently, he added molecular dynamics simulations of proteins to his repertoire through a sabbatical appointment at the University of Florida with Kennie M. Merz. His research uses an amalgam of these computational and experimental approaches to explore the bonding and reactivity of inorganic systems. His contributions include mechanistic studies of bioactive selenium and sulfur compounds; theoretical studies of the active sites of molybdenum and tungsten enzymes; determination of the electronic structure of luminescent coinage metal materials; and bonding models of π-stacking interactions. He is currently Professor of Chemistry and Biochemistry at Old Dominion University.

Julia L. Brumaghim

Julia L. Brumaghim received her Ph.D. degree from the University of Illinois at Urbana-Champaign and completed postdoctoral research at the University of California at Berkeley in both Bioinorganic Chemistry and Cellular and Molecular Biology. She is currently an Associate Professor in the Department of Chemistry at Clemson University in Clemson, SC. Recent work primarily focuses on metal-mediated oxidative DNA damage and the mechanisms by which sulfur, selenium, and polyphenolic antioxidant compounds prevent this damage. In addition, she is investigating the ability of nanomaterials to generate reactive oxygen species and promote oxidative damage. She has published over 40 papers, has presented her research at over 80 scientific conferences and universities, and has been a recipient of the ACS PROGRESS/Dreyfus Lectureship Award from the American Chemical Society and Camille and Henry Dreyfus Foundation, a CAREER award from the National Science Foundation, and the 2008 Award for the Best Paper from A Young Investigator, *Journal of Inorganic Biochemistry* and Elsevier Publishers.

Indexes

Author Index

Subject Index